I0014968

Security of Information and Networks

Editors

Atilla Elçi
Eastern Mediterranean University, Gazimagusa (TRNC), North Cyprus

S. Berna Ors
Istanbul Technical University, Istanbul, Turkey

Bart Preneel
Katholieke Universiteit Leuven, Leuven, Belgium

Proceedings of the First International Conference on Security of Information and Networks (SIN 2007), 7-10 May 2007, Gazimagusa (TRNC), North Cyprus

Order this book online at www.trafford.com/07-1689
or email orders@trafford.com

Most Trafford titles are also available at major online book retailers.

Other Contributors
Formatting By Behnam Rahnama, EMU, North Cyprus

Cover Graphics By Artiss & Ecem Akarra, Turin Design, USA

Note for Librarians: A cataloguing record for this book is available from Library
and Archives Canada at www.collectionscanada.ca/amicus/index-e.html

ISBN: 978-1-4251-4109-7

*We at Trafford believe that it is the responsibility of us all, as both individuals
and corporations, to make choices that are environmentally and socially sound.
You, in turn, are supporting this responsible conduct each time you purchase a
Trafford book, or make use of our publishing services. To find out how you are
helping, please visit www.trafford.com/responsiblepublishing.html*

*Our mission is to efficiently provide the world's finest, most comprehensive
book publishing service, enabling every author to experience success.
To find out how to publish your book, your way, and have it available
worldwide, visit us online at www.trafford.com/10510*

www.trafford.com
North America & international
toll-free: 1 888 232 4444 (USA & Canada)
phone: 250 383 6864 ♦ fax: 250 383 6804
email: info@trafford.com
The United Kingdom & Europe
phone: +44 (0)1865 722 113 ♦ local rate: 0845 230 9601
facsimile: +44 (0)1865 722 868 ♦ email: info.uk@trafford.com

10 9 8 7 6 5 4 3 2

Preface

The International Conference on Security of Information and Networks (SIN 2007) was hosted by the Eastern Mediterranean University in Gazimagusa, North Cyprus and co-organized by the Istanbul Technical University, Turkey.

SIN 2007 followed a national conference held in Turkey. It is the intention of the SIN steering committee that SIN will become an annual series of international conferences in the area of Security of Information and Networks. The conference provided a relaxed atmosphere in which researchers could present and discuss new work on selected areas of current interest.

While SIN 2007 covered all areas of information and network security, the papers presented focused on the following topics:

- cryptology: design and analysis of cryptographic algorithms, hardware and software implementations of cryptographic algorithms, and steganography;
- network security: authentication, authorization and access control, privacy, intrusion detection, grid security, and mobile and personal area networks;
- IT governance: information security management systems, risk and threat analysis, and information security policies.

They represented an interesting mix of innovative academic research and experience reports from practitioners. A total of 80 papers were submitted. After an extensive reviewing process, the program committee accepted 24 papers and 16 short papers for presentation at the conference.

The program was complemented by a number of invited speakers, who delivered excellent overview talks:

- Elisabeth Oswald, University of Bristol, Bristol, UK: Power Analysis Attack -- A Very Brief Introduction;
- Marc Joye, Thomson R&D, France: On White-Box Cryptography;
- Çetin Kaya Koç, Istanbul Commerce University, Turkey and Oregon State University, USA: Micro-Architectural Side-Channel Attacks and Branch Prediction Attack;
- Karthik Bhargawan, Microsoft Research Cambridge, UK: Web Services Security: Protocols, Implementations, and Proofs;
- Bart Preneel, Katholieke Universiteit Leuven, Leuven, Belgium: Research Challenges in Cryptology;
- Mehmet Ufuk Çağlayan, Boğaziçi University, Turkey: Secure Routing in Ad Hoc Networks and Model Checking.

On the days before the conference five tutorials were presented on a broad range of topics.

First and foremost we would like to thank the members of the program committee for their efforts on reviewing the papers.

We would also like to thank the numerous external reviewers for their assistance. We are also indebted to North Cyprus Turkcell, TRNC Ministry of Education and Culture, Gazimagusa Teknopark, Microsoft Research Cambridge, Microsoft Turkey, Argela Technologies, Nethouse Broadband Internet and EC-Council for financial support of the workshop. We would like to thank the following organizations for their support of the conference: National Research Institute of Electronics and Cryptology (UEKAE), IEEE Turkey Section, IEEE Computer Society Turkey Branch, Chamber of Computer Engineers, TRNC, North Cyprus, Internet Technologies Association (INETD), Turkey and Linux Users Association (LKD), Turkey.

We thank our organizers, namely the Faculty of Engineering at the Eastern Mediterranean University (EMU), and the Faculty of Electrical and Electronics Engineering of the Istanbul Technical University.

The extensive hosting facilitation provided by the Department of Computer Engineering of EMU in terms of faculty, staff, and equipment has been our steady base in preparing the conference. As such, we gratefully acknowledge the support accorded to SIN 2007 Conference by the Chairman of the Department, Assoc. Prof. Dr. Hasan Kömürcügil, by the Dean, Prof. İzzet Kale, and by the Vice Rector (Research), Prof. Mehmet Altınay.

We also wish to thank Internet Technology Research Center of EMU for administrative support, Mr. Behnam Rahnama for developing and maintaining the SIN 2007 website, and, Microsoft Research for providing the MSR Conference Management Tool, the web-based review site.

We are gratefull to quite a few people for their unfaltering services during the conference: Ms. Fatma Tansu and İdil Candan at the Registration Desk, Mr. Osman Veysal and Mr. Hurol Mear for technical services, Mr. Erdal Altun for all provisioning relations, volunteers of the ACM EMU Student Chapter under the leadership of Mr. Özcan İlikhan for conference hosting services, Dr. Yıltan Bitirim for his ever-ready & willing to produce results management in local arrangements, and Asst. Prof. Dr. Ahmet Ünveren for his able and timely organizing in order to faciliate the conference site.

We also thank Kre-Com Travel Agent and the Salamis Bay Conti Resort Hotel management for their support of the conference.

Camera-ready formatting of this book was done by Mr. Behnam Rahnama who painstakingly dealt with every detail to render it right. The elusive background image of the front cover is the creation of Mr. Artiss Akarra and Ms. Ecem Akarra, both of Turin Design, USA. Incidentally, they are also to receive kudos for the attractive SIN 2007 posters. We are greatly indebted to these fine people.

Finally we would like to thank all the participants, submitters, authors and presenters who all together made SIN 2007 a great success. We hope that the development of an international conference in this area of the world will bring together researchers in information security, strengthen their cooperation and stimulate their research.

June 2007   Atilla Elci, Sıddıka Berna Örs, Bart Preneel

1st International Conference on Security of Information and Networks

May 7-10, 2007, Gazimagusa, North Cyprus
Conference Website URL http://www.sinconf.org/

Co-Chairs

Atilla Elçi..................Eastern Mediterranean University, TRNC, North Cyprus
Berna Örs...Istanbul Technical University, Turkey

Program Committee

Bülent Örencik.. TÜBİTAK MAM, Turkey
Atilla Elçi....................................Eastern Mediterranean University, TRNC
Berna Örs...Istanbul Technical University, Turkey
Ahmet Hasan Koltuksuz........................Izmir Institute of Technology, Turkey
Attila Özgit...................................Middle East Technical University, Turkey
Alexander Chefranov.......................Eastern Mediterranean University, TRNC
Şaban Eren...Maltepe University, Turkey
Sefer Kurnaz..Air Force Academy, Turkey
Justin Zhan...Carnegie Mellon University, USA

Local Arrangements Committee

Atilla Elçi....................................Eastern Mediterranean University, TRNC
Alexander Chefranov.......................Eastern Mediterranean University, TRNC
Muhammed Salamah.......................Eastern Mediterranean University, TRNC
Süha Bayındır...............................Eastern Mediterranean University, TRNC
Ender Yüksel....................................Istanbul Technical University, Turkey
Alev Elçi....................................Eastern Mediterranean University, TRNC
Yıltan Bitirim..............................Eastern Mediterranean University, TRNC
Ahmet Ünveren............................Eastern Mediterranean University, TRNC
Behnam Rahnama..........................Eastern Mediterranean University, TRNC
Nurten Kara..................................Eastern Mediterranean University, TRNC
Süleyman İrvan..............................Eastern Mediterranean University, TRNC

External Referees

Khalil Abuosba   Sheikh Ahamed   Ersan Akyıldız

Taner Altınok	Fuat İnce	Mesut Razbonyalı
Serap Atay	Hai Jin	Şeref Sağıroğlu
Lyudmila Babenko	Tai-Hoon Kim	Erkay Savaş
Arkadiy Barskiy	Il Seok Ko	Ashutosh Saxena
Süha Bayındır	Çetin Koç	Bruce Schneier
Zeki Bayram	Özgül Küçük	Ali Selçuk
Ayşe Bener	Klaus Kursawe	Stefaan Seys
Jun Bi	Geuk Lee	Siraj Shaikh
Müslim Bozyiğit	Tiberiu Letia	Sushil Sharma
Ufuk Cağlayan	Albert Levi	Dave Singelee
Jan Cappaert	Chuchang Liu	Gennadiy Slyusarev
Alexander Chefranov	Robert Maier	Boris Sobol
Anton Chuvakin	Oleg Makarevich	Eugene Spafford
Danny De Cock	Nele Mentens	Kaile Su
Mina Deng	Jorge Nakahara	Cengiz Tavukçuoğlu
Claudia Diaz	Gregory Neven	Dongvu Tonien
Tharam Dillon	Svetla Nikova	Tugkan Tuğlular
Rüyal Ergül	Jose Onieva	Ion Tutanescu
Ana Ferreres	Mehmet Orgun	Vladimir Vasilyev
Dieter Gollmann	Valeriy Osipyan	Huaxiong Wang
Johann Groszchaedl	Elisabeth Oswald	Dai Watanabe
Song Han	Pascal Paillier	Thomas Wollinger
Selim Hendrickson	Josef Pieprzyk	Karel Wouters
Edward Humphreys	Norbert Pramstaller	Brecht Wyseur
David Hwang	Vasant Raval	Mohammad Zulkernine

List of tutorials

- Ahmet Koltuksuz & Selma Tekir, Izmir Institute of Technology, Turkey: Open-Source Intelligence and Analysis
- Ersan Akyıldız, Middle East Technical University, Turkey: Elliptic Curve Cryptography
- Mustafa Başak, TÜBİTAK-UEKAE, Turkey: AKiS, Smart Card Operating System
- Hüseyin Hışıl, Queensland University of Technology, Australia and Serap Atay, Izmir Institute of Technology, Turkey: Making of a Multiprecision Cryptographic Software Library: Experiences with CRYMPIX .
- Murat Lostar, Lostar Information Security, Turkey: Creating/Using a Management System to Manage Information Security and ISO 27001 Certification

Sponsoring Institutions

North Cyprus Turkcell,
TRNC

Ministry of Education and
Culture, TRNC

Gazimagusa Teknopark,
TRNC

Microsoft Research
Cambridge, UK

Microsoft Turkey

Argela Technologies,
TRNC & Turkey

Nethouse Broadband
Internet, TRNC

EC-Council, USA

National Research
Institute of Electronics and
Cryptology (UEKAE),
Turkey

IEEE Turkey Section &
IEEE Computer Society
Turkey Branch

Chamber of Computer
Engineers, TRNC

Turkey Linux Users
Association (LKD),
Turkey

Internet Technologies Association (INETD), Turkey

Organized by

**Faculty of Engineering
Eastern Mediterranean University
TRNC**

**Faculty of Electrical and Electronics
Engineering
Istanbul Technical University Turkey**

Hosted by

**Department of
Computer Engineering**

**Internet Technologies Research Center
(ITRC)**

**Eastern Mediterranean University
Gazimagusa (TRNC), North Cyprus**

Table of Contents

Attacks, Intrusion Detection, and Security Recommendations

Keynote Speech

Keynote Speech

Security Software, Performance, and Experience

Access Control and Security Assurance

Keynote Speech

Security Software, Performance, and Experience

Access Control and Security Assurance

Keynote Speech

Power Analysis Attacks—A Very Brief Introduction

Elisabeth Oswald

Department of Computer Science,
University of Bristol,
Merchant Venturers Building,
Woodland Road,
Bristol, BS8 1UB,
United Kingdom.
eoswald@cs.bris.ac.uk

Abstract. Power analysis attacks are a fascinating field of contemporary research in security. They are interesting because understanding them requires bringing together knowledge from various disciplines such as cryptography, electrical engineering and statistics. Furthermore, counteracting them has proven to be a very challenging task. This article aims to give a rather brief introduction to various types of power analysis attacks and some of the techniques that can be used to counteract them.

1 Introduction to Power Analysis

Power analysis attacks [KJJ99] work because the power consumption of the predominant technology that is used for building cryptographic devices, CMOS, depends on the data. When a cryptographic device computes a cryptographic algorithm, it does so step-by-step. This means that in each computation step a certain intermediate value of the cryptographic algorithm is calculated. Obviously, these intermediate values depend on the cryptographic key. Furthermore, in typical CMOS logic styles, the logical values are represented by voltage levels in the cryptographic circuit. The reason why power analysis attacks work should now be apparent: there is a link between the cryptographic key, the intermediate values and their representation in the circuit and the power consumption of the device. This link is exploited in all power analysis attacks.

In general power analysis attacks can be conceptualized in the following way. The attacker attempts to reveal information about some unknown secret value that is stored inside a cryptographic device. This value can have any meaning, but we will refer to it as the *secret key* in the remainder of this article. The attacker may access the device in certain ways. For instance, the attacker knows the inputs or outputs (plaintexts, ciphertexts) of the device, and the attacker can invoke the execution of a certain cryptographic algorithm, using a certain (but unknown) secret key. The attacker then also has a model of the cryptographic device. This model can be very crude, if very little is known about the cryptographic device, or it can be very sophisticated, if the attacker has for instance full access to all specifications or even low-level descriptions (net lists) of the device. The attacker uses the model to predict the values of a certain intermediate result that she assumes to occur in the computation of the cryptographic algorithm. These intermediate values depend on the unknown secret key. Hence, she needs to guess (a part) of the secret key for this prediction. Next, she maps the intermediate values to hypothetical power consumption values. Hence, she needs to make an assumption about the power consumption characteristics of the device. Last, she compares her hypothetical power consumption values to the real power consumption of the device. If all her assumptions were correct and her key guess was correct too, then the hypothetical power consumption values correlate to the real power consumption values. If this correlation is high enough, then she accepts the key guess as correct. If there is very little correlation, then one or several of her assumptions were incorrect.

It should be clear that many assumptions are made. In particular, the mapping from intermediate values to hypothetical power consumption values is a rather critical one. In many publications, so-called standard power consumption models are used. Standard power consumption models are the bit-model, the Hamming-weight model or the Hamming-distance model. They often work well if intermediate values that are stored in registers are attacked, but they work not so well in many other cases. However, it could be that the attacker has detailed knowledge about the power consumption characteristics of a device. Such knowledge can be obtained through a characterization of the device before the actual attack. Power analysis attacks which use such a characterization are often called *template attacks* [CRR03].

2 Power Analysis Attacks without Characterization of Device

In this section we assume for all attacks that the attacker uses standard power models. Depending on how the analysis of the power traces actually proceeds, we distinguish between different types of attacks: simple power analysis (SPA) attacks, differential power analysis (DPA) attacks and higher-order (HO) DPA attacks.

2.1 Simple Power Analysis

The goal of SPA attacks is to reveal the key when given only a small number of power traces (for a small number of plaintexts). In the most extreme case, this means that the attacker attempts to reveal the key based on one single power trace.

SPA attacks are useful in practice if only one or very few traces are available for a given set of inputs. Consider for example a scenario where a consumer uses a smart card to pay for gas at a gas station. The customer has to refill the gas tank of the car on a regular basis and always buys a similar amount of gas. A malicious smart card reader could record the power consumption of the card. In this way, the attacker could gather a couple of traces for similar plaintexts.

The principle of SPA attacks is the following. The attacker needs to be able to monitor the power consumption of the device under attack. In the attacked device, the key must have (directly or indirectly) a significant impact on the power consumption. SPA attacks exploit key-dependent differences (patterns) within a trace.

2.2   Differential Power Analysis

In contrast to SPA attacks, DPA attacks require a large number of power traces. It is therefore usually necessary to physically possess a cryptographic device for some time in order to mount a DPA attack on it. Consider for example an owner of an electronic purse. This person can record a large number of power traces by transferring small amounts of money to and from the purse. These traces could then be used to reveal the cryptographic key that is used by the purse.

Another important difference between the two kinds of attacks is that the recorded traces are analyzed in a different way. In SPA attacks, the power consumption of a device is mainly analyzed along the time axis. The attacker tries to find patterns in a single trace. In case of DPA attacks, the shape of the traces along the time axis is not so important. DPA attacks analyze how the power consumption at fixed moments of time depends on the processed data. Hence, DPA attacks focus exclusively on the data dependency of the power traces.

2.3   Higher-Order Differential Power Analysis

DPA attacks, as described in the previous section, have the property that one intermediate value is predicted and used in the attack. Because only one intermediate value is used, these DPA attacks are also referred to as *first-order* DPA attacks. If several intermediate values are used to formulate the hypotheses, then the corresponding DPA attacks are called *higher-order* DPA attacks. We continue to write DPA attacks in order to refer to first-order DPA attacks in the remainder of this article. The so-far main application of higher-order DPA attacks is to attack implementations that are protected by masking.

Higher-order DPA attacks exploit the joint leakage of several intermediate values that occur inside the cryptographic device. Note that due to performance reasons, typical implementations of masking schemes conceal several intermediate values by the same mask. However, even if several masks are used throughout the algorithm, they are generated before the algorithm starts, they are applied to the data and (or) the key, and they are altered by the operations of the algorithm. Consequently, in an implementation where efficiency (memory, speed) is needed, it is always the case that a mask (or a combination of masks) and an intermediate value that is concealed by this mask (or a combination of masks) occur in the device. Hence, in practice it is typically not necessary to study higher-order DPA attacks in general. In practice, it is sufficient to concentrate on higher-order DPA attacks that exploit the leakage that is related to *two* intermediate values. These attacks are called *second-order* DPA attacks. The two intermediate values can either be two values that are concealed by the same mask, or, they can be a mask and a value that is concealed by this mask.

3   Power Analysis Attacks with Characterization of Device

In template attacks, we characterize power traces by a multivariate normal distribution, which is fully defined by a mean vector and a covariance matrix $(\mathbf{m}, \mathbf{C})$. This pair is then called a *template*. In contrast to other types of power analysis attacks, template attacks usually consist of two phases: A first phase, in which the characterization takes place, and a second phase, in which the characterization is used for an attack.

As said before, in a template attack, we assume that we can characterize the device under attack. This means, we can determine templates for certain sequences of instructions. For example, we might possess another device of the same type as the attacked one that we can fully control. On this device, we execute these sequences of instructions with different data d_i and keys k_j in order to record the resulting power consumption. Then, we group together the traces that correspond to a pair of (d_i, k_j), and estimate the mean vector and the covariance matrix of the multivariate normal distribution. As a result, we obtain a template for every pair of data and key (d_i, k_j): $h_{d_i, k_j} = (\mathbf{m}, \mathbf{C})_{d_i, k_j}$.

Later on, we use the characterization together with a power trace from the device under attack to determine the key. This means, we evaluate the probability density function of the multivariate normal distribution with $(\mathbf{m}, \mathbf{C})_{d_i, k_j}$ and the power trace of the device under attack. In other words, given a power trace $\mathbf{t}'$ of the device under attack, and a template $h_{d_i, k_j} = (\mathbf{m}, \mathbf{C})_{d_i, k_j}$, we compute the probability $p(\mathbf{t}'|(\mathbf{m}, \mathbf{C})_{d_i, k_j})$. This probability is often written more concise as $p(\mathbf{t}'|k_j)$.

3.1   Template-Based Simple Power Analysis

In a template-based SPA attack, the question of the attacker is basically the following: given a power trace $\mathbf{t}'_i$, what is the probability that the key of the device equals k_j? This conditional probability $p(k_j|\mathbf{t}'_i)$ can be calculated using *Bayes' theorem* using the prior probability $p(k_l)$ and the probability $p(\mathbf{t}'_i|k_l)$.

Given just one trace $\mathbf{t}'_i$, the best guess for the key that is used by the device is the key k_j that leads to the highest probability $p(k_j|\mathbf{t}'_i)$. Guessing the key of the device based on this strategy is called maximum-likelihood approach. This approach is optimal in the sense that it minimizes the probability of error.

3.2   Template-Based Differential Power Analysis

In a template-based DPA attack we have a matrix of power consumption values $\mathbf{T}$. We are hence interested in determining the probability $p(k_j|\mathbf{T})$. This is the probability that, given a matrix $\mathbf{T}$ of power consumption values, the device uses the key k_j.

The extension from $p(k_j|\mathbf{t}_i')$ to $p(k_j|\mathbf{T})$ is not very difficult. Since the power traces are statistically independent, we can multiply the probabilities that correspond to different traces.

4   Countermeasures

DPA attacks work because the power consumption of cryptographic devices depends on the intermediate values of the executed cryptographic algorithms. Hence, the goal of countermeasures against DPA attacks is to make the power consumption of the cryptographic device independent of the intermediate values of the executed cryptographic algorithm.

The countermeasures against DPA attacks that have been published so far can essentially be categorized into different groups: protocols, hiding, and masking.

4.1   Protocols

Keys that are only used for a few cryptographic operations are usually called session keys. Using session keys can make power analysis attacks significantly more difficult. This idea is used for example in the EMV standard. Session keys should be used whenever it is possible.

However, there are many scenarios where it is not practical or not possible to update the secret keys frequently enough to prevent power analysis attacks. This is why hiding and masking countermeasures are generally the first line of defense for cryptographic devices in practice.

4.2   Hiding

The basic idea of hiding is to remove the data dependency of the power consumption. This means that either the execution of the algorithm is randomized or the power consumption characteristics of the device are changed in such a way that an attacker cannot easily find a data dependency. The power consumption can be changed in two ways to achieve this goal: the device can be built in such a way that every operation requires approximately the same amount of energy, or it can be built in such a way that the power consumption is more or less random. In both cases, the data dependency of the power consumption is reduced significantly. However, in practice the data dependency cannot be removed completely—there always remains a certain amount of data dependency.

It is important to point out that implementations protected by hiding countermeasures process the same intermediate results as unprotected implementations. Resistance against power analysis attacks is solely achieved by altering the power consumption characteristics of the cryptographic device.

4.3   Masking

The basic idea of masking is to randomize the intermediate values that are processed by the cryptographic device. The motivation behind this approach is that the power that is needed to process randomized intermediate values is independent of the actual intermediate values. A big advantage of masking is that the power consumption characteristics of the device do not need to be changed. The power consumption can still be data dependent. Attacks are prevented because the device processes randomized intermediate values only.

5   Summary

Power analysis attacks exploit the fact that the instantaneous power consumption of a cryptographic device depends on the data it processes and on the operation it performs. Based on this dependency it is possible to extract the secret key of a cryptographic device. Power analysis attacks can be prevented by adequate countermeasures. Such countermeasures break the link between the intermediate values and the operations of a cryptographic algorithm and the power consumption of the device that executes the algorithm.

This article has provided a rather brief introduction to power analysis attacks. The text used in this article is taken from the book called *Power Analysis Attacks: Revealing the Secrets of Smart Cards* [MOP07].

References

[CRR03] Suresh Chari, Josyula R. Rao, and Pankaj Rohatgi. Template Attacks. In Burton S. Kaliski Jr., Çetin Kaya Koç, and Christof Paar, editors, *Cryptographic Hardware and Embedded Systems – CHES 2002, 4th International Workshop, Redwood Shores, CA, USA, August 13-15, 2002, Revised Papers*, volume 2523 of *Lecture Notes in Computer Science*, pages 13–28. Springer, 2003.

[KJJ99] Paul C. Kocher, Joshua Jaffe, and Benjamin Jun. Differential Power Analysis. In Michael Wiener, editor, *Advances in Cryptology - CRYPTO '99, 19th Annual International Cryptology Conference, Santa Barbara, California, USA, August 15-19, 1999, Proceedings*, volume 1666 of *Lecture Notes in Computer Science*, pages 388–397. Springer, 1999.

[MOP07] Stefan Mangard, Elisabeth Oswald, and Thomas Popp. *Power Analysis Attacks – Revealing the Secrets of Smart Cards*. ADIS Book Series. Springer, 2007. ISBN 0-387-30857-1.

On White-Box Cryptography

Marc Joye

Thomson R&D France
Technology Group, Corporate Research, Security Laboratory
1 avenue de Belle Fontaine, 35576 Cesson-Sévigné Cedex, France
marc.joye@thomson.net

Abstract. White-box cryptography techniques are aimed at protecting software implementations of cryptographic algorithms against key recovery. They are primarily used in DRM-like applications as a cost-effective alternative to token-based protections. This paper discusses the relevance of white-box implementations in such contexts as a series of questions and answers.

Q1: What is white-box cryptography?

A major issue when dealing with security programs is the protection of "sensitive" (secret, confidential or private) data embedded in the code. The usual solution consists in encrypting the data but the legitimate user needs to get access to the decryption key, which also needs to be protected. This is even more challenging in a software-*only* solution, running on a non-trusted host.

White-box cryptography is aimed at protecting secret keys from being disclosed in a *software* implementation. In such a context, it is assumed that the attacker (usually a "legitimate" user or malicious software) may also control the execution environment. This is in contrast with the more traditional security model where the attacker is only given a black-box access (i.e., inputs/outputs) to the cryptographic algorithm under consideration.

Q2: What is the difference with code obfuscation?

Related and complementary techniques for protecting software implementations but with *different* security goals include code obfuscation and software tamper-resistance. *Code obfuscation* is aimed at protecting against the reverse engineering of a (cryptographic) algorithm while *software tamper-resistance* is aimed at protecting against modifications of the code.

All these techniques have however in common that the resulting implementation must remain directly executable.

Q3: How realistic is the white-box threat model?

The traditional (i.e., black-box) threat models for encryption schemes are the chosen-plaintext attack (CPA) model and the chosen-ciphertext attack (CCA) model. In the *CPA model*, the adversary chooses plaintexts and is given the corresponding ciphertexts; in the *CCA model*, the adversary chooses ciphertexts and is given the corresponding plaintexts.

In the white-box threat model, the adversary can get access to the same resources as in the black-box model *plus* full control of the encryption/decryption software. The goal of the adversary is to extract the key. One may wonder why such a scenario makes sense since an adversary controlling the encryption/decryption software can make use of it to encrypt or decrypt arbitrary data without needing to extract the keys. We note that a white-box implementation can be useful as it forces the user to use the software at hand.[1] Furthermore, other security measures can be used concurrently.

If an adversary could recover the decryption key, then the data could be decrypted and used with *any* software on *any* host (cf. BORE attacks — break once, run everywhere). This would allow a global crack with more severe damages.

Q4: What are the applications of white-box cryptography?

The main application of white-box cryptography is the secure distribution of "valuable" content such as in *digital rights management* (DRM) applications. Here the main goal is to prevent the unauthorized use of bulk data processed by software, like music or movies.

More surprisingly, white-box techniques also allow the development of "lightweight" cryptography — note that only the private operation is "light", for example:

CONVERTING A SECRET-KEY ENCRYPTION INTO A PUBLIC-KEY ENCRYPTION
It is easy (in principle) to construct a *public-key encryption* scheme from a white-box implementation of a secret-key encryption algorithm E_K, say $\mathrm{WB}(E_K)$. Anyone in possession of $\mathrm{WB}(E_K)$ can encrypt messages while only one possessing secret key K is able to decrypt using decryption algorithm E_K^{-1}.

Note: To be valid, this transformation requires that E_K and E_K^{-1} are different. One should also ensure that the release of $\mathrm{WB}(E_K)$ does not contradict the usual security properties, like one-wayness or better, semantic security.

TRANSFORMING A MAC INTO A DIGITAL SIGNATURE
Dual to above, a *(keyed) message authentication code* (MAC_K) scheme can be used to produce a digital signature. Being given secret-key K, one can compute a signature using MAC_K on any message; further, using a public (and certified) white-box implementation of the "verification" algorithm, say $\mathrm{WB}(\mathrm{MAC}_K^{-1})$, anyone can verify the validity of signatures. Contrary to

---

[1] We also note that external encodings can be used so that the encryption (respectively, decryption) software requires encoded inputs and produces encoded outputs. As a result, the white-box implementation cannot be used for evaluating the encryption algorithm (respectively, decryption algorithm) in isolation. See Q5.

traditional MAC constructs, the verification does not require the knowledge of secret K.

Note: Again, in order to satisfy the non-repudiation property, one must assume that the operations of computing and verifying a "MAC" are different. Using a cryptographic hash function h, if E_K and $E_K{}^{-1}$ are different, a signature on message m can e.g. be produced as $S = E_K(h(m))$; its correctness can then verified with $\text{WB}(E_K{}^{-1})$ by checking whether $\text{WB}(E_K{}^{-1})(S)$ is equal to $h(m)$.

Q5: What is the general methodology behind white-box implementations?

The main idea of white-box implementations is to rewrite a *key-instantiated* version so that all information related to the key is "hidden". In other words, for each secret key, a key-customized software is implemented so that the key input is unnecessary.

Most symmetric block-ciphers, including the AES and the DES, are implemented using substitution boxes and linear transformations. Imagine that such a cipher is white-box implemented as a huge lookup table taking on input any plaintext and returning the corresponding ciphertext for a given key. Observe that this white-box implementation has exactly the same security as the same cipher in the black-box context: the adversary learns nothing more than pairs of matching plaintexts/ciphertexts. Typical plaintexts being 64-bit or 128-bit values, such an ideal approach cannot be implemented in practice.

Current white-box implementations apply the above basic idea to smaller components. They represent each component as a series of lookup tables and insert random input- and output bijective encodings to introduce ambiguity, so that the resulting algorithm appears as the composition of a series of lookup tables with randomized values.

To add further protection, external (key-independent) encodings may be used by replacing the encryption function E_K (respectively, decryption function $E_K{}^{-1}$) with the composition $E_K' = G \circ E_K \circ F^{-1}$ (respectively, $E_K'{}^{-1} = F \circ E_K{}^{-1} \circ G^{-1}$). Input encoding function F and output decoding function G^{-1} (respectively, G and F^{-1}) should not be made available on the platform that computes E_K' (respectively, $E_K'{}^{-1}$) so that the white-box implementation cannot be used to compute E_K (respectively, $E_K{}^{-1}$). Although the resulting implementation is not standard, such an approach is reasonable for many DRM applications.

Q6: Are there alternatives to white-box cryptography?

Tamper-resistant tokens (e.g., smart cards) also help in preventing key recovery attacks. Typically, cryptographic keys are stored in the non-volatile memory of the token and cryptographic computations take place *inside* the token. Such a token may therefore be viewed as a black-box device. Unfortunately, things are not so easy: they are susceptible to the so-called *side-channel attacks*.

The corresponding threat model is sometimes referred to as *grey-box cryptography*. The adversary has access to the inputs and outputs of the crypto-algorithm *plus* extra side-channel information. The fact that the adversary may have the device in his possession means that he can run a series of experiments and collect information like the running time, the power consumption or the electromagnetic radiation, from which he may infer the secret key. Those attacks are now well understood and efficient (hardware/software) countermeasures are available in recent smart-card implementations.

Grey-box cryptography also encompasses physical attacks (e.g., probing) and fault attacks. In the latter case, an adversary may induce faults and try to recover secret information from the faulty output. Again, there are known protections against those attacks, including sensors (hardware level) and space/time redundancy (hardware or software level).

Note that all the attacks available in the grey-box context are readily applicable — and easier to mount — in the white-box context.

Q7: What are the pros and cons?

Without considering security issues, we list below the advantages of white-box solutions compared to hardware-based solutions. Security aspects are discussed in the next section.

Advantages White-box solutions
  - are *cost-efficient*: they are easy to distribute and to install;
  - are easily *renewable*: if a security flaw is discovered, updating the software or distributing patches can be done remotely.

Disadvantages White-box solutions
  - are orders of magnitude *slower* and require more resources (e.g., memory, processing power, etc);
  - are restricted to *symmetric-key* cryptography: there are no known white-box implementations of public-key algorithms.

Q8: How secure are white-box implementations?

There is no complete system that is absolutely secure (and that will never be the case). A system is said secure *relatively* to a security model: one defines the adversary goal (e.g., recovering a key) and the resources the adversary has access to (e.g., oracle decrypting chosen ciphertexts — CCA model).

In the white-box context, it is much more difficult to define the resources of an attacker as they are virtually endless. The best we can hope is to prevent all *known* relevant threats, in an *effective* way. The security is highly dependent on the implementation: there is no need to use strong cryptographic algorithms if they are poorly implemented. Furthermore, white-box implementations are also more sensitive to known attacks; in particular, they are prone to fault attacks.

Concluding Remarks & Open Problems

Most reported attacks exploit *software* security flaws and not weaknesses in cryptographic algorithms. This implies that software protection deserves a higher consideration: *the threats related to the white-box context should be addressed carefully in the design process of secure applications.*

There are currently no fully satisfying solutions for implementing a standard block-cipher in the white-box context (all known proposals are more or less broken). Progresses should be made to provide higher resistance. In some cases, the situation is even worse; we quote two open problems in white-box cryptography:

- the *dynamic* case: when keys are frequently changing over time; and
- the *public-key* case: when public-key techniques are used (e.g., for key exchange, digital signatures, ...).

In those two cases, there are no known solutions; only token-based schemes are available for building security in the white-box context.

White-box implementations cannot be used *alone* as a protection against key recovery attacks, they should be used in conjunction with other techniques (including non-technical ones). White-box cryptography is still in its early days and requires further research before being widely adopted in commercial products. Grey-box implementations (i.e., token-based solutions) should be preferred —when relevant— as they are more mature: they have a longer history and have undertaken peer evaluation and public scrutiny. Another point against the widespread use of white-box techniques is the penalty cost they may incur: they are orders of magnitude slower and require more resources.[2] On the plus side, we remark that certain techniques of white-box cryptography could be used to improve the security of grey-box implementations against active implementation attacks (e.g., fault attacks). Finally, we remind that security comes at cost and, as a corollary, cannot (should not) be perfect. The *appropriate* security level is dictated by the application (the value of what needs to be protected), the environment (i.e., the threat model) and the costs to develop the corresponding security solution.

Acknowledgments

I am grateful to B. Preneel and B. Wyseur for sending a copy of [7]. I am also grateful to O. Billet, E. Diehl, and C. Salmon-Legagneur for comments.

References

1. J. Algesheimer, C. Cachin, J. Camenisch, and G. Karjoth. Cryptographic security for mobile code. In *2001 IEEE Symposium on Security & Privacy*, pages 2–11, IEEE Press, 2001.

---

[2] This should however be of lesser concern as technologies are evolving. We also admit that the impacts on performance are not that dramatic in certain contexts.

2. J. Bringer, H. Chabanne, and E. Dottax. White-box cryptography: Another attempt. Cryptology ePrint Archive, Report 2006/468, December 2006. Available at URL http://eprint.iacr.org/2006/468.

3. O. Billet, H. Gilbert, and C. Ech-Chatbi. Cryptanalysis of a white-box AES implementation. In *Selected Areas in Cryptography − SAC 2004*, volume 3357 of *Lecture Notes in Computer Science*, pages 227–240. Springer, 2004.

4. B. Barak, O. Goldreich, R. Impagliazzo, S. Rudich, A. Sahai, S. Vadhan, and K. Yang. On the (im)possibility of obfuscating programs. In *Advances in Cryptology − CRYPTO 2001*, volume 2139 of *Lecture Notes in Computer Science*, pages 1–18. Springer, 2001.

5. S. Chow, P. Eisen, H. Johnson, and P.C. van Oorschot. White-box cryptography and an AES implementation. In *Selected Areas in Cryptography − SAC 2002*, volume 2595 of *Lecture Notes in Computer Science*, pages 250–270. Springer, 2003.

6. S. Chow, P. Eisen, H. Johnson, and P.C. van Oorschot. A white-box DES implementation for DRM applications. In *Digital Rights Management − DRM 2002*, volume 2696 of *Lecture Notes in Computer Science*, pages 1–15. Springer, 2003.

7. J. Cappaert, B. Wyseur, and B. Preneel. Software security techniques. Internal report, COSIC, Katholieke Universiteit Leuven, October 2004.

8. Louis Goubin, Jean-Michel Masereel , and Michael Quisquater. Cryptanalysis of white box DES implementations. Cryptology ePrint Archive, Report 2007/035, February 2007. Available at URL http://eprint.iacr.org/2007/035.

9. M. Jacob, D. Boneh, and E.W. Felten. Attacking an obfuscated cipher by injecting faults. In *Digital Rights Management − DRM 2002*, volume 2696 of *Lecture Notes in Computer Science*, pages 16–31. Springer, 2003.

10. H.E. Link and W.D. Neumann. Clarifying obfuscation: Improving the security of white-box DES. In *International Conference on Information Technology: Coding and Computing − ITCC 2005*, volume 1, pages 679–684. IEEE Press, 2005. Also available as Cryptology ePrint Archive, Report 2004/025, January 2004 at URL http://eprint.iacr.org/2004/025.

11. A. Main and P.C. van Oorschot. Software protection and application security: Understanding the battleground. International Course on State of the Art and Evolution of Computer Security and Industrial Cryptography, Heverlee, Belgium, June 2003.

12. A.J. Menezes, P.C. van Oorschot, and S.A.Vanstone. Handbook of Applied Cryptography. CRC Press, 1997.

13. T. Sander and C.F. Tschudin. Towards mobile cryptography. In *1998 IEEE Symposium on Security & Privacy*, pages 215–224, IEEE Press, 1998.

14. P.C. van Oorschot. Revisiting software protection. In *Information Security − ISC 2003*, volume 2851 of *Lecture Notes in Computer Science*, pages 1–13. Springer, 2003.

15. Brecht Wyseur, Wil Michiels, Paul Gorissen, and Bart Preneel. Cryptanalysis of white-box DES implementations with arbitrary external encodings. Cryptology ePrint Archive, Report 2007/104, March 2007. Available at URL http://eprint.iacr.org/2007/104.

ECSC-128: New Stream Cipher Based on Elliptic Curve Discrete Logarithm Problem[*]

Khaled Suwais[1], Azman Samsudin[1]

[1] School of Computer Sciences, Universiti Sains Malaysia (USM)
Minden, 11800, P. Penang / Malaysia
{khaled, azman}@cs.usm.my

Abstract. ECSC-128 is a new stream cipher based on the intractability of the Elliptic Curve discrete logarithm problem. The design of ECSC-128 is divided into three important stages: Initialization Stage, Keystream Generation Stage, and the Encryption Stage. The design goal of ECSC-128 is to come up with a secure stream cipher for data encryption. ECSC-128 was designed based on some hard mathematical problems instead of using simple logical operations. In terms of performance and security, ECSC-128 was slower, but it provided high level of security against all possible cryptanalysis attacks.

Keywords: Encryption, Cryptography, Stream Cipher, Elliptic Curve, Elliptic Curve Discrete Logarithm Problem, Non-Supersingular Curve.

1 Introduction

Data security is an issue that affects everyone in the world. The use of unsecured channels for communication opens the door for more eavesdroppers and attackers to attack our information. Therefore, current efforts tend to develop new methods and protocols in order to provide secure communication over public and unsecured communication channels. In this paper we are interested in one class of the Symmetric Key cryptosystems, which is known as Stream cipher. The basic design is derived from the One-Time-Pad cipher, which XOR a sequence stream of plaintext with a keystream to generate a sequence of ciphertext. Currently, most of the stream ciphers are based on some logical and mathematical operations (e.g. XOR, shift, rotation, etc), while the others are based on the use of Linear Feedback Shift Registers (LFSR). Both types of ciphers are proven to provide high performance stream ciphers according to the results presented in [1], [2]. However, generally these ciphers are not able to reach the required highest level of security [3], [4], [5]. Various attacks managed to manifest the weaknesses of existing ciphers. Therefore, the direction of using simple logic and mathematical operations has shown the imperfection of some known stream ciphers.

[*] The authors would like to express their thanks to Universiti Sains Malaysia for supporting this study.

The idea behind our proposed stream cipher is based on the use of Elliptic Curve Discrete Logarithm Problem (ECDLP). The ECDLP along with other two intractable problems are considered the three most important pivots in current secure public-key cryptosystems. The other two problems are the Integer Factorization Problem (IFP), and the Discrete Logarithm Problem (DLP). The first use of ECDLP was in 1985 by Koblitz [6] and Victor Miller [7] independently. They proposed a new cryptosystem known as Elliptic Curve Cryptosystem (ECC), whose security depends on the discrete logarithm problem over the points on an Elliptic Curve.

The general idea of ECDLP rests on the difficulty of finding integer k given points kG and G, on Elliptic Curve E defined over a finite field with q elements (Fq), where in common, q is a large prime number or a field of characteristic two. In fact, the reason of choosing ECDLP as a core for our proposed stream cipher rests on the belief that ECDLP is harder than IFP and DLP [8]. The best known algorithm to solve the ECDLP has exponential complexity compared to the algorithm used to solve *IFP* which has sub-exponential complexity [9], [10]. In addition, the National Security Agency (NSA) announced *Suit B* at the *RSA* conference in 2005, which exclusively uses ECC for Digital Signature and Key Exchange [11]. Therefore, ECDLP with no doubt will enable our proposed stream cipher to reach the high level of security.

2   Related Work

Currently, many state-of-the-art stream ciphers seem to have appeared. The obvious relation between current ciphers enabled us to classify them into two important distinguishable categories. The first category is called the *logical and bit manipulation-based stream ciphers*. This category includes those ciphers whose construction design rests on some logical operations (Shift, XOR, etc), and uses some other substitution techniques. Examples of ciphers that belong to this category are: RC4 [12] and SEAL [2]. The second category includes all ciphers whose design rests on the use of the LFSRs. This category is labeled as *LFSR-based stream ciphers*, and it includes SNOW [13], LILI-128 [4], and other stream ciphers.

2.1   Logical and Bit-Manipulation-Based Stream Cipher

The concept of constructing a stream cipher aims to deliver a secure algorithm against all kind of attacks. In 1987, Ron Rivest of RSA Security designed RC4 [12], which later became the most widely used stream cipher. RC4, as any other stream ciphers, generates a stream of bits (keystream) which will be XOR'ed with a stream of plaintext bits to produce the ciphertext bits. Generating the keystream in RC4 requires a permutation array S of all 256 possible byte using the RC4 Key Scheduling Algorithm (*KSA*). At the present time, RC4 is not recommended for use in new application. As appears in [14], several weaknesses of the *KSA* algorithm of RC4 can be summarized in two points. The first weakness is the existence of massive classes of weak keys. These classes enabled the attackers to determine a large number of bits of *KSA* output by using a small part of the secret key. Thus, the initial outputs of the weak keys are disproportionally affected by a small portion of key bits. The second weakness rests on a related key vulnerability. Therefore, RC4 falls off the standards for secure cipher.

Another example that belongs to this category is SEAL. It was introduced by Rogaway and Coppersmith as a fast software encryption algorithm [2]. The notion of SEAL is based on a mapping of a 32-bit string n to an L-bit string under the control of a 160-bit length secret key of. In general, the length of the output, L, is limited to 64 KB. A few years later, a cryptanalysis attack presented by Handschuh and Gilbert [15] showed that an attack is capable of distinguishing SEAL from a random function by using 2^{30} computations. In addition, the attack has sufficient power to derive some bits from SEAL secret tables.

2.2 LFSR-Based Stream Ciphers

A LFSR is a hardware device made up by registers, which is capable to hold one value at a time [16]. The values are elements from a chosen field Fq (for binary field q = 2, or $q = 2^W$ for extension field of the binary field). The purpose of using LFSR was to deliver a stream cipher with high performance property. In most cases, the immediate output of LFSR is not acceptable as a keystream generator since the value production process is done in linear fashion. The linearity of the operations makes the direct output of LFSR easy to predict. Thus it requires a combination of other techniques to reach a reasonable level of security.

SNOW and LILI-128 are two examples of stream ciphers based on the used of LFSRs. The design of SNOW rests on the combination between LFSR and a Finite State Machine (FSM) with a recurrence defined over the Galois Field $GF(2^{32})$. It accepts both 128-bit and 256-bit keys. The keystream is generated by combining the values from LFSR with values in FSM in time t:

$$Z_t = \text{FSM}_{out\_t} \oplus \text{LFSR}_{out\_t} \qquad (2.1)$$

SNOW was broken by two attacks known as Guess-and-Determine (GD) attack [5]. These two attacks utilize the relationships between internal values and relationship employed to form the keystream symbols from the internal values. The GD attack determines many internal values based on its guessing of some internal values using the above relationship (Equation 2.1) with time complexity $O(2^{256})$.

LILI-128 is another example of LFSR based stream ciphers designed by Dawson et al [4]. The key length used in LILI-128 is 128 bits. It uses two binary LFSRs labeled by LFSR_c (39 bits), LFSR_d (89 bits), and two functions to generate a pseudorandom binary keystream sequence. The keystream generator includes two components: *Clock Controller* and *Data Generator*. The clock controller uses LFSR to generate an integer sequence which controls the clocking of the LFSR within the data generator component. Time-Memory-Tradeoff attack has been used to attack the design of

LILI-128. This attack required 2^{40} observed keystream bits and 2^{36} memory records to be stored in the dictionary. The time required for this attack is 2^{104}, which is considered faster than the exhaustive search. The full attack is described in [17].

Both categories of stream cipher discussed earlier in section 2.1 and 2.2 show some weakness in one form or another. Therefore, we believe that most of the stream ciphers are fast in terms of the execution time but are not fully satisfy the best level of security requirements against cryptanalysis attacks. For that reason, we are proposing a new stream cipher based on the intractability of the ECDLP to come up with a new fully secure stream cipher.

3   ECSC-128: Design and Implementation

The use of Elliptic Curves (EC) in Cryptography is promising since these curves had showed their resistance against all types of attacks. Elliptic Curve Cryptography (ECC) provides more "*security per bit*" than other types of asymmetric cryptography. The security of the proposed ECSC-128 rests on the intractability of ECDLP, which is defined as follow:

Definition:
Given an Elliptic Curve E defined over a finite field Fq $(E(Fq))$, a point $P_1 \in E(Fq)$, a point $P_2 \in E(Fq)$, find the integer k such that $P_2 = k\,P_1$.

The general equation of E is known as the *Weierstrass* form, and has the following formula:

$$y^2 + a_1xy + a_3y = x^3 + a_2x^2 + a_4x + a_6 \qquad (3.1)$$

The equation in (3.1) has two simplified versions:

$$y^2 = x^3 + ax^2 + b \qquad (3.2)$$

$$y^2 + y = x^3 + ax + b \qquad (3.3)$$

The simplified version (3.3) is called a *Supersingular* curve, and it is not suitable for Cryptography. Supersingular curves are forbidden in all standards of EC systems such as IEEE P1363 and ANSI (X9.62 and X9.63). Therefore, we avoid using these curves in our stream cipher, and we used the *non-Supersingular* curves (Equation 3.2) instead, for ECSC-128.

3.1   The Design of ECSC-128

The proposed design works as follows: an input key of 128-bit length generates a keystream. The keystream length is in 320-bit block. The keystream is divided into10-

word (sub-keystream). These sub-keystream bits are XOR'ed with the plaintext bits (one word of plaintext at a time) to generate a stream of ciphertext. The design of this cipher is divided into three main stages: Initialization Stage (IS), Keystream Generation Stage (KGS), and Encryption Stage (ES).

- Initialization Stage (IS):

At this stage, the input key ℓ (128-bit) determines the initial value of point P_1 on the curve E, and the initial value of the key k (128-bit length). The value of k is generated by:

$$k = (\ell \oplus V) << r \qquad (3.4)$$

where V is constant value, and r is variable value based on the second half of the bits from $\ell \oplus V$. The symbol ($<<$) refers to the left shift operation by r position. The generated key k will be later multiplied by point P_1 to produce new point P_2 on the curve E. The Elliptic Curve E is formed by computing function (3.5) defined over Fp (where p is a publicly known large irreducible polynomial).

$$E: y^2 = x^3 + a x^2 + b \quad \text{(where } a, b \text{ elements in } Fp) \qquad (3.5)$$

Given the curve E and the value of ℓ, the point generation on E is straightforward. The process of generating P_1 includes embedding ℓ on E with the correct (x,y) coordinates. The embedding process involves solving Equation (3.5) over an irreducible (prime) polynomial field. The solution (as implemented in [18]) requires two matrices (*Tmatrix, Smatrix*) and one vector (*Trace_Vector*). The *Smatrix* used to compute the square root of a polynomial modulo irreducible polynomial. With the assumption that the solution to Equation (3.5) exists, the *Tmatrix* is formed from the set of basis vectors which are summed together. The *Trace_Vector* is used to determine if the solution exists.

- Keystream Generation Stage (KGS):

The generated point P_1 from the previous stage (IS) is used as base point to generate a keystream K_s of 320-bit length. The idea of generating K_s is based on multiplying the key k by the point P_1 in a process known as *Point Multiplication*. Point multiplication is the idea behind the ECDLP. The rule of multiplying P_1 by k over Fq is stated as follow [19]:

Given point $P_1 = (x_1, y_1)$ and integer k, then $P_2 = kP_1 = (x_2, y_2)$ is new point on the curve E generated by multiplying P_1 by itself k times.

The point multiplication process consists of a finite set of point additions and doubling operations ($kP_1 = P_1 + P_1 + P_1 + \ldots + P_1$). Adding P_1 to itself (doubling operation) computed in equation (3.6).

$$\Theta = x + \frac{x}{y}$$

$$X_2 = \Theta^2 + \Theta + b \tag{3.6}$$

$$Y_2 = x_1 + (\Theta + 1)\, x_1$$

The addition will be performed k times, and the success of the multiplication operation means we have a new point P_2 with new coordinates (x_2, y_2) on E. The procedure of generating K_S based on the points generated from the previous stage is illustrated in **Fig 1**.

- Encryption Stage (ES):

The process of encrypting the input stream of plaintext involves some actions to control the number of times we need to generate new keystreams. The length of the keystream K_S generated from (KGS) is 320-bit. K_S is divided into ten words (32-bit for each sub-keystream). Each sub-keystream is XOR'ed with 32-bit of plaintext (Pt) to produce a stream of ciphertext (Ct). The encryption routine will check the status of the unused bits of K_S to determine if the encryption process needs more keystream bits. When the routine returns zero unused bits, new value for integer k will be generated by adding 1 to the previous values k. In turn, k will be converted into new point P_1 as discussed in the first stage (IS). Subsequently, the encryption routine will invoke the keystream generator (in KGS) to generate new 320-bit keystream based on the multiplication of the new values of P_1 and k. The generated keystream can be XOR'ed with new plaintext bits to generate new stream of ciphertext. The structure of Encryption Stage (ES) is portrayed in **Fig 2.**

Given P_1, P_2 on $E(Fp)$ and Input ℓ

- Compute $k = (\ell \oplus V) << r$
- $P_1 = $ embed (ℓ) on E
- $P_2 = k P_1$ on E
- Transform P_2 into K_s

Fig. 1. The Functionality of *IS* and *KGS* stages

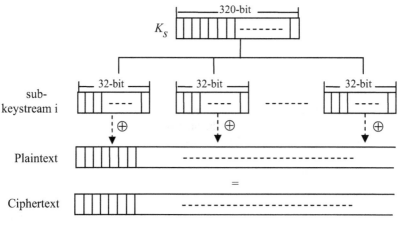

$i = 1, 2, 3 \dots 10$

$\oplus$ Refer to exclusive-or

Fig. 2. The Design of the Encryption Stage (ES).

As described above, the Initialization Stage (IS) will control the keystream generator in (KGS) by feeding it with the point P_1 and the integer k. These two parameters will be multiplied in (KGS) to create a new point P_2, which forms the new keystream K_s after transformation. The encryption process is controlled by the first two stages (IS and KGS) in terms of generating new points and later generating new keystream to perform the encryption on the given stream of plaintext. The overall structure of our proposed stream cipher is divided into three related stages as appears in **Fig 3**.

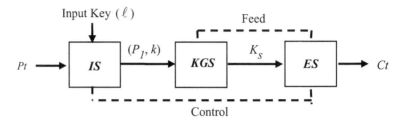

Fig. 3. Overall Design of ECSC-128.

3.2 The Implementation of ECSC-128

ECSC-128 is a word-oriented stream cipher. All operations are done on 32-bit word, which makes ECSC-128 applicable for 32-bit platforms. Unlike other stream ciphers, the key setup stage (IS) requires one XOR operation on one 128-bit variable which can be executed in less than 1 μs on Pentium 4 2.00 GHz.

ECSC-128 encrypts 27108 byte/second for a given input key (128-bit) on Pentium 4 CPU (2.00 GHz). Compared to other stream ciphers, the encryption rate of ECSC-128 is considered slow. The reason is the time that is required to calculate Θ when we perform point addition operation. The calculation of Θ requires an inversion operation over Fp, which is the most time consuming among EC operations. However, there is a high possibility of increasing the speed of ECSC-128 by improving Θ calculation which requires inversion operation. Point Multiplication on E is the core of ECSC-128, since point multiplication is an ECDL problem in EC. For that reason, ECSC-128 performance can be improved by making more efforts to enhance the operations involved in calculating the value of Θ.

4   Security Analysis

The cryptographic strength of ECSC-128 lies in the difficulty of solving the intractability of ECDLP. In this section we analyze ECSC-128 in terms of the security design properties as well as its strength against the cryptanalysis attacks.

4.1 Brute-Force Attack

ECSC-128 uses *non-Supersingular* curves instead of using supersingular curves. The reason is that supersingular curves are prohibited by all standards as mentioned previously. For a given curve E, the total number of points that satisfy equation E

(which is known as the order of the curve #E) must contain big prime number. Hasse's Theory [18] stated that the range of #E lies in the range:

$$2^n - 2\sqrt{2^n} + 1 \leq \#E \leq 2^n + 2\sqrt{2^n} + 1 \qquad (4.1)$$

where n is the number of bits in the field. For ECSC-128, n=160, and therefore the range is between:

$1.4615016373309029182036836237902 \times 10^{48}$

and

$1.4615016373309029182036860416418 \times 10^{48}$

The quantity of numbers in the range of the above equation is $2^{68} = 295147905179352825856 \times 10^{20}$. Therefore, brute force search for the largest prime within this range is infeasible, since matching such range to a particular curve is non-trivial. From another perspective, the length of the input key for ECSC-128 is 128-bit, which is secure against brute force search as well. The length of the key used in ECSC-128 was chosen to achieve the expected security against brute force search attack and to satisfy the required balance of the difficulty between the forward and the inverse operations. The forward operation of ECSC-128 is the point multiplication $(P_2 = kP_1)$ on E, which is fast enough compared to its inverse operation (retrieve k given P_1 and P_2). In general, the difficulty of the inverse operation against the key length must be exponential for secure cryptographic primitives as portrayed in **Fig 4** [20].

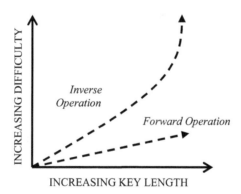

Fig. 4. Difficulty of forward, inverse operation against key length

According to the intractability of ECDLP, applying brute force is infeasible. Logically, finding k from $P_2 = kP_1$ means performing repeated addition operation starting from P_1, $2P_1$, until we find kP_1. Then, we double P_1 then we add P_1 to the doubled point $2P_1$ to find $3P_1$, then $3P_1$ to P_1 finding $4P_1$ and so on. This attack is avoided in ECSC-128 by using a large prime field. This means that; the number of possible values of k is large. Searching through the bearable values of k will take hundreds of years using current available processors.

4.2 Baby-Step-Giant-Step Attack

The baby-step-giant-step algorithm is used to solve the discrete logarithm problem in arbitrary groups [21]. Let n be the order of a point P (order(P): is the number of times we can add the point to itself until we find the point at infinity) on E. This attack requires a memory for $\sqrt{n}$ points and time complexity of $O\left(\sqrt{n}\right)$ [22], which is infeasible due to the use of a large prime number n in ECSC-128.

4.3 Pollard's rho Attack

Pollard's rho is a collision-based algorithm which was first proposed by J. Pollard in 1978 [23]. It relies on finding collisions between elements of a cyclic group. Thus, Pollard's algorithm can be applied to a group of points on E. The idea of Pollard's attack is to reveal the solution to ECDLP by computing the trail of points on E. This computation is known as a *walk*, and it is achieved by generating a sequence of points using an iterating function. The success ratio of this attack is related to calculating the square root of the size of the cyclic group. In fact, according to the time complexity of Pollard's algorithm $O\left(\sqrt{\#E}\right)$, ECSC-128 is secured against this attack. However, ECSC-128 uses a large enough value of $\#E$ (see section 3), which make Pollards rho attack infeasible.

5 Conclusion

In this paper we present a new stream cipher (ECSC-128) based on the intractability of the Elliptic Curve discrete logarithm problem (ECDLP). ECSC-128 is divided into three main stages: Initialization Stage (IS), Keystream Generation Stage (KGS), and the Encryption Stage (ES). Complete descriptions of ECSC-128 stages, implementation, and an overview of possible attacks have been discussed. ECSC-128 has changed the current direction of stream ciphers from using simple logic and mathematical operations into a new direction of using complex mathematical problems that are computationally infeasible.

References

1. M. Boesgaard, M. Vesterager, T. Pedersen, J. Christiansen, and O. Scavenius: Rabbit: A New High-Performance Stream Cipher, In Proc. Fast Software Encryption 2003, volume 2887 of LNCS. Springer, 2003.
2. P. Rogaway, D. Coppersmith: A Software-Optimized Encryption Algorithm, Journal of Cryptology, vol. 11, num, 4, pp. 273-287, 1998.
3. Scott R. Fluhrer, I. Mantin and A. Shamir: Weaknesses in the Key Scheduling Algorithm of RC4. Selected Areas in Cryptography. pp1 – 24, 2001.
4. E. Dawson, A. Clark, J. Golic, W. Millan, L. Penna, L. Simpson: The LILI-128 Keystream Generator, NESSIE submission, In Proc. of the first Open NESSIE Workshop 2000.
5. Christophe De Canniere: Guess and Determine Attack on SNOW. Nessie public report. NES/DOC/KUL/WP5/011/a. www.cryptonessie.org. 2001
6. N. Koblitz: Elliptic curve cryptosystems, in Mathematics of Computation, Vol. 48, pp. 203–209. 1987.
7. V. Miller, Use of elliptic curves in cryptography, CRYPTO 85, 1985.
8. Advanced Technology Center: Elliptic Curve Cryptography. TATA Public Report. http://www.atc.tcs.co.in/esecurity/research.html. 2006.
9. N. Jha and I. Sengupta: A generalized scheme for multiparty secure access using elliptic curve cryptography, In Proc. of the 6th Intl. Conference on Information Technology, pp. 119-124. Bhubaneswar, India. December 2003.
10. Zhu, H: Survey of Computational Assumptions Used in Cryptography Broken or Not by Shor's Algorithm. School of Computer Science. Montreal, McGill University. Master in Science. 2001
11. NSA: NSA Suite B Cryptography, http://www.nsa.gov/ia/industry/crypto_suite_b.cfm. 2005.
12. R. Rivest: The RC4 Encryption Algorithm. RSA Data Security Inc. Document No. 003-013005-100-000000. Mar. 1992.
13. P. Ekdahl, and T. Johansson: SNOW-A New Stream Cipher, In Proc. of First NESSIE Workshop, Heverlee, Belgium. 2000.
14. S. Fluhrer, I. Mantin, and A. Shamir: Weaknesses in the key scheduling algorith of RC4. In Proc. 8th workshop on selected Areas in cryptography, LNCS 2259. Springer-Verlag. 2001.
15. H. Handschuh and H. Gilbert: X^2 cryptanalysis of the SEAL encryption algorithm. In Proc. of the 4th Workshop on Fast Software Encryption, volume 1267 of Lecture Notes in Computer Science, pages 1-12. Springer-Verlag, 1997.
16. P. Ekdahl: On LFSR based Stream Ciphers Analysis and Design. Information Technology. Lund, Lund University. Ph.D. 2003.
17. S. Babbage: Cryptanalysis of LILI-128. NESSIE Public Report, www.cosic.esat.kuleuven.ac.be/nessie/reports. 2001.
18. M. Rosing: Implementing elliptic curve cryptography. Manning Publications Co., Greenwich, CT, USA. 1999.
19. A. J. Menezes: Elliptic Curve Public Key Cryptosystems. Boston: Kluwer Academic Publishers. 1993.
20. Certicom: An Elliptic Curve Cryptography (ECC) Primer. Catch the Curve Series. http://www.certicom.com/download/aid-317/WP-ECCprimer.pdf. 2005
21. D. Shanks: Class number, a theory of factorization and genera. In Proc. Symp. Pure Math. 20, pages 415--440. AMS, Providence, R.I. 1971.
22 T.E. Gueneysu, C. Paar, J. Pelzl: On the security of Elliptic Curve Cryptosystems against Attacks with Special-Purpose Hardware, 2nd Workshop on special-purpose Hardware for Attacking Cryptographic Systems - SHARCS 2006, April 3-4. 2006
23. J. M. Pollard: Monte Carlo methods for index computation mod p. mathematics of computation, 32(143):918-924. July 1978.

Hardware Implementation of Elliptic Curve Cryptosystem over $GF(p^m)$

Ilker Yavuz[1,2], Berna Ors[2],

[1]TUBITAK National Research Instute of Electronics and Cryptography, Gebze, Kocaeli, Turkey

[2]Istanbul Technical University, Faculty of Electrical & Electronic Eng., Istanbul, Turkey

Siddika.Ors@ehb.itu.edu.tr, ilkery@uekae.tubitak.gov.tr

Abstract. This paper describes an efficient FPGA implementation for modular multiplication operation in the field $GF\left(p^m\right)$ that is suitable for implementing Elliptic Curve Cryptosystems. We have developed a systolic array implementation of a Montgomery modular multiplication (MMM). This implementation is efficient for large finite fields ($m = 160 - 193$ bit) that offer a high security level. Also our implementation is independent of reduction polynomial because of MMM technique. We have specialized to fields $GF\left(3^m\right)$ and provide implementation results. We choose $m = 97$ to compare our results with previous elliptic curve implementations that use different multiplication techniques. In contrast to earlier works, Montgomery multiplication technique has never been used in the field $GF\left(p^m\right)$.

Keywords: Elliptic Curve, Montgomery, FPGA, Systolic Array

1 Introduction

In 1976, Diffie and Hellman introduced the idea of public key cryptography [1]. They showed that private communication is possible even when one only has an authenticated channel. Moreover, they have introduced the concept of a digital signature, which allows to uniquely bind a message to its sender. Since then, numerous public-key cryptosystems (PKCs) have been proposed and all these systems based their security on the difficulty of some mathematical problem. Elliptic Curve Cryptosystems (ECC), which were proposed by Koblitz [2] and Miller [3], are examples of PKCs. The basic operation for ECC is a point multiplication which relies on an efficient finite field multiplication. Commonly used finite fields for ECC are $GF(p)$ and $GF\left(2^m\right)$. Recently, there has been interest on applications based on fields $GF\left(p^m\right)$. As a consequence, a substantial amount of research is focused on efficient and secure implementation of modular multiplication in hardware.

In 1985 Montgomery has introduced a new method for modular multiplication [4]. The approach of Montgomery avoids the time consuming trial division that is a common bottleneck of other algorithms. His method has been proved to be very efficient and is the basis of many implementations of modular multiplication, both in software and hardware [5,6,7,8,9,10].

In 1998 Koç and Acar introduced a method to use Montgomery multiplication over $GF(2^m)$ [11]. They showed that the multiplication operation in the field $GF(2^m)$ can be implemented significantly faster than the standard multiplication.

In 2003 Bertoni, Guajardo, Kumar, Orlando, Paar and Wollinger introduced an efficient $GF(p^m)$ arithmetic for cryptographic applications [12]. They specialized their application to fields $GF(3^{97})$ but they did not use MMM structure. Their application consists of a special multiplier and modulo reduction operation.

In this paper we look at an efficient hardware implementation of a MMM over $GF(p^m)$ in a Field Programmable Gate Array (FPGA). We then specialize to fields $GF(3^m)$ and choose $m = 97$. Our contribution consists of an FPGA implementation of the MMM in a systolic array, which allows pipelining and makes the clock frequency of the design independent of the field size m.

There are MMM implementations for $GF(p)$ and $GF(2^m)$ [13, 14] but there are no practical ASIC or FPGA implementation results on previously designed systolic array architectures for Montgomery multiplication over $GF(p^m)$.

2 Montgomery Modular Multiplication over $GF(p^m)$

The Montgomery multiplication of a and b is defined as follows:

$$c = abr(x)^{-1} \bmod N \qquad (1)$$

Where $r(x) = x^m \bmod N$. In this equation, if field is chosen $GF(p^m)$; $a = (a_{l-1}a_{l-2}\cdots a_1a_0)_p$, $b = (b_{l-1}b_{l-2}\cdots b_1b_0)_p$.

This Montgomery representation allows very efficient modular arithmetic especially for multiplication. Montgomery's method for multiplying two numbers that is defined in modulo p domain avoids division which is the most expensive operation in hardware. The method requires some conversions from normal to Montgomery representation and back conversion. The procedure is as follows; to compute $c = ab \bmod N$, first of all, Montgomery representation of a' must be calculated. These presentations are obtained by computing the Montgomery multiplication of a and r^2 that is $a' = \text{Mont}(a, r^2) = ar \bmod N$. When this Montgomery representation of a and normal representation of b are used as inputs to Montgomery multiplication, the desired value is obtained. That is, $c = \text{Mont}(ar, b) = arbr^{-1} \bmod N = ab \bmod N$.

In the original notation of Montgomery after each multiplication a reduction is needed (Step 7 in Algorithm 1). The input had the restriction $a, b < N$ and the output c was bounded by $c < 2N$. As a consequence, if $c > N$, N must be subtracted so that the output can be used as input to the next multiplication. To avoid this subtraction a bound for r is known [17] such that for the inputs $a, b < 2N$ the output is also bounded by $c < 2N$.

In [17], the need of avoiding reduction after each multiplication is addressed. In practice this means that the output of multiplication can be directly used as an input to the next Montgomery multiplication. We want to find a bound on r such that with $a, b < 2N$ and the output of the Montgomery multiplication $c < 2N$. In [16, 17], Walter explain that $r = b^{l+2}$. Algorithm 2 gives the MMM without final subtraction.

Algorithm 1: Montgomery Modular Multiplication with final subtraction
Require: $N = (n_{l-1}n_{l-2\ldots}n_1n_0)_b$, $x = (x_{l-1}x_{l-2\ldots}x_1x_0)_b$, $y = (y_{l-1}y_{l-2\ldots}y_1y_0)_b$ with x and $y\varepsilon[0, N-1]$, $r = b^l \bmod N$ with $\gcd(N,b) = 1$ and $N' = -N^{-1} \bmod b$.
Ensure : $x.y.r^{-1} \bmod N$

1:      $T \leftarrow 0$
2:      for i from 0 to $l-1$ do
3:          $u_i \leftarrow (t_0 + x_i.y_0)N' \bmod b$
4:          $T \leftarrow (T + x_i.y_0 + u_i.N)/b$
5:      end for
6:      if $T \geq m$ then
7:          $T \leftarrow T - m$
8:      end if
9:      Return (T)

Algorithm 2: Montgomery Modular Multiplication without final subtraction
Require: $N = (n_{l-1}n_{l-2\ldots}n_1n_0)_b$, $x = (x_{l-1}x_{l-2\ldots}x_1x_0)_b$, $y = (y_{l-1}y_{l-2\ldots}y_1y_0)_b$ with x and $y\varepsilon[0, N-1]$, $R = b^{l+2}$ with $\gcd(N,b) = 1$ and $N' = -N^{-1} \bmod b$
Ensure : $x.y.r^{-1} \bmod 2.N$

1:      $T \leftarrow 0$
2:      for i from 0 to $l+1$ do
3:          $u_i \leftarrow (t_0 + x_i.y_0)N' \bmod b$

4:      $T \leftarrow \left(T + x_i.y_0 + u_i.N\right)/b$

5:      end for

6:      Return (T)

3 Hardware Implementation

3.1 Design Overview

Montgomery Modular Multiplication circuit (MMMC) over $GF\left(p^m\right)$ can be divided hierarchically into 3 levels [14]:

1. Systolic Array Cell: Computes 1 step (a couple of bits that defines a value in modulo 3) of T in Step 4 of Algorithm 2.
2. Systolic Array: Computes one iteration of T in Step 2 of Algorithm 2.
3. Montgomery Modular Multiplication Circuit (MMMC): Computes complete Algorithm 2.

3.2   Systolic Array Cells

The ith iteration of Step 2 in Algorithm 2 computes the temporary results T_i as it is shown in equation (2)

$$T_i = \left(T_i + x_i.y + u_i.N\right)/b \tag{2}$$

Where $i = 0,1,...,l+1$ and $T_{-1} = 0$ [15]. The jth value of T_i is obtained using the recurrence relation.

$$t_{i,j} = \left(t_{i-1,j} + x_i y + u_i N\right) \tag{3}$$

In equation (3), all values are in modulo 3 domain and also all operations are made in modulo 3 domain. Therefore, all values can be defined using 2-bit length in binary representation.

The rightmost cell just computes the intermediate value U_i. The equation of U_i is as follows;

$$U_i = \left(t_0 + x_i y_0\right)N' \bmod 3 \tag{4}$$

N' is related to input. A circuit can be used to the calculate N' or it must be calculated and given to the circuit as a parameter. In our implementation, we choose p= 3.Using this equation $N' = -N^{-1} \bmod b$, $N' = 2$. It is obvious that, multiplying N' is shift left operation.

According to the equation of the U_i, top level design of the rightmost cell is as seen in Fig.1. (a). It can be seen in Fig.1 (a), there are two multipliers, 1adder and 3 modulo blocks.

Each regular cell computes a step (a couple of bits) of T_i. The equation of T_i is as follows;

$$t_{i,j} = \left(t_{i-1,j} + x_i y_i + u_i n_i\right) \bmod 3 \tag{5}$$

According to equation of the T_i, top level design of the regular cell is as shown in Fig.1. (b). As shown in Fig.1. (b), there are two multipliers, 1adder and 3 modulo 3 blocks.

The leftmost cell is different from regular cells. Because the leftmost cell does not use m as an input this value is equal to zero. So, the multiplication $u_i n_i$ equals to zero. Therefore, the leftmost cell does not consist of $u_i n_i$ multiplier. Also leftmost cell does not use $t_{i,j}$ as an input. As a result, it just consists of one multiplier, one adder and two mod 3 blocks. Fig. 1. (c)

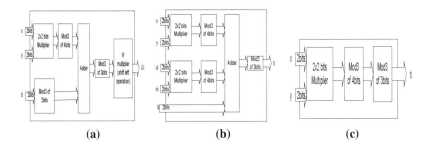

Fig. 1 Schematic view of systolic array cells; (a) Rightmost, (b) Regular and (c) Leftmost

3.3   Systolic Array

To obtain a linear pipelined modular multiplier, all these cells are arranged as shown in Fig.2.

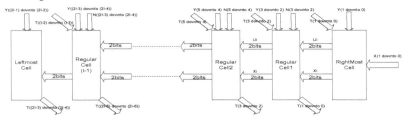

Fig. 2 Schematic view of complete systolic array.

The jth cell behaves like cell (i, j) computing $t_{i,j}$ at time $2i + j$ for $i = 0, 1, ..., l + 1$

X register stores the constant value of X and $X(0)$ denotes the least significant 2 bits of the X register. T denotes the intermediate value register. Y and N registers store the constant values of Y and N. All these registers are $2l$-bit long (l is number of steps) because every step is 2-bit width.

Because all arithmetic operations are done in $\mod p$ domain in $GF\left(p^m\right)$, none of these cells generate carry.

3.4 Systolic Array

The MMMC has three $2l$-bit data inputs X, Y and N, one START instruction input to define the beginning of process, one DONE output to define the end of process and $2l$-bit RESULT output. The circuit works $2i + j$ clock cycles to obtain result where "i" is the iteration number and j is the number of cells.

The algorithmic state machine (ASM) approach is used to design MMMC. For reference of ASM [18] is a good reference. The circuit has a top level block to control all data flow between sub blocks.

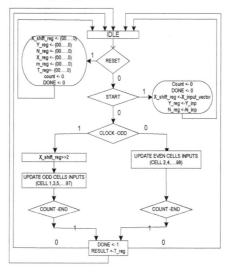

Fig. 3 Algorithmic State machine of Montgomery modular multiplier.

Fig. shows the algorithmic state machine chart of the MMMC. The circuit waits in IDLE state. If the RESET signal arrives, all internal and external registers are reset. If the START signal arrives, all input data is loaded to the internal registers. That is, X_input_vector is loaded to X_shift_reg, Y_inp is loaded to the Y_reg and N_inp is loaded to the N_reg registers. After START signal trigged the circuit, it state to the low state and the circuit becomes ready to process all inputs. The X_shift_reg is a 2 bit right shift register and shifted in only odd clocks that is, it is shifted 2 bits per 2 clocks. Using this input registers, all cells in systolic array are updated regularly. Odd cells inputs $(1,3,5...)$ are updated in odd clocks and even cells inputs are updated in even clocks $(2,4,6...)$ This process goes on until the last step (last 2 bits) of X_shift_reg are evaluated by the last cell of systolic array. When all X_shift_reg values are evaluated by all systolic array cells, T_reg value is assigned to RESULT register.

4   Obtaining Elliptic Curve Processor Using Multiplication Circuit

The core of an elliptic curve processor is the multiplication circuit. All other sub-blocks are obtained by calling multiplication block with different inputs with different times. The block diagram of the elliptic curve point multiplier circuit and EC Processor are given in Fig. 4.

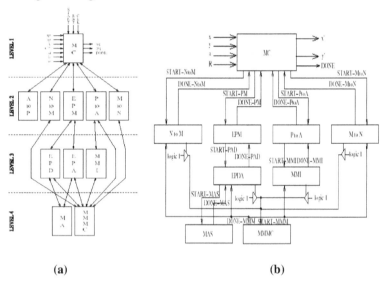

(a)   **(b)**

Fig. 4 Block diagram of (a) EC point multiplier circuit and (b) EC Processor

The operation blocks at each level from top to bottom are as follows:
- Level 1: Main Controller (MC) : A state machine with 5 state
- Level 2:
    1. Affine to projective coordinates converter (AtoP)
    $$(x, y)\longrightarrow(X,Y,Z,aZ^4)\ \text{such that}\ X =x, Y =y, Z =1, aZ^4 =a$$
    2. Normal to Montgomery representation converter (NtoM)
    $$Mont(x, R^2) =xR^2R^{-1} \bmod M =xR \bmod M$$
    3. EC point multiplier (ECPM) : Use double-and-add algorithm [2]
    4. Projective to affinene coordinates converter (PtoA)

$(X, Y, Z, aZ^4) \rightarrow (x, y)$ such that $x = XZ^{-2}$ and $y = YZ^{-3}$

    4. Montgomery to normal representation converter (MtoN)

$x = Mont(xR, 1) = xRR^{-1}, y = Mont(yR, 1) = yRR^{-1}$

- Level 3:
  - 1. EC point doubling circuit (ECPDC) : Use double-and-add algorithm [2]
  - 2. EC point addition circuit (ECPAC) : Use double-and-add algorithm [2]
  - 3. Modular multiplicative inverter (MMI) : Fermat's Theorem [19]
- Level 4:
  - 1. Modular addition (MA)
  - 2. Montgomery modular multiplication circuit (MMMC)

As it is given above, all these sub-blocks consist of multiplication operation and use MMMC. The only operation which doesn't call MMM is MA and it is a simple operation. Usage of sub-blocks and configuration of MC are given in Fig. 4.(b).

5 Implementation Results

We present an implementation of Montgomery Modular Multiplication circuit over $GF(p^m)$. The most important difference of our implementation is using a different finite field that is $GF(p^m)$ and using MMM technique. Also our work is a hardware implementation of MMM Circuit over $GF(p^m)$. There is not enough research about $GF(p^m)$. Some researches for $GF(p^m)$ are software implementation. For instance if we compare our design with [12], in [12], a hardware implementation of $GF(p^m)$ is introduced but it consists of a multiplication block and reduction block separately. A modular reduction block is very expensive in hardware implementation.

    Because there is no previous hardware implementation which uses MMM in $GF(p^m)$ and all other designs use different platforms or software implementation, it is hard to compare the area and performance.

The implementation results of the MMMC on a Xilinx Virtex1000E FPGA are as follows. M, number of clock Cycle, number of slices, number of External IOBs, maximum clock Delay(ns), minimum clock period T_p and total MMM latency for one MMM (T_{MMM}) are given in Table 1.

Table 1. Implementation Results of MMM Circuit.

Parameters	Implementation Results
m	97
# of Clock Cycle ($2i + j$)	296
Number of Slices	1902 out of 12288 (%15 of Virtex 1000E)

Number of External IOBs	18 out of 158 (%11 of Virtex 1000E)
Max.Clock Delay(ns)	1.016
T_p minimum Clock Period (ns)	10.262
Total MMM Latency(μs)	3.038

Our Montgomery implementation is efficient and secure as it is proven. Namely, the optimal bound is achieved which, with some saving in hardware, omits completely all reduction steps that are presumed to be vulnerable to side-channel attacks.

This implementation can be used for Elliptic Curve Cryptosystem (ECC). For ECC, the multiplication circuit is core of the system. Also any reduction block is used to obtain modulo of multiplication results. The advantage of MMM is that if MMM circuit is used for point multiplication, there is no need to use any separate reduction circuit which is very expensive in hardware implementation. MMM circuit consists of both these operations. Also, MMM circuit performance is independent of reduction polynomial

In addition, our implementation is reusable and can be extended in $GF\left(3^m\right)$.

Repeating regular cells extends the degree m . This extension does not affect clock frequency. The MMM circuit is independent of the clock frequency because of systolic array structure. This property gives our circuit the advantage of suitability to various applications different bit lengths like ECC 163 bits, ECC 193 bits, ECC 233 bits.

6 Future Works

With this implementation, we show that Montgomery modular multiplication over $GF\left(3^m\right)$ can be implemented in hardware. Using this implementation and some additional sub blocks such as modular addition/ subtraction and a main controller to control MMM circuit, an elliptic curve cryptosystem can be achieved easily. With the usage of MMM circuit flexibility, a clock frequency independent ECC can be easily extended to larger bits to obtain more security.

References

1.   Diffie, W., Hellman, M., E.: New directions in cryptography. IEEE Transactions on Information Theory, Vol. 22 (1976), pp. 644-654.
2.   Koblitz, N.: Elliptic curve cryptosystem. Math. Comp., Vol. 48 (1987) pp. 203-209.
3.   Miller, V.: Uses of elliptic curves in cryptography. In Advances in Cryptology: Proceedings of CRYPTO'85, H. C. Williams, Ed. (1985) number 218 in Lecture Notes in Computer Science, pp. 417-426, Springer-Verlag.
4.   Montgomery, P.: Modular multiplication without trial division. Mathematics of Computation, Vol. 44 (1985) pp. 519-521.

5.   Ors, S.B., Batina, L., Preneel B., Vandewalle J.: Hardware Implementation of a Montgomery Multiplier in a Systolic Array, Proceedings of the 10th Reconfigurable Architectures Workshop (RAW), Nice, France (2003)

6.   Batina, L., Muurling, G.: Montgomery in practice: How to do it more efficiently in hardware. In: Proceedings of RSA 2002 Cryptographers' Track, B. Preneel, Ed., San Jose, USA, February 18-22 (2002) number 2271 in Lecture Notes in Computer Science, pp. 40-52, Springer-Verlag.

7.   Trichina, E., Tiountchik, A.: Scalable algorithm for Montgomery multiplication and its implementation on the coarse-grain reconfigurable chip. In: Proceedings of Topics in Cryptology – CT-RSA 2001, D. Naccache, Ed. (2001) number 2020 in Lecture Notes in Computer Science, pp. 235-249, Springer-Verlag.

8.   Freking, W., L., Parhi, K., K.: Performance-scalable array architectures for modular multiplication. In: Proceedings of the IEEE International Conference on Application-Specific Systems, Architectures and Processors (2002) pp. 149-160, IEEE.

9.   Tsai, W., -C., Shung, C., B., Wang, S., -J.: Two systolic architectures for modular multiplication. IEEE Transactions on Very Large Scale Integration (VLSI) Systems, Vol. 8, number 1, pp. 103-107, February (2000)

10.   Eldridge, S., E., Walter, C., D.: Hardware Implementation of Montgomery's Modular Multiplication Algorithm. IEEE Transactions on Computers, Vol. 24, number 6, pp. 693-699, June (1993).

11.   Koç, C., K., Acar, T.: Montgomery multiplication in $GF(2^k)$. Designs, Codes and Cryptography, Vol. 14, pp. 57-69 (1998)

12.   Bertoni, G., Guajardo, J., Kumar, S., Orlando, G., Christof, P. and Wollinger, T.: Efficient $GF(p^m)$ Arithmetic Architecture for Cryptographic Applications, In Marc Joye (Ed.): The Cryptographers' Track at the RSA Conference - CT-RSA 2003, volume LNCS 2612, pp. 158-175, San Francisco, CA, USA, April 13-17 (2003)

13.   Ors, S.B., Batina, L., Preneel B., Vandewalle J.: Hardware Implementation of an Elliptic Curve Processor over $GF(p)$, The Proceedings of the IEEE 14th International Conference on Application-specific Systems, Architectures and Processors (ASAP), pp. 433-443, The Hague, The Netherlands (2003)

14.   Mentens, N., Ors, S. B., Preneel , B., and Vandewalle, J.: An FPGA Implementation of a Montgomery multiplier over $GF(2^k)$, The Proceedings of the 7th IEEE Workshop on Design & Diagnostics of Electronic Circuits & Systems (DDECS), pp. 121-128 (2004)

15.   Walter, C. D.: Montgomery's multiplication technique: How to make it smaller and faster. In Ç. K. Koç, and C. Paar, editors, Proceedings of Cryptographic Hardware and Embedded Systems (CHES1999), number 1717 in Lecture Notes in Computer Science, Vol. 1717. Springer-Verlag, Berlin Heidelberg New York (1999)

16.   Walter, C.D.: Precise bounds for Montgomery modular multiplication and some potentially insecure RSA moduli. In B. Preneel, editor, Proceedings of Topics in Cryptology- CTRSA 2002, number 2271 in Lecture Notes in Computer Science, pages 30–39(2002)

17.   Walter, C.D.: Montgomery exponentiation needs no final subtraction. *Electronic letters*, 35(21):1831–1832, October (1999)

18.   Mano, M. M. and Kime, C. R.: Logic and Computer Design Fundamentals., second edition, Prentice Hall, Upper Saddle River, New Jersey 07458 (2001)

19.   A. Menezes, P.van Oorschot, and S. VanStone. Handbook of Applied Cryptography. CRC Press, 1997

Secure Hill Cipher Modification SHC-M

Eastern Mediterranean University, North Cyprus,
Taganrog Institute of Technology, Federal University of South Russia, Russia

Abstract. The secure Hill cipher strength is improved by avoiding a
permutation transfer via network channels; the number of the permutation in the
pseudo-random sequence is transferred instead. Pseudo-random permutation
generator is used by a receiver to restore the permutation.

Key words: Hill cipher, permutation, pseudo-random permutation generator

1 Introduction

Hill cipher (HC) is a well known computationally simple cipher; it is easily broken in
the plaintext-cipher-text attack [1]. Secure Hill cipher (SHC) was proposed in [2]. As
in the original Hill cipher, encryption is performed by multiplication of a plaintext, P,
with a key matrix, K_t, but in SHC [3], the key matrix changes by a sender, S, for
each new plaintext. A receiver, R, gets from S a permutation, t, which is generated by
the latter and used to calculate K_t from the shared by the both parties key matrix, K.
The permutation is send encrypted by HC with the key matrix K. It is the reason for
SHC weakness that was specified in [4]. Here, we propose a way of SHC strength
enhancing which is based on elimination of encrypted permutation transfers. In
Section 2, we give more details on SHC. Proposed algorithm, SHC-M, is described in
Section 3. Section 4 contains conclusion.

2 Secure Hill cipher SHC

SHC works as follows [3].
Initial settings. All matrices considered below are $m \times m$ sized over Z_m, $m > 0$
(e.g., $m = 26$ for English language texts). A sender, S, and a receiver, R, share an
invertible key matrix K. A plaintext, P, and a cipher-text, C, are also matrices.
Encryption. The sender S chooses a permutation, t, over Z_m, and, using it, builds a
permutation matrix, M_t. For example, if $t = (3,0,2,1)$, then

$$M_t = \begin{vmatrix} 0 & 0 & 0 & 1 \\ 1 & 0 & 0 & 0 \\ 0 & 0 & 1 & 0 \\ 0 & 1 & 0 & 0 \end{vmatrix},$$

it has just one 1 in each row and column, and placement of ones in rows is dictated by the permutation t. Multiplication of M_t with any matrix does not change the latter excepting the order of the rows that may be illustrated by the following:

$$M_t \cdot P = \begin{vmatrix} 0 & 0 & 0 & 1 \\ 1 & 0 & 0 & 0 \\ 0 & 0 & 1 & 0 \\ 0 & 1 & 0 & 0 \end{vmatrix} \cdot \begin{vmatrix} 0 & 1 & 2 & 3 \\ 4 & 5 & 6 & 7 \\ 8 & 9 & 10 & 11 \\ 12 & 13 & 14 & 15 \end{vmatrix} = \begin{vmatrix} 12 & 13 & 14 & 15 \\ 0 & 1 & 2 & 3 \\ 8 & 9 & 10 & 11 \\ 4 & 5 & 6 & 7 \end{vmatrix}.$$

We see that after such a multiplication, rows of the matrix P are permuted according to t. Inverse of M_t, M_t^{-1}, exists because determinant of M_t is not equal to zero. For our example,

$$M_t^{-1} = \begin{vmatrix} 0 & 1 & 0 & 0 \\ 0 & 0 & 0 & 1 \\ 0 & 0 & 1 & 0 \\ 1 & 0 & 0 & 0 \end{vmatrix},$$

and, in general case, inverse is obtained just by matrix transposition, or directly from the permutation t (it dictates placements of ones, now in columns). It is easy to see that right multiplication by M_t^{-1} leads to permutation of columns in the multiplied matrix:

$$P \cdot M_t^{-1} = \begin{vmatrix} 0 & 1 & 2 & 3 \\ 4 & 5 & 6 & 7 \\ 8 & 9 & 10 & 11 \\ 12 & 13 & 14 & 15 \end{vmatrix} \cdot \begin{vmatrix} 0 & 1 & 0 & 0 \\ 0 & 0 & 0 & 1 \\ 0 & 0 & 1 & 0 \\ 1 & 0 & 0 & 0 \end{vmatrix} = \begin{vmatrix} 3 & 0 & 2 & 1 \\ 7 & 4 & 6 & 5 \\ 11 & 8 & 10 & 9 \\ 15 & 12 & 14 & 13 \end{vmatrix}.$$

The matrix K_t is defined as follows:

$$K_t = M_t \cdot K \cdot M_t^{-1}, \tag{1}$$

i.e. currently used for encryption key matrix is obtained from the master key matrix K by permutations of its rows and columns, and, hence, it is also invertible. Cipher-text, C, is obtained as follows:

$$C = K_t \cdot P. \tag{2}$$

Sender S calculates also a vector

$$u = K \cdot t,$$

and sends it to R together with the cipher-text:

$$S\text{->}R:\ C,\ u.$$

Decryption. Receiver, R, calculates:

$$t = K^{-1} \cdot u . \tag{3}$$

Using calculated in (3) permutation t, R builds matrices M_t, M_t^{-1}, and uses them to calculate K_t according to (1). Having K_t, R calculates its inverse, and decrypts obtained cipher-text as follows:

$$P = K_t^{-1} \cdot C . \tag{4}$$

3   Modification of the secure Hill cipher SHC-M

The weakness of SHC is in the transfer of the vector u via network channels. We avoid it in the proposed SHC-M. We assume the same initial settings and the same encryption-decryption as in SHC. Additionally, we assume that the parties share a secret value, *SEED*, that is used to generate pseudo-randomly a sequence of permutations. Such sequence of permutations can be obtained, for example, in the frame of algorithm RC4 that uses pseudo-random permutations to generate pseudo-random numbers [1]. Let's denote such generator returning n-th permutation in the sequence as *PRPG(SEED, n)*. Then, the sender, S, specifies the number of the permutation in such a sequence and sends it to the receiver, R, instead of permutation itself as it is made in SHC. If an opponent knows the permutation number, he can't get permutation itself because no secret information about the permutation is transferred. The receiver, R, knowing *SEED* and the permutation number, generates it and uses it to build the permutation matrices, the current key matrix, and to decrypt the cipher-text. Hence, SHC-M is as follows.

Initial settings. Shared data: K, *SEED*

Encryption. Sender, S, selects a number, n, and calculates

$$t = PRPG(SEED, n) . \tag{5}$$

Then S calculates K_t according to (1), gets a cipher-text, C, according to (2), and sends it to R together with n:

$$S\text{->}R:\ C,\ n.$$

Decryption. Receiver, R, calculates t according to (5), calculates K_t according to (1), and gets the plaintext, P, according to (4).

4   Conclusion

Thus far, we proposed a modification to SHC which gives more strength to it due to the avoidance of secret information transfer via network channels. In our approach, only the number of a permutation is transferred (together with a cipher-text), and it is restored by a receiver by feeding the permutation number and shared with a sender secret seed value into some pseudo-random permutation generator, for example, the one used in RC4. All the rest in SHC-M is the same as in SHC.

References

1. Stallings, W.: Cryptography and Network Security. 4th edn, Prentice Hall, Upper Saddle River (2006)
2. Saeednia, S.: How to Make the Hill Cipher Secure. Cryptologia, Vol. 24. 4 (2000) 353-360
3. Overbey,   J.,   Wojdylo,   J.:   The   Secure   Hill   Cipher. http://jeff.over.bz/presentations/undergrad/ma464-secure-hill-cipher.ppt, last access date 28.02.2007
4. http://cat.inist.fr/?aModele=afficheN&cpsidt=16282094, last access date 28.02.2007

Keyed Blind Multiresolution Watermarking Algorithm For Digital Images.

Mahmoud Hassan [1)]Sarah Alkuhlani [2], Wasan Talhouni [1] and Laith Smadi [1]
[1] Electronics Engineering Department, Princess Sumaya University, Amman ,
Jordan, m.hassan@psut.edu.jo
[2] King AbdullahII School for Information Technology, University of Jordan,
Amman, Jordan

Abstract. We are introducing a keyed blind multiresolution watermarking algorithm for digital images. This algorithm uses discrete wavelets transform. With this method it is possible to make invisible watermark that has robustness against image compression such as JPEG2000. The proposed algorithm is an additive watermarking algorithm that adds a static number to the selected coefficient at the high and middle frequency of transform domain that is generated by discrete wavelets transform. A blind watermarking algorithm allows detecting the watermark from the target without original watermark image.

Keywords: Blind image watermarking, discrete wavelet transform, compression attack, noise attack.

1 Introduction

With the widespread of Internet and the distribution of digital media, the protection of the intellectual property rights has become increasing important. One of the current copyright protection methods is digital watermarking [1].

Watermarking is the process of embedding a digital signature or a digital watermark in a document to assert its ownership. The insertion step may be performed in spatial, frequency or multiresolution domains. The watermark can then be extracted from the watermarked media to identify the owner [2]. This step can be developed in two different ways: using the original image (non-blind) [3], [4], [5] or without using the original image (blind) [6], [7]. Each digital watermarking technique must satisfy three essential requirements: perceptual transparency, robustness to attacks and capacity of the embedded signature.

Watermarking techniques can be classified into various categories in various ways. Such as can be classified according to working domain (spatial, frequency, multiresolution), or according to type of document (text, image, audio, video) or according to human perception (invisible, visible, dual).

Also the invisible watermarking can be divided into: blind, non-blind and semi-blind.

The proposed algorithm in this paper is a blind watermarking algorithm that doesn't use the original image in the extraction stage of watermarking.

2   Watermarking in the Wavelet Transform (DWT) domain

The idea of the DWT can be explained briefly as follows. An image is decomposed into a pyramid structure with various bands as shown in Fig. 1.

The discrete wavelet transform has a number of advantages over other transforms such as discrete cosine transform (DCT). The discrete wavelet transform is a multi-resolution description of an image, i.e., the decoding can be processed sequentially from a low resolution to the higher resolutions [8],[9]. Also the discrete wavelet transform has better match to the human visual system than the DCT because the artifacts introduced by the wavelet domain coding with high compression radio are less annoying than those introduced at the same bit rate by the DCT.

The discrete wavelet transform is computationally efficient and can be implemented in a variety of ways and wavelet-based watermarking has the advantage of matching the emerging image/video compression standards (JPEG 2000) [10].

Watermarking in the DWT domain consists of two parts:
- Encoding (embedding the watermark) stage and
- Decoding (extraction) stage.

As an introduction, Xia et al [3] encoding/decoding algorithms can be summarized as follows.

Fig.2 shows the encoding as per Xia et al algorithm.

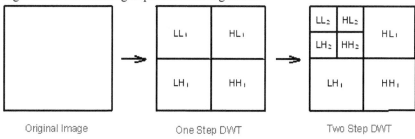

Original Image   One Step DWT   Two Step DWT

Fig. 1. Wavelet transforms (DWT) image decomposition of an image.

An image is decomposed into several bands using the DWT.A pseudo-random sequence (Gaussian noise) watermark is inserted. This watermark is added to the large coefficients which are not located in the lowest resolution. Finally, the inverse discrete wavelet transform (IDWT) is applied on the modified DWT coefficients at the lowest resolution.

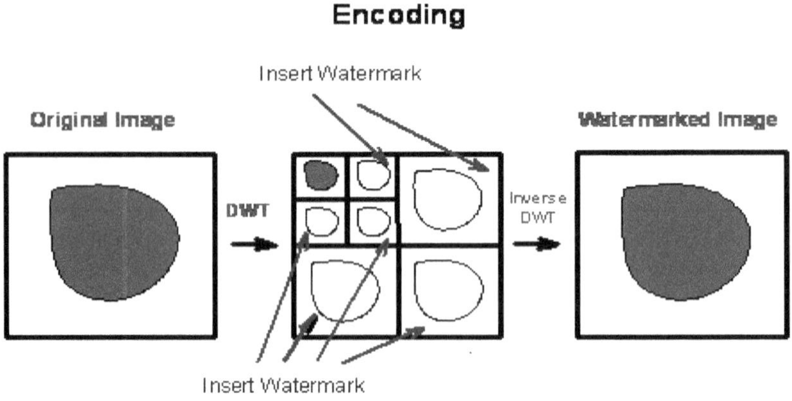

Encoding

Insert Watermark

Original Image   Watermarked Image

Insert Watermark

Fig.3 shows the decoding process as per Xia et al algorithm.

Fig.2: Encoding in Xia's Algorithm
The watermarked and the original images are decomposed using DWT into four bands, LL1,LH1,HL1,HH1, respectively. The difference of the DWT coefficients in HH1 bands of the received and the original images are taken. Then the cross correlation of the original watermarked of the difference is calculated. Signature is only detected if the result of the cross correlation is a peak. Otherwise repeat above process using HH1 and LH1 bands and check for a peak. If there is still no peak use the LL2, LH2, HL2 and HH2 bands until a peak appears, otherwise, the signature can not be detected.

3   The Proposed Algorithm

The algorithm is a blind watermarking which means that the original image is not needed in the detection process. Also a key is used to add the watermark in some selected coefficients.

In a system where no key-based mechanism is employed, a potential attacker would first have to find out the specifications of the scheme (which might not be of great difficulty since they almost share common principles) and then, without leaving any evidences, she/he could remove, extract or change the data of an embedded watermark.

Therefore, the use of a key-based mechanism to protect the scheme becomes an urgent requirement for the development of a secure authentication.

Fig. 4 shows encoding the watermark. The steps are similar to Xia et al algorithm.

The only difference is the step on inserting the watermark. The watermark is embedded into the detail wavelet coefficients of the host image with the use of a key. This key is randomly generated based on a seed. This key consists of zeros and ones and is used to select the exact locations in the wavelet domain in which to embed the watermark. For each coefficient within the wavelet domain, the key has a corresponding value of one or zero to indicate if the coefficient is to be marked or not, respectively . The number of ones in the key must be greater or equal to the size of the watermark.

The encoding steps can be summarized as follows:
- Two levels DWT decomposition is performed on the original image.
- Generate the watermark: the watermark is a pseudo-random sequence (Gaussian noise) W of length N which is seeded with a given value.
- Generate the key K, according to a given seed S.
- Consider each coefficient location (m,n) which are not located in the lowest resolution. If the associated value of the key ki is one, insert the watermark wi to the coefficient. Otherwise do nothing.
- Finally, the inverse transformation IDWT is applied on the modified DWT coefficients and the unchanged.

DWT coefficient at the lowest resolution.

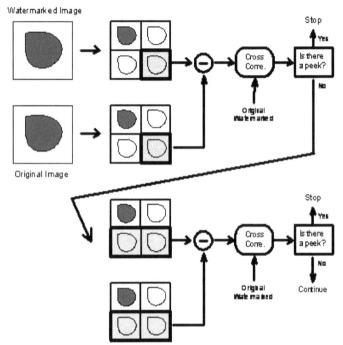

Fig.3. Decoding as per Xia algorithm

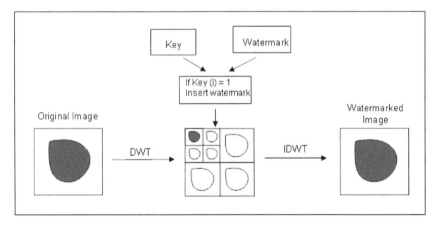

Fig.4. Inserting the watermark as per our proposed algorithm

Fig.5 shows the watermark detection. Here, the original image is not needed for the detection process. The equation used for watermark detection is similar to the one used in Dugad et al [6] . Decoding steps can be summarized as follows:

- Two-level DWT decomposition is performed on the original image.
- Generating the watermark and the key using the given seeds.
- Consider each coefficient location (m,n) which are not located in the lowest resolution if the associated value of the key ki is one, otherwise do nothing.
- Take the correlation δ between the DWT coefficient of the watermarked image yi and the watermark wi.
- The correlation is computed according to the following formula:

$$\delta = \frac{1}{M}\sum yiwi \qquad (1)$$

M: is the number of coefficients where the watermark is inserted.
y: is the DWT coefficient of the watermarked image
w: is the watermark

- The threshold S is defined as follows:

$$S = \frac{\alpha}{2M}\sum (yi) \qquad (2)$$

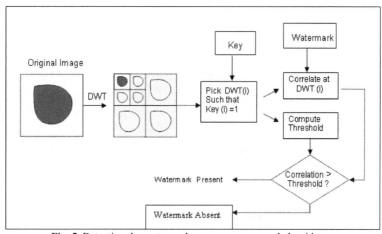

Fig. 5. Detecting the watermark as per our proposed algorithm

α: is a scale factor used in the watermark insertion.
- If the correlation is greater than the threshold the watermark is detected (present) otherwise, the watermark is not detected (absent).

4   Experimental Results

For estimating the performance of the proposed watermarking algorithm, we experiment on a test image. The test image is Lena image having 256x256 size. The Haar DWT is used. Two step DWT is implemented and images are decomposed into 7 subbands. Watermarks and Gaussian noise are added into all 6 subbands but not in the lowest subband (the lowest frequency components). This method makes the watermark invisible. Fig.6 shows the original image and Fig.7 shows the watermarked image. As can be seen, it is difficult to distinguish between the two images.

The algorithm was tested with two kinds of image attacks: additive noise and JPEG2000 compression. Results show that the algorithm is robust and capable of detecting the watermark without the need of the original image.

Correlation and threshold values are shown in Table1. In both cases, compression and additive noise, the correlation value is greater than the threshold value, which means that the watermark is successfully detected.

The cross-correlations of the original and detected watermarks were computed using the Matlab corr2 function. Fig.8 shows the cross-correlation between the original and detected watermark after two kinds of image distortions.

Fig.7. Watermark image   **Fig.6.** Original image

In Fig. 8(a), compression was applied on the watermark image. A peak can be clearly seen in the middle which indicates that the watermark is robust against

JPEG compression. In Fig. 8(b), additive noise was added to the watermarked image. A peak can be also seen clearly in the middle which indicates that the watermarked image is robust against additive noise.

Above results show that the proposed blind algorithm is robust against additive noise and JPEG compression.

Distortion	Correlation	Threshold	Watermark Detected?
JPEG Compression	0.2054	0.1987	Yes
Additive Noise	0.3651	0.2832	Yes

Table1.Correlation compared with the threshold. In both compression and additive noise, the correlation values were greater than the threshold values, which mean that the watermark was successfully detected.

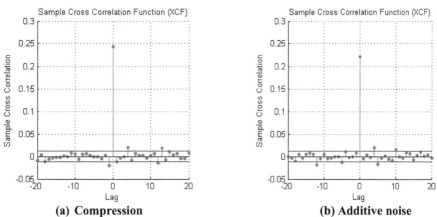

(a) Compression   (b) Additive noise

Fig.8. Cross-correlation between the original watermark and the extracted watermark. In both compression and additive noise, a peak can be clearly seen in the middle which indicates that the watermark is robust against JPEG compression.

5   Conclusion

A blind watermarking algorithm is introduced. This algorithm is additive and blind. It is based on the discrete wavelet transform. Since we embed the watermark in the wavelet transform domain of an image, the watermark's energy is concentrated in the edge components of an image. This method makes the watermark invisible. Watermark was added as a Gaussian noise to the DWT large coefficient based on a key. The use of a key based mechanism prevents attackers from removing, extracting or changing the data of an embedded watermark.

To detect the watermark the original image is not needed.

The robustness of this algorithm was examined against JPEG compression and additive noise. Results show that this algorithm is robust against these image distortions, while the DCT technique is not.

References

1. Bae, Sung-Ho: Copyright Protection of Digital Images. Artech House Books (2006)
2. Gonzalez, F., Hernandez, J.: A   Tutorial on Digital Watermarking. Security Technology, 1999 Proceedings, IEEE 33 rd 1999 International Carnahan Conference .
3   X. Xia, C. Boncelet and G. Arce: 'A multiresolution watermark for digital images', Proc. IEEE, ICIP97, 1997, Vol.1, 548-551.
4   D. Kundur and D. Hatzinak: 'A robust digital image watermarking method using wavelet – based fusion', Proc. IEEE ICIP97, 1997, Vol.1, 544-547.
5   R. Wolfgang and E. Delp : 'A watermark for digital images', Proc. IEEE ICIP96, 1996, Vol.3, 219-222.
6   R. Dugad, K. Ratakonda and N. Ahuja: 'A new wavelet-based scheme for watermarking images', Proc. IEEE ICIP98, Chicago, IL, USA, October 1998.
7   M. Barni, F. Bartolini and A. Piva: 'Improved wavelet-based watermarking thrugh Pixel, wise masking', IEEE Trans. Image Processing 2001, Vol.1, 783-791.
8   Wang, Y., Doherty, J., Dyck, R.: A wavelet based watermarking algorithm for ownership verification of digital images, IEEEt Trans Image Processing, Vol. 11, no.2(2000) 77-88
9   Meerwald, P., Uhl, A.: A survet osf wavelet- domain watermarking algorithms.Proceedings of SPIE , Security and Watermarking of Multimedia Contents III, Vol 4314 (2001)
10   A. Skodras, C. Christopoulos and T. Ebrahim: 'The JPEG2000 still image compression standard', Signal Processing Magazine, IEEE, 2001, Vol.18, 36-58.

Covert Channel Communication in RFID (Short Paper)

Md. Sadek Ferdous, Farida Chowdhury[1]

Department of Information & Communication Technology, Metropolitan University, Sylhet, Bangladesh
[1]Department of Computer Science & Engineering, Shah Jalal University of Science & Technology, Sylhet, Bangladesh
sferdous@metrouni.edu.bd, [1]farida-cse@sust.edu

Abstract. Radio Frequency Identification (RFID) is being considered as one of the most pervasive computing technologies in history. It is believed that tags based on RFID technology will be attached to every consumer products replacing the bar codes as RFID tags provide some excellent facilities such as location tracking by which several ubiquitous services can be facilitated. It is assumed that even every person will carry RFID tag in future for obtaining those ubiquitous services. But this provision has opened up a new arena of security concern in which this tracking facility can be abused and personal privacy can be compromised. Different mechanisms have been proposed and applied to get around this situation. This paper presents a novel approach to increase security in RFID by introducing covert channel communication in RFID. Covert Channel communication in Computer Network has been a topic of both discussion and active research for more than three decades. Here in this paper, we've outlined the mechanism for covert channel communication in RFID and proposed how Covert Channel communication can be used to enhance security and protect privacy.

1   Introduction

Radio-Frequency Identification (RFID) is a pervasive technology that provides automatic identification facility of objects. In a few ways it can be considered as 'Radio Barcode', but unlike barcodes, RIFD provides some excellent additional facilities, eliminating the need of human involvement [1]. It is believed that in near future RFID will enable us, humans, to interact with computing environment automatically and subconsciously and thus providing 'Ubiquitous Computing' capability [2]. These unique features have led many concerned peoples to adopt RFID for many intelligent application scenarios such as counterfeiting of goods, automatic checkout system, infectious animal disease tracking, managing supply chain and super market, access control, pet identification, automatic toll collection, remote keyless entry for automobiles, etc [2], [3]. But due to its wireless nature, information revealed during the RFID communications can be abused by any knowledgeable party and individual privacy can be threatened. In this paper we've introduced a unique approach to fight back the 'Violation of Privacy' problem by proposing Covert Channel communication in RFID for the first time.

Following this introduction, this paper is organized as follows: Section 2 introduces the basic concept of RFID system, threat model in RFID and depicts the concept of Covert Channel communication in terms of wired network. Section 3 outlines the previous works. We've proposed the algorithms for Covert Channel communication for RFID in Section 4, exemplified how Covert Channel can be used to increase security and analyzed other performance issue and we conclude in Section 5.

2   RFID Concept, Threat Model & Covert Channel Communication

An RFID System has two basic components: RFID Tag or Transponder, RFID Reader or Transceiver and the system is supported by a backend database. RFID tag is attached to any object to provide its identification and usually contains a microchip with small computation and storage capability and a coupling antenna to communicate through Radio-Frequency [4], [5]. Transponder usually is the data provider in RFID system. On the other hand transceiver/reader is usually the data seeker. As the RFID system uses RF for communication, data can be read by the reader from the transponder within a limited distance automatically and thus eliminating the need of any human interaction for data collection.

The general format of a packet for RFID response by the transponder derived from the EPC (Electronic Product Code) Network is given in fig. 1 [6].

Header	EPC Manager	Object Class	Serial Number
8 bits	28 bits	24 bits	36 bits
Version (00110101)	Code of Manufacturer	Article Classification	

Fig. 1. The General EPC Format Specified for Retail

EPC is meant to replace UPC (Universal Product Code) used in bar code to identify an object. For different possible situations, there are different packet formats for EPC and usually consist of 64 or 96 bits [7].

Due to its RF Communication RFID system is open in nature. That means traffic generated by the RFID Transponder and reader can be read and analyzed. Several attack methods on security and privacy have been brought into existence. Some of those methods are Sniffing/Eavesdropping [8], Tracking [2], Spoofing/Cloning [2], [3], Replay/Relay Attack [9], Denial of Service [2], etc.

Covert Channel Communication has been in security paradigm since 1973 when B. W. Lampson published a paper titled "A Note on the Confinement Problem" published by ACM in which the term Covert Channel was introduced for the first time. [10], [11]. But the widely accepted definition for Covert Channel comes from the US Department of Defense Trusted Computer System Evaluation Criteria [12]. But different kinds of schemes classifying the Covert Channel communication exist today. Here we're adopting the classification model proposed in [13]. According to that, Covert Channel can be classified into: Value Based Spatial Channel, Transition Based Spatial Channel, Value Based Temporal Channel & Transition Based Temporal Channel [13], [14].

3   Related Works

Increasing security in RFID by using Convert Channel communication is a novel approach and to the best of the knowledge of the authors of this paper is not implemented elsewhere.

4   Proposed Approach

Assumption: One of the main features of the RFID Communication is that information provided by the tag/transponder is static. But this is not true in wired or wireless LAN where two consecutive packets in the same communication are never same. So covert communication based on spatial channel (both value based and transition based) can be implemented in the wired or wireless LAN by changing one of the header data on the packets [15]. But such covert channel can't be implemented in RFID System. That's why we've adopted Temporal Channel Based Covert Communication. That is, our proposition will not alter any bit of the packet, rather it will depend either on the frequency of event (Value based) or on the delay in-between the frequency of events (Transition Based). Our proposition will assume packet arrival time as the frequency of event.

Proposition: Our proposed algorithm goes below:
- **Value Based Temporal Channel:**
– At the Transponder's End.
Algorithm SendingVBTCCommunication
1.   Wait for any interrogation request initiated by the reader.
2.   If a request is initiated by the reader:
3.   Start Covert Channel communication. Repeat lines 4 to 7 until communication ends.
4.   Build a false EPC packet of 96 bits with any value.
5.   Build an original EPC packet of 96 bits with original value.
6.   To transmit a bit with value 1 of the original packet, send the false packet within a defined time frame.
7.   To transmit a bit with value 0 of the original packet, don't send the false packet within a defined time frame.
8.   End If
9.   Exit.
– At the Reader's End.
Algorithm ReceivingVBTCCommunication
1.   Start Covert Channel communication by initiating an interrogation request towards the transponder. Repeat lines 2 to 5 until communication ends.
2.   If a response packet arrives within a defined time:

3.   Ignore the value of the EPC packet as it contains false value regarding the transponder, interpret the whole packet as a single bit with value 1 and stores it into a queue.
4.   Else interpret the absence of packet within that defined time period as a single bit with value 0 and stores it into a queue.
5.   End If.
6.   At the end of communication, the queue will hold the valid EPC packet containing original EPC data.
7.   Exit.

- **Transition Based Temporal Channel.:**
− At the transponder's End.

Algorithm SendingTBTCCommunication
1.   Wait for any interrogation request initiated by the reader.
2.   If a request is initiated by the reader:
3.   Start Covert Channel communication. Repeat lines 4 to 8 until the communication ends.
4.   Build a false EPC packet of 96 bits with any value.
5.   Build an original EPC packet of 96 bits with original value.
6.   At first send a false EPC packet.
7.   To transmit a bit with value 1 of the original packet, send the next packet in such a way that interval time between two packets exceeds a threshold time.
8.   To transmit a bit with value 0 of the original packet, send the next packet in such a way that interval time between two packets remains below the threshold time.
9.   End If.
10.   Exit.

− At the Reader's End.

Algorithm ReceivingTBTCCommunication

1.   Start Covert Channel communication by initiating an interrogation request towards the transponder. Repeat lines 2 to 6 until communication ends.
2.   Ignore the value of the EPC packets as it contains false value regarding the transponder. If the interval of arrival time between two consecutive packets exceeds a certain threshold value:
3.   Interpret that interval time as a single bit with value 1 and stores it into a queue.
4.   Else if the interval of arrival time between two consecutive packets does not exceed a certain threshold value:
5.   Interpret that interval time as a single bit with value 0 and stores it into a queue.
6.   End If.
7.   At the end of communication, the queue will hold the valid EPC packet containing original EPC data.
8.   Exit.

Illustration: Let's illustrate the algorithm by an example. Suppose in a RFID system, the transponder wants to send a data packet (Assumed) like Fig. 2:

00110101	1101010···1011

Fig. 2. A Valid EPC response packet (Assumed)

At first the transponder will build a false EPC packet. A false EPC packet may look like Fig. 3 (Assumed):

| 00110101 | 0011011···0101 |

Fig. 3. A False EPC response packet (Assumed)

So according to the Value Based Covert Channel algorithm, to send a bit with value 1 of the original packet (Fig. 2) the transponder will send the false packet (Fig. 3) at defined time interval and to send a bit with value 0 the transponder will not send the false packet at that interval. So, to send the 96 bits of the EPC packet, the transponder will go though 96 transition periods.

Now according to the Transition Based Covert Channel algorithm, to send a bit with value 1 of the original packet (Fig. 2) the transponder will send two consecutive false packets (Fig. 3) in such a way that the time interval between two packets exceeds a threshold value and to send a bit with value 0 of the original packet the transponder will send two consecutive false packets in such a way that the time interval between two packets does not exceed the threshold limit.

Performance Analysis: It is obligatory in our proposed approach that synchronization between the transponder and the reader should be set up at first, that is they should be synchronized by defining a definite time interval for Value Based algorithm and a threshold time for Transition Based algorithm. Some of the attack methods in RFID such as Sniffing, Tracking, Spoofing, etc. require the interaction with the direct RFID response. In our proposed approach, valid RFID response is never transmitted, only the false response is transmitted. So if any attack tries to build an attack model based on RFID response, he will build a false model and won't be able intercept the legitimate data until he knows/assumes the underlined approach and required synchronization. According to our algorithms, QoS will mainly depend on synchronization. As long as synchronization is maintained, QoS can be ensured. As RFID operates normally in small range, the packet will have to travel a short distance. This property will ensure two things: Minimal chance for a packet to be lost and short time for a packet to reach the destination which almost eradicates the possibility of losing a RFID packet or taking longer time for a RFID packet to reach the reader, for any or both of which synchronization could be lost between the reader and the transponder. So if synchronization is set, it is most likely to prevail during the whole communication and thus maintaining QoS.

5   Conclusions

Lack of security in RFID system can threaten a whole group of invasive computer applications. So it is very important to enhance the security aspects of RFID. Different approaches have been proposed, illustrated and implemented by researchers around the world. This paper presents a novel approach to apply security in RFID by proposing the possibility of Covert Channel communication in RFID which is first of its kind. The authors believe that this approach will establish a new paradigm of security in RFID system.

References

1. Ari Juels, RFID PRIVACY: A TECHNICAL PRIMER FOR THE NON-TECHNICAL READER, Chapter From the book: Privacy and Technologies of Identity, A Cross-Disciplinary Conversation (Eds K. Strandburg and D.Stan Raicu), Springer-Verlag, 2005.
2. Melanie R. Rieback Bruno Crispo Andrew S. Tanenbaum , Is Your Cat Infected with a Computer Virus?, In Pervasive Computing and Communications, Pisa, Italy, March 2006. IEEE, IEEE Computer Society Press.
3. Simson L. Garfinkel, Ari Jules, Ravi Pappu, RFID Privacy: An Overview of Problems and Proposed Solutions, In IEEE Security and Privacy, 3(3):34–43, May-June 2005.
4. Pedro Peris-Lopez, Julio Cesar Hernandez-Castro, Juan M. Estevez-Tapiador, and Arturo Ribagorda, RFID Systems: A Survey on Security Threats and Proposed Solutions, In 11th IFIP International Conference on Personal Wireless Communications – PWC06, volume 4217 of Lecture Notes in Computer Science, pages 159–170.Springer-Verlag, September 2006.
5. Sanjay E. Sarma, Stephen A. Weis, and Daniel W. Engels, RFID Systems and Security and Privacy Implications, In Burton Kaliski, C¸ etin Kaya ¸co, and Christof Paar, editors, Cryptographic Hardware and Embedded Systems –CHES 2002, volume 2523 of Lecture Notes in Computer Science, pages 454–469, Redwood Shores, CA, USA, August 2002. Springer-Verlag.
6. Thomas Hjorth, Supporting Privacy in RFID Systems, Master thesis, Technical University of Denmark, Lyngby, Denmark, December 2004.
7. EPC Standard Speci_cation, version 1.1 rev. 1.24. April 1, 2004.
    http://www.epcglobalinc.org/standards technology/ EPCTagDataSpeci_cation11rev124.pdf
8. Biometrics deployment of machine readable travel documents, May 2004.
    http://www.icao.int/mrtd/download/documents/Biometrics   deployment   of   Machine Readable Travel Documents 2004.pdf
9. Gerhard Hancke, A Practical Relay Attack on ISO 14443 Proximity Cards, Manuscript, February 2005.
10.Lampson, Butler: A Note on the Confinement Problem,
    http://www.cis.upenn.edu/ KeyKOS/Confinement.html
11.Pukhraj Singh, Whispers On The Wire: Network Based Covert Channels Exploitation & Detection (BETA Draft),
    http://gray-world.net/papers/pukhrajsingh_covert.doc
12.US DoD: Trusted Computer System Evaluation Criteria, 1985,
    http://csrc.nist.gov/publications/history/dod85.pdf
13.Wang, Zhenghong, New Constructive Approach to Covert Channel Modeling and Channel Capacity Estimation, In ISC 2005, LNCS 3650, pp. 498-505, 2005.
14.Marc Smeets, Matthijs Koot, Research Report: Covert Channels, 2006
    http://www.os3.nl/~mrkoot/courses/RP1/researchreport_2006-02-15_final2.pdf
15.Steven J. Murdoch and Stephen Lewis, Embedding Covert Channels into TCP/IP, Draft for Information Hiding Workshop 2005 proceedings (Revision 1159: July 29, 2005)

Privacy-Preserving Decision Tree Classification in Horizontal Collaboration

Justin Zhan

Carnegie Mellon University
justinzh@andrew.cmu.edu

Abstract. In this paper, we discuss privacy preserving decision tree classification problem. We then design a solution for such a problem based on homomorphic encryption. We show a way to prove the privacy of our solution.

Key Words: Privacy, Data Mining, Decision Tree Classification.

1   Introduction

Data privacy is becoming an important issue for may applications. In this paper, we would like to address the privacy problem within data mining area. Specifically, we consider the privacy-preserving collaborative decision tree classification problem.

The decision tree is one of the classifiers. The induction of decision trees [3] from attribute vectors is an important and fairly explored machine learning paradigm.

According to ID3 algorithm, each non-leaf node of the tree contains a splitting point, and the main task for building a decision tree is to identify an attribute for the splitting point based on the information gain. Information gain can be computed using *entropy*. In the following, we assume there are nc number of classes in the whole training data set. $Entropy(S)$ is defined as follows:

$$Entropy(S) = -\sum_{j=1}^{nc} Q_j \log Q_j, \tag{1}$$

where Q_j is the relative frequency of class j in S. Based on the entropy, we can compute the information gain for any candidate attribute A if it is used to partition S:

$$Gain(S, A) = Entropy(S) - \sum_{v \in A} (\frac{|S_v|}{|S|} Entropy(S_v)), \tag{2}$$

where v represents any possible values of attribute A; S_v is the subset of S for which attribute A has value v; $|S_v|$ is the number of elements in S_v; $|S|$ is the number of elements in S. To find the best split for a tree node, we compute information gain for each attribute. We then use the attribute with the largest information gain to split the node.

There are two types of collaborative models. In the vertical collaboration, diverse features of the same set of data are collected by different parties. In the horizontal collaboration, diverse sets of data, all sharing the same features, are gathered by different parties. The collaborative model that we consider is the *horizontal collaboration*.

Next, we introduce homomorphic encryption and digital envelope technique.

2   Tools

Our solution is based on homomorphic encryption [2] and digital envelope technique [1]. Specifically, we utilize the following property of the homomorphic encryption functions: $e(m_1) \times e(m_2) = e(m_1 + m_2)$ where m_1 and m_2 are the data to be encrypted. Because of the property of associativity, $e(m_1 + m_2 + .. + m_n)$ can be computed as $e(m_1) \times e(m_2) \times \cdots \times e(m_n)$ where $e(m_i) \neq 0$. That is

$$e(m_1 + m_2 + \cdots + m_n) = e(m_1) \times e(m_2) \times \cdots \times e(m_n). \tag{3}$$

$$d(e(m_1)^{m_2}) = d(e(m_1 \times m_2)), \tag{4}$$

where $\times$ denotes multiplication.

To build a decision tree classifier, we have to decide and assign an attribute for each node. In order to determine each node for the tree, we need to conduct the following steps:

1. To compute $Entropy(S_v)$.
2. To compute $\frac{|S_v|}{|S|}$.
3. To compute $\frac{|S_v|}{|S|} Entropy(S_v)$.
4. To compute information gain for each candidate attribute.
5. To compute the attribute with the largest information gain.

Next, we will provide privacy-oriented protocols to conduct each step in the scenarios of horizontal collaboration. Due to the page limit, we cannot show the entire solution. We only show the first step and the extended version will provide the whole solution.

3   Privacy-Preserving Protocols for Horizontal Collaboration

Problem 1. Assume that P_1 has a private data set DS_1, P_2 has a private data set DS_2, $\cdots$ and P_n has a private data set DS_n. The goal is to compute $e(Entropy(S_v))$, $e(Q_j log(Q_j))$, $\frac{|S_v|}{|S|} Entropy(S_v)$ and the attribute with the largest information gain for horizontal collaboration involving DS_1, $\cdots$, and DS_n without compromising data privacy.

We first select a key generator who produces the encryption and decryption key pairs. Let us assume that P_n is the key generator who generates a homomorphic encryption key pair (e, d). Next, we will show how to conduct each step.

3.1 To Compute $e(Entropy(S_v))$

Highlight of Protocol 1: The protocol contains three steps. In step I, each party computes $|S_v|$ based on their own datasets. Assuming P_1 gets c_1, P_2 gets c_2, $\cdots$, P_n gets c_n. P_n sends $e(c_n)$ to P_1 who computes $e(c_n) \times e(c_1) = e(c_1 + c_n)$ and sends it to P_2. Repeat until P_{n-1} obtains $e(c_1 + c_2 + \cdots + c_n)$. P_{n-1} computes $e(\sum_{i=1}^{n} c_i)^{\frac{1}{N}} = e(Q_j)$.

In step II, P_{n-1} generates a set of random numbers R_1, R_2, $\cdots$, and R_t. P_{n-1} sends the sequence of $e(Q_j)$, $e(R_1)$, $e(R_2)$, $\cdots$, $e(R_t)$ to P_n in a random order. P_n decrypts each element in the sequence, and sends $log(Q_j)$, $log(R_1)$, $log(R_2)$, $\cdots$, $log(R_t)$ to P_1 in the same order as P_{n-1} did. Next, P_1 and P_{n-1} computes $e(Q_j log(Q_j) + RQ_j) \times e(-RQ_j) = e(Q_j log(Q_j))$. Finally, P_{n-1} computes $e(Entropy(S_v))$.

We present the protocol as follows:

Protocol 1 .

1. *Step I: To compute $e(Q_j)$*
   (a) *Each party computes $|S_v|$ based on their own datasets. Assume P_1 gets c_1, P_2 gets c_2, $\cdots$, P_n gets c_n.*
   (b) *P_n sends $e(c_n)$ to P_1.*
   (c) *P_1 computes $e(c_n) \times e(c_1) = e(c_1 + c_n)$ and sends it to P_2.*
   (d) *Repeat until P_{n-1} obtains $e(c_1 + c_2 + \cdots + c_n)$.*
   (e) *P_{n-1} computes $e(\sum_{i=1}^{n} c_i)^{\frac{1}{N}} = e(Q_j)$.*
2. *Step II: To compute $e(Q_j log(Q_j))$*
   (a) *P_{n-1} generates a set of random numbers R_1, R_2, $\cdots$, and R_t.*
   (b) *P_{n-1} sends the sequence of $e(Q_j)$, $e(R_1)$, $e(R_2)$, $\cdots$, $e(R_t)$ to P_n in a random order.*
   (c) *P_n decrypts each element in the sequence, and sends $log(Q_j)$, $log(R_1)$, $log(R_2)$, $\cdots$, $log(R_t)$ to P_1 in the same order as P_{n-1} did.*
   (d) *P_1 adds a random number R to each of the elements, then sends them to P_{n-1}.*
   (e) *P_{n-1} obtains $log(Q_j)+R$ and computes $e(Q_j)^{(log(Q_j)+R)} = e(Q_j log(Q_j) + RQ_j)$.*
   (f) *P_{n-1} sends $e(Q_j)$ to P_1.*
   (g) *P_1 computes $e(Q_j)^{-R} = e(-RQ_j)$ and sends it to P_{n-1}.*
   (h) *P_{n-1} computes $e(Q_j log(Q_j) + RQ_j) \times e(-RQ_j) = e(Q_j log(Q_j))$.*
3. *Step III: To compute $e(Entropy(S_v))$*
   (a) *Repeat protocol step I and step II to compute $e(Q_j log(Q_j))$ for all j's.*
   (b) *P_{n-1} computes $e(Entropy(S_v)) = \prod_j e(Q_j log(Q_j)) = e(\sum_j Q_j log(Q_j))$.*

The Correctness Analysis of Protocol 1: In step I, P_{n-1} obtains $e(Q_j)$. In step II, P_{n-1} gets $e(Q_j log(Q_j))$. The two steps repeatedly used until P_{n-1} obtains $e(Q_j log(Q_j))$ for all j's. In step III, P_{n-1} computes the entropy by all the terms previously obtained. Notice that although we use $Entropy(S_v)$ to illustrate, $Entropy(S)$ can be computed following the above protocols with different input attributes.

The Complexity Analysis of Protocol 1: The communication cost is linear in the number of parties n. The computation cost is linear in the total number of records N.

Theorem 1. *Protocol 1 preserves data privacy at a level equal to* ADV_{P_n}.

Proof. According to the definition of privacy in [5, 6], we need identify the value of ϵ such that

$$|Pr(T|CP) - Pr(T)| \le \epsilon$$

holds for $T = T_{P_i}$, $i \in [1, n]$, and CP = Protocol 1.

According to the notations defined in [5, 6],

$$ADV_{P_n} = Pr(T_{P_i}|VIEW_{P_n}, Protocol 1) - Pr(T_{P_i}|VIEW_{P_n}), i \ne n,$$

and

$$ADV_{P_i} = Pr(T_{P_j}|VIEW_{P_i}, Protocol 1) - Pr(T_{P_j}|VIEW_{P_i}), i \ne n, i \ne j.$$

The information that P_i where $i \ne n$ obtains from other parties is encrypted by e which is semantically secure, therefore,

$$ADV_{P_i} = ADV_S.$$

In order to show that privacy is preserved according to Definition **??**, we need to know the value of the privacy level ϵ. We set

$$\epsilon = max(ADV_{P_n}, ADV_{P_i}) = max(ADV_{P_n}, ADV_S) = ADV_{P_n}.$$

Then

$$Pr(T_{P_i}|VIEW_{P_n}, Protocol 1) - Pr(T_{P_i}|VIEW_{P_n}) \le ADV_{P_n}, i \ne n,$$

and

$$Pr(T_{P_j}|VIEW_{P_i}, Protocol 1) - Pr(T_{P_j}|VIEW_{P_i}) \le ADV_{P_n}, i \ne n, i \ne j,$$

which completes the proof [1].

---

[1] Note that the information that P_n obtains from P_{n-1} is $e(Q_j)$, $e(R_1)$, $e(R_2)$, $\cdots$, $e(R_t)$ in a random order.

4   Discussion

Encryption is a well-known technique for preserving the confidentiality of sensitive information. To date, there are many such systems. Homomorphic encryption [2] is a very powerful cryptographic tool and has been applied in several research areas such as electronic voting, on-line auction, etc. [4] is mainly based on homomorphic encryption where Wright and Yang applied homomorphic encryption to the Bayesian networks induction for the case of *two* parties. Zhan et. al. [7] proposed a cryptographic approach to tackle collaborative association rule mining among multiple parties. In this paper, we have used homomorphic encryption and digital envelope technique to achieve collaborative decision tree classification without sharing the private data among the collaborative parties. Specifically, we provide a solution for decision tree classification with horizontal collaboration. Due to page limit, we cannot show our entire solution. In the extended version of this paper, we will show further details and add more related works.

References

1. D. Chaum. Security without identification. In *Communication of the ACM, 28(10): 1030–1044, October*, 1985.
2. P. Paillier. Public-key cryptosystems based on composite degree residuosity classes. In *Advances in Cryptography - EUROCRYPT '99, pp 223-238, Prague, Czech Republic*, 1999.
3. J. Quinlan. Induction of decision trees. In *Machine Learning, 1:81 106*, 1986.
4. R. Wright and Z. Yang. Privacy-preserving bayesian network structure computation on distributed heterogeneous data. In *Proceedings of the 10th ACM SIGKDD International Conference on Knowledge Discovery and Data Mining (KDD)*, 2004.
5. J. Zhan. *Privacy Preserving Collaborative Data Mining*. PhD thesis, Department of Computer Science, University of Ottawa, 2006.
6. J. Zhan and S. Matwin. A crypto-based approach to privacy preserving collaborative data mining. In *Workshop on Privacy Aspect of Data Mining (PADM'06) in conjunction with the IEEE International Conference on Data Mining (ICDM'06), HongKong, December 1*, 2006.
7. J. Zhan, S. Matwin, and L. Chang. Privacy-preserving collaborative association rule mining. In *19th Annual IFIP WG 11.3 Working Conference on Data and Applications Security, Nathan Hale Inn, University of Connecticut, Storrs, CT, U.S.A., August 7-10*, 2005.

Generalization of open key knapsack cryptosystems

V. O. Osipyan

Kuban State University, Krasnodar, Russia, RRWO@mail.ru

Abstract. The class of cryptosystems with the open-key in the basis of which there is nonstandard NP-full problem about a knapsack is considered. The problem of stacking such nonstandard (generalized) knapsack is offered. New concepts the generalized knapsack, the generalized supergrowing knapsack are introduced. Then the algorithm of construction of the generalized knapsack cryptosystems with the open-key is given. As an example of such cryptosystems the cryptosystem on the basis of which lies Warshamov's W_n p-nary code with generalized knapsack function $F_p(x)$.

1 Introduction

As it is known [1–3, 5–8] in the basis of all standard knapsack security information systems (**KSIS**) there is NP-full problem about stacking a knapsack. In order to construct an easy subclass standard **KSIS** on the basis of supergrowing knapsack R. Merkle and M. Hellman proposed "to disguise" a knapsack with the help of linear transformation of the latter by means of strong modular multiplication.

In report Chor-Rivest's [1] report unlike Merkle-Hellman report the knapsack presents a set of logarithms in multiplicate group of the expanded field and is characterized by higher density.

Although after opening Merkle-Hellman's original schemes by means of the algorithm of polynomial complexity **LLL** [4] many experts began to treat the cryptostability of such systems skeptically. Simultaneously a set of other complicated variants **KSIS** [3, 7, 8] were offered namely Grem-Shamere's knapsack based on principle of stacking a standard knapsack among which there still exist not opened **KSIS** [8] ones.

Unlike the existing **KSIS** [1–8] based on standard knapsacks in the given paper first an attempt is made to generalized a classical problem about a knapsack secondly the security system information (**SSI**) based on the generalized knapsack is offered.

2 A problem about the generalized knapsack K_G

According to mathematical model about standard knapsack K_S [3, 7] it is required for set vector $A = (a_1, a_2, ..., a_n)$ with natural components a_i and natural number v to establish the existence or absence of binary vector $W =$

$(\alpha_1, \alpha_2, ..., \alpha_n)$, the one for which equality takes place: $v = A * W^T$. If for the given v there is such a binary vector W then it is accepted that the input of a standard knapsack (A, v) is possibly or has a solution and accordingly v refers is referred to as an possibly value for A formulate.

Let's the generalized (non-standard) problem about knapsack K_G which generalizes the standard one K_S.

A problem about the generalized knapsack K_G. Let there be a set n subjects with set volumes a_i, $i = 1..n$ ($1..n$ – from 1 up to n), and also a knapsack with the volume v, it is necessary to find such set of subjects which will completely fill the knapsack. Unlike a standard problem about knapsack K_S where i-th object is either put in the knapsack or is not, in the given problem (we shall call it generalized) - it is possible to set for each i-th object its maximum quantity of p inputted in the knapsack.

2.1   The solution of the generalized knapsack K_G problem

We will examine the class of open key cryptosystems based on the well-known task of knapsack. We will study the question of Knapsack Cryptosystems generalization based on the given knapsack in case of applying preliminary p-nary coding of the elements of message, and we reveal the necessary and sufficient conditions of solution existence for (A, v) entrance. Unlike the conventional knapsack systems in which while defining v one or another component of knapsack vector may be either present or not we consider the case when they may be replicated by the given number of times.

Let $A = (a_1, a_2, ..., a_n)$ be n's dimensional knapsack vector consisting of n ($n \geq 3$) different components a_i, $i = 1..n$. And (A, v) is an entrance of a knapsack task, where v is also a natural number. For the sake of simplicity let's suppose that values of knapsack vector components are sorted in ascending order. And let $Z_p = \{0, 1, ..., p - 1\}$, $p \geq 2$ be a set of replication factors of knapsack vector A. At first let's consider the question of presenting the given number v as a sum of vector's A components if its components can't be repeated as it is demanded in some sources on the construction of the open key knapsack cryptosystem or can be repeated as it is suggested in this article. It's obvious that such presentation for any entrance (A, v) is either impossible or has more than one way of presentation and at last has the only unique one. First of all we will be interested in the last case when there is the unique presentation for the synthesis of cryptosystem which is based on the given knapsack.

The solution without replication of components for the entrance (A, v) is called a subset of elements A, the sum of them equals v, i.e.

$$\sum_{i=1}^{N} \alpha_i a_i = v, \quad \alpha_i \in 0, 1 \ . \tag{1}$$

But if $\alpha_i \in \{0, 1, ..., p - 1\}$, $p > 2$ and at least one of the factors $\alpha_i \geq 2$ in (1) then we call it the solution with replications with corresponding replication factor components of Z_p. This definition means that vector $W_v = (\alpha_1, \alpha_2, ..., \alpha_n)$

exists. This vector is the spectrum of replications for number v, where $\alpha_i \in \{0, 1, ..., p-1\}$ and $v = A * W_v^T$, moreover the solution will be with replications if $p = 2$ or without if $p > 2$.

It should be especially stressed that all entrances (A, v) of knapsack vectors, for any valid meaning must have unique solution with replications or without them. And all entrances (A, v) for any arbitrary v have no more than one solution. Let's call them injective. If the solutions with replication factor components of vector A from Z_p are considered for this vector then this injective vector should be called generalized knapsack vector and named as $\tilde{A}_p$.

In particular if $p = 2$ then it means that $\tilde{A}_2 = A$, i.e. the number of components $\tilde{A}_2$ replications for entrance $(\tilde{A}_2, v)$ is equal to 0 or 1.

The knapsack vector $A = (a_1, a_2, ..., a_n)$ is called increasing if for any $j = 2..n$ there is an inequality: $a_j > a_{j-1}$ and respectively it is called overgrowing without replications [1] if $a_j > \sum_{k=1}^{j-1} a_k$.

The knapsack vector $\tilde{A}_p = (a_1, a_2, ..., a_n)$ is called generally overgrowing if for any $j = 2..n$ there is an inequality: $a_j > \sum_{k=1}^{j-1}(p-1)a_k$, where $p > 2$ is a natural number. So for example $\tilde{A}_{13} = (6, 74, 965, 12556, 163145)$ is generally overgrowing knapsack vector of the length 5. It is evident that if knapsack vector is overgrowing then it is injective and increasing simultaneously. Though it is possible that knapsack vector is not overgrowing, however it is injective, for example for vector: $A = (2, 3, 4, 8, 16)$. The injective knapsack vector A is called dense if the difference of its two arbitrary neighboring components is minimal. Knapsack is considered to be increasing too if it is dense. Let's consider the following theorems about the existence of generalized entrance $(\tilde{A}_p, v)$ solutions, without any proof.

Theorem 1. *The generalized knapsack vector $\tilde{A}_p = (a_1, a_2, ..., a_n)$ of n dimension, $n \geq 3$, is dense and injective if $a_1 = c$, $a_j = p^{j-2}((p-1)c + 1)$, $j = 2..n$, where c is some integer positive constant.*

Corollary 1. *In particular, if $c = 1$ then the injective knapsack vector $A = (1, p, p^2, ..., p^{n-1})$ is also dense.*

So, for example if $p = 3$, $c = 2$, $n = 8$ then $\tilde{A}_3 = (2, 5, 15, 45, 135, 405, 1215, 3645)$ is dense and injective generalized knapsack vector.

Theorem 2. *The generalized knapsack vector $\tilde{A}_p = (a_1, a_2, ..., a_n)$ of n dimension, $n \geq 3$, is injective if it is generally overgrowing, i.e. for any $j = 2..n$ there is inequality $a_j > \sum_{k=1}^{j-1}(p-1)a_k$ where $a_1 = const$, $p > 2$ is some natural number.*

One should notice, that the condition of injectiveness is fulfilled for all general knapsacks with lower density in case when the inequality is fulfilled: $a_j > \sum_{k=1}^{j-1} \alpha a_k$, $j = 2..n$, where $\alpha > p - 1$ – and it represents a special direction of research.

3   Nonstandart knapsack information security systems

Now we are going to create cipher algorithm using an open key based, on the task about generalized knapsack, using Warshamov's p-code of and generalized function of cipher $F_p(x)$. If letters are elementary messages but not individual sub-task of some given length (the so called l-grammes or blocks with length l), then as their digital equivalents you can take special meanings of the given segment, otherwise, it's necessary to choose one from p^n – meanings of function $F_p(x)$, as the equivalent for block with length l of the given alphabet, consisting of k-letters. It's obvious for the conditions to be performed: $n \geq \lceil l * \log_p k \rceil$, where n – knapsack definition. In order to simplify the understanding here let's match every letter of the open text with one cipher word from W_n – Warshamov's code of the according length. Let $\tilde{A}_p = (a_1, a_2, ..., a_n)$ be a generalized injective knapsack vector of n's dimension, where $n \geq 3$. It's obvious, that any possible meaning for v can be presented as: $v = \alpha_1 * a_1 + ... + \alpha_n * a_n$, $\alpha_i \in Z_p$ or $v = A * W_v^T$, where $W = (\alpha_1, \alpha_2, ..., \alpha_n)$ – spectrum of number v, the interval of its meaning can be defined by the inequality: $0 \leq v \leq (p-1)(a_1 + ... + a_n)$.

For every possible value v from p^n we define generalized encryption function in the following way:
$$F_p(0) = F_p(00...00) = 0, F_p(1) = F_p(00...00) = a_1, ...,$$
$$F_p(p^n - 1) = F_p(p-1, p-1, ..., p-1) = (p-1)a_1 + (p-1)a_2 + ... + (p-1)a_n.$$

So let any elementary message (in particular a letter) of open text be in accordance with its spectrum. $W_* \leftrightarrow (\alpha_1^*, \alpha_2^*, ..., \alpha_n^*)$ and define cipher E of such message as: $E = F_p(\alpha_1^*, \alpha_2^*, ..., \alpha_n^*) = \tilde{A}_p^* (\alpha_1^*, \alpha_2^*, ..., \alpha_n^*)^T$. Then let's present the open text T as strings of p-nary using its representation corresponding Warshamov's code, then divide a resulting string into n-length segments. Afterward encode an open text by segment, using encryption function $F_p(x)$. Let's examine the example of Warshamov's code application in mono-alphabet cryptosystem. For simplicity sake we'll use only a plain substitution method.

Let's assume $p = 3$, $n = 6$, $a = 0$ and for selected p, n and a build W_6 code, consisting of 105 words. Further we select of random 26 words from W_6 for initial letter-coding method of the English alphabet. Each selected word will correspond only to one letter.

Let's assume that we have following the accordance's:

$A - 000000$ $\quad F - 001212$ $\quad K - 010010$ $\quad P - 011222$ $\quad U - 020020$
$B - 000120$ $\quad G - 002022$ $\quad L - 010211$ $\quad Q - 012121$ $\quad V - 020221$
$C - 000201$ $\quad H - 002111$ $\quad M - 011021$ $\quad R - 012202$ $\quad W - 021000$
$D - 001011$ $\quad I - 002200$ $\quad N - 011102$ $\quad S - 012210$ $\quad X - 021120$
$E - 001100$ $\quad J - 010002$ $\quad O - 011110$ $\quad T - 020012$ $\quad Y - 021120$
$\qquad\qquad\qquad\qquad\qquad\qquad\qquad\qquad -100001 \qquad Z - 021201$

Lets open text look like this $T = $ CODE RRW, and generalized injective knapsack vector equals $\tilde{A}_3 = (2, 5, 15, 45, 135, 405)$. Let's define the values of $F_p(x)$ function for all the letters in the open text: $T = $ CODE RRW.

As far as that letters of alphabet corresponds to ciphers received by crypting function $F_3(x) = \tilde{A}_3 * W_x^T$ where W_x is a Warshamov's cipher of x we get

$$F_3(\_) = (2, 5, 15, 45, 135, 405) * (100001)^T = 407$$
$$F_3(C) = (2, 5, 15, 45, 135, 405) * (000201)^T = 495$$
$$F_3(D) = (2, 5, 15, 45, 135, 405) * (001011)^T = 555$$
$$F_3(E) = (2, 5, 15, 45, 135, 405) * (001100)^T = 60$$
$$F_3(O) = (2, 5, 15, 45, 135, 405) * (011110)^T = 200$$
$$F_3(R) = (2, 5, 15, 45, 135, 405) * (012202)^T = 935$$
$$F_3(W) = (2, 5, 15, 45, 135, 405) * (021000)^T = 25$$

In this case the initial text $T = $ CODE RRW is encrypted as follows: $E = 495, 200, 555, 60, 407, 935, 935, 25$.

It's obvious, that code W_n where the number of words is less than number of letters in initial alphabet is unacceptable. And unlike the standard knapsack, the number of $\tilde{A}_3$ vector components that may be repeated here is equal to 2.

The procedure of cryptogram E restoration is the same as in binary case (standard case of knapsack [1]). The difference is that the choosing regular component of knapsack it's necessary to consider its possible repetition. For example just for one cipher 935 in this case we have: $935 = 0*2 + 1*5 + 2*15 + 2*45 + 0*135 + 2*405$, therefore, spectrum of the code is equal to $W = (012202)$, that's why the character, corresponding to it is R. The uniqueness of the cryptogram E reconstruction is guaranteed by **Theorem 1**.

Thus defined cryptosystem on the basis of a generalized backpack function $F_p(x)$, represents a symmetric cryptosystem. With the help of such cryptosystem if is equally difficult to decrypt the cryptogram on a generalized backpack both for the cryptanalyst and for the legal receiver.

As in binary **KSIS**, as a secret loophole for the legal user, we will define a secret pair (m, e) with a condition: $(m, e) = 1$, $m > \max\{a_i\}$. Then according to the principle of a backpack system the illegal user should analyze a crypto text, using other generalized backpack vector – open key of encoding: $\tilde{B}_p = (b_1, ..., b_n)$, where $b_i = e * a_i (\bmod\ m)$ for all indexes i. Vector $\tilde{B}_p$ is said to be received from vector $\tilde{A}_p$ due to strong modular multiplication. The legal receiver of the initial report can analyze a crypto text, having initial permutations and the simplest version of the backpack vector, which are a loophole known only to the legal users of a cryptosystem. In this case (m, e) are secret keys and a way to represent letters with the help of Warshamov's code. It will take the cryptanalyst a lot of time to decode it, even if he knows an encoding system and cryptogram, but doesn't know the loophole.

He will have to solve a hard task about knapsack. This task for the cryptanalyst belongs to NP-complete class.

Theorem 3. Let $\tilde{A}_3 = (a_1, a_2, ..., a_n)$ – injective generalized knapsack vector of dimension n, $n \geq 3$, and vector $\tilde{B}_p = (b_1, ..., b_n)$, $b_i = e * a_i (\bmod\ m)$, $(m, e) = 1$ = 1 can be calculated from $\tilde{A}_p$ by strong module multiplication relating to m and e. Then the solution of knapsack task for input $(\tilde{B}_p, v * e)$ is congruent with unique solution for input $(\tilde{A}_p, v)$.

We must note that we can examine cryptosystems not based on p-nary Warshamov's codes, but p-nary representations of message elements with base p or any other code with corresponding modification. It is necessary to note, that open text reconstruction procedure, doesn't depend on knapsack vector components themselves, it depends only on the size of knapsack and open original letter encoding method.

4 Conclusions

A new model **KSIS** with open key on the basis of the generalized (non-standard) problem about knapsack K_G is offered. In order to stack the knapsack an algorithm with be time required $(p/2)^n$-th time complexity with be twice as which much than in the similar model of a standard problem about knapsack K_S the particular case of which is the case when $p = 2$.

References

1. Arto Salomaa. Public-Key Cryptography. (1995)
2. Ginzburg B.D.// Problems of cybernetics **19**(1967) 249–252
3. Merkle R., Hellman M. Hiding information and signatures in trapdoor knapsacks // IEEE Transactions on Information Theory. **IT** − **24** (1978) 525–530
4. Merkle R., Hellman M. On the security of multiple encryption // Communications of the ACM. **24** (1981) 465–467
5. Lenstra A.K., Lenstra H.W., Lovasz L. Factoring polynomials with rational coefficients // Mathematische Annalen. **261** (1982) 515–534
6. Chor B., Rivest R. A knapsack-type public key cryptoystem based on arithmetic in finite fields//IEEE Transactions on Information Theory. **IT** − **34** (1988) 901–909.
7. Salomaa A. Public-Key Cryptography Springer-Verlag Berlin Heidelberg New York London Paris Tokyo Hong Kong Barcelona.
8. Koblitz N. A Course in Number Theory and Cryptography. Springer-Verlag New York. (1987)
9. Schneir B. Applied Cryptography: Protocols, Algorithms and Source Code in C, 2nd edition. New York: J. Wiley & Sons, (1996)

Practical Public Key Solution in Mobile Ad Hoc Networks

Wen-Jung Hsin
Information and Computer Science
Park University, Parkville, MO 64152

Lein Harn
School of Computing and Engineering
University of Missouri-Kansas City, Kansas City, MO 64111

Abstract. A mobile ad hoc network is a distributed system in which nodes are self-organizing and rapidly deployable without relying on existing infrastructure. In this highly complex and flexible network, public key cryptography can be used to bring strong security features into a mobile ad hoc network. However, due to the cost associated with the public key computation and public key infrastructure, one can not directly bring public key cryptography into resource limited mobile devices. In this paper, we provide practical public key solution to mobile ad hoc networks with the special consideration of the resource limitation imposed by a mobile device. Our key idea is to utilize an on-line/off-line framework where a strong public key based share key is established off-line and used on-line. The off-line computation is performed in preparation for the on-line fast interaction. For secure on-line communications, we propose light-weight authentication and key agreement without public key computation.

Keywords: Mobile Ad Hoc Network; Wireless Communications; Authentication and Key Agreement; Public Key Cryptography

1 Introduction

In recent years, mobile ad hoc networks have gained tremendous amount of attention due to their self-organizing and rapidly deployable features. The use of such a network may arise as part of an emergency response to a natural disaster or terrorist attack, or military communications in a battle field, or communication among attendees at a conference. Due to its dynamic topology without relying on existing infrastructure, the feasibility of such a network is already a challenging task even without considering the security issues. In considering security issues such as authentication and key agreement, many papers such as [2, 6, 7, 9, 10, 14] in the existing literature have considered the use of public key based cryptography. However due to the resource limitation such as limited power supply and bandwidth in the mobile devices in a small isolated ad hoc network, some of the approaches adopted in these papers are not practical. In this paper, we aim to bring strong public key security services into the mobile ad hoc networks with the consideration of resource limitation in the mobile devices.

Public key cryptography is the trend in modern communications. It not only solves many problems in symmetric key cryptography, but also brings in many features that can not be achieved in symmetric key cryptography. First and foremost, public key cryptography simplifies key distribution and key storage requirement as one only needs a public and private key pair and a digital certificate to establish a shared key with any communicating party on the fly. In contrast, in symmetric key cryptography, each pair of communicating parties needs to pre-establish a secret key using some out-of-band mechanism such as a phone call, an e-mail, or a secure messenger. Second, the most well-known assumptions of public key cryptographic algorithms are based on solving difficult computational problems such as solving discrete logarithms or factoring; whereas the security of the conventional cryptographical algorithms, such as one-way hash functions and block ciphers, are based on the complexity of analyzing simple iterated functions of multiple rounds. To summarize, as the contribution of the public key cryptography is so great, many literature considered it as the true revolution in the entire cryptography history [11].

Since public key cryptography offers many superior properties, it would be nice to bring it to mobile ad hoc networks. Public key cryptography has some major costs nonetheless. For starter, to make public key cryptographic systems work, a universally acceptable public key infrastructure (PKI) must be established. PKI includes digital certificates, public key computations, and certificate authorities (CAs). The CAs are needed to certify public keys so that verifiers are assured the authenticity of the public keys. This infrastructure is mostly available via today's Internet, but it would be too costly to replicate for every small, isolated network as in mobile ad hoc environment. Second, public key certificates bring in the requirement of the real-time validation of Certificate Revocation List (CRL) so that verifiers do not use the revoked public key certificates. The limited communication bandwidth in mobile devices makes real-time certificate validation impossible. Third, even though the security base of public key cryptographic algorithms is strong, the computation of them can be costly as compared to symmetric key cryptography. Limited computational/battery power in mobile devices makes public key computations infeasible. With the consideration of above costs, one must overcome many challenges before bringing public key cryptography directly into resource limited mobile devices and mobile ad hoc networks.

Nevertheless, as the applications for mobile ad hoc networks are getting immensely popular, the need for the strong security is imminent. While one can not directly bring the public key cryptography into mobile devices at the present time, one can change the way public key cryptography put into the mobile devices and still maintain the strong security features in the services provided by the public key cryptography. These services include user authentication, data confidentiality, and data integrity. These services can be achieved through Authentication and Key Agreement (AKA). This paper provides several practical approaches to bring these services into mobile ad hoc networks.

1.1 Model

Our main idea is to utilize an on-line/off-line framework where a strong public key based share key is established off-line and used on-line. The key established off-line is a master key. During the on-line phase, a session key is established for each session. Each session key can only be used one time. This is akin to modern cellular communications where a session key is derived from a master key and used one-time for each communication session.

Specifically, in the off-line environment, we assume that abundant computation power and bandwidth with accessibility to wireline network are available to a mobile device, see Figure 1. Thus, in the off-line phase when timing is not critical, a mobile device can perform sophisticate public key computations such as key agreement with other mobile devices and store the results for use in the on-line phase. (For example, SSL, commonly used in our daily life, can be used for key exchange.) However, under ample resource condition, if a mobile device itself still can not perform sophisticate public key computation, we assume other devices can perform the computation on behalf of the mobile device and load the result into the mobile device. In the on-line phase when timing is very critical, all interactions between the mobile devices are light-weight (i.e., no public key computation), utilizing the public key based share key established in the off-line phase.

We provide three AKA models. First, we assume that the set of mobile devices to be deployed to a field is known in advance. These mobile devices can perform off-line pair-wise key agreement in advance. Second, assume that in a dynamic environment the mobile devices that will be deployed are not known ahead of time. A key registration agent R can be used. A mobile device can first register with agent R to obtain a public key based share key with agent R in the off-line phase, and then perform light-weight AKA with other mobile devices through agent R in the on-line phase. A key registration agent is common for mobile ad hoc networks. For example, an emergency response team for disaster relief usually has a command center which can serve as a key registration agent. Third, a hybrid of the previous two models can be used for the mobile devices to be deployed, and the mobile devices which wish to join in after an ad hoc network is established.

1.2 Advantages

There are several advantages in our proposed models for mobile ad hoc networks.

- Provide strong public key agreement in mobile devices during off-line phase.
- Provide strong yet light-weight authentication between the resource limited mobile devices during on-line handshaking.

- Allow for group member formulation statically (i.e., the set of mobile devices to be deployed is known in the off-line phase) and dynamically (i.e., mobile devices can join in after a mobile ad hoc network is established.)

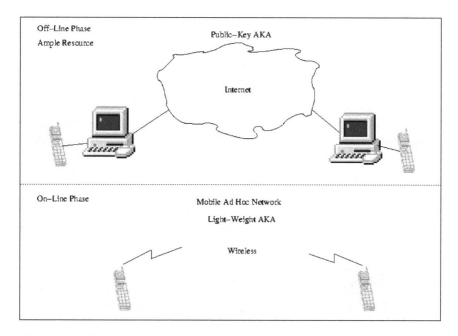

Figure 1: Mobile Ad Hoc Network: On-line/Off-line AKA

1.3 Motivation

Currently, a lot of research in this area centers around making mobile ad hoc networks feasible, and relatively less attention is given to the security related issues. Compared to other wireless networks, a mobile ad hoc network is especially vulnerable to security attacks (such as eavesdropping, spoofing, and denial-of-service attacks) due to its dynamically changing topology. In fact, at the present time, there is no security model that is applicable to the mobile ad hoc network. When security is lacking (e.g., relief teammates can not trust each other, or no communication privacy among the rescuers), the response to the disaster itself may contribute to an even larger problem. For example, a relief worker must be able to verify the legitimacy of a message received in his mobile device commanding him to move to another part of disaster area. Otherwise, it could be an impersonator trying to purposely disrupt the relief effort.

1.4 Related Work

In the literature, Aboudagga et al. [1] provided an excellent review of the authentication protocols with a considerable number of references for ad hoc networks. The authentication protocols used in the sensor networks can also be used for ad hoc networks [4].

Among the papers that use public key in mobile ad hoc networking, Capkun et al. [2] considered the public key management system by having the mobile users themselves generate the keys, issue certificates, and perform authentication. Zhou et al. [14] used threshold cryptography for distributed public key management. Yi et al. [13] used threshold cryptography for establishing the certificate authority. In general, threshold cryptography requires a fair amount of computation and communications as multiple entities are involved to produce a key. Therefore, it is not appropriate for mobile ad hoc networks. Lehane et al. [9] proposed the use of distributed shared RSA key generation techniques. Many of these papers however do not consider the computation consumption and bandwidth limitation in the mobile devices, thus they are not practical for mobile ad hoc networks.

Tseng [12] proposed to use heterogeneous networks, such as satellite and cellular networks, for public-key management in the mobile ad hoc networks. An excellent survey paper in various aspects of mobile ad hoc networks can be found in [3].

1.5 Main Contributions

The following lists the main contributions of this paper.

- We utilize the idea of on-line/off-line framework in mobile ad hoc networks to overcome resource limitation in the mobile devices. In particular, strong public key based share key is established during the off-line phase and used during the on-line phase.

- We propose light-weight authentication and key agreement during the on-line phase without public key computation.

1.6 Organizations

The rest of the paper is organized as follows. Section 2 describes the practical on-line/off-line public-key based AKA models. Section 3 describes the peer-to-peer On-line Light-weight AKA procedure, and section 4 describes the on-line light-weight AKA procedure via a registration agent. Section 5 provides a summary and conclusion.

2   Practical On-Line/Off-Line Public-Key Based AKA

In this section, we describe a practical way of using public key in establishing AKA in mobile ad hoc networks. In particular, three models are proposed depending on whether the set of mobile devices is known before deployment. These models utilize the on-line light-weight AKA techniques described in sections 3 and 4.

For establishing a mobile ad hoc network dynamically, there can be an assumption as to whether the set of mobile devices is known before the network is established. The simplest case is when the set of the mobile devices is known before deployment. These mobile devices can establish common shared keys ahead of time. If the set of mobile devices is not known before deployment, a registration agent can be used. The following lists three models.

Model 1: The set of mobile devices is known before deployment

> This is the case for a small emergency relief team in which the teammates are trained together and know each other before deployment. In such case, during the off-line phase in which ample resource is available, the mobile devices can perform in advance pair-wise authenticated public key agreement using Harn et al.'s DH (Diffie-Hellman) key exchange integrated with DSA (Digital Signature Algorithm) scheme in [5] or IEEE standard P1363 [8]. If a mobile device itself is not capable of performing sophisticate public key computation in the off-line phase, one can utilize other devices for performing the computation on behalf of the mobile device and load the result into the mobile device. Once a shared key is established between each pair of mobile devices, immediately after deployment, each pair of mobile devices can execute the peer-to-peer on-line light-weight AKA as described in section 3 to authenticate each other.

Model 2: The set of mobile devices is not known before deployment

> In this dynamic setting, a key registration agent R can be used. A mobile device can first register with agent R to obtain a public key based share key between the mobile device and agent R, and then perform light-weight AKA with other mobile devices via agent R using on-line light-weight AKA described in section 4. A key registration agent is common for mobile ad hoc networks. For example, a conference usually has a registration desk which can serve as a key registration agent.

Model 3: A hybrid model between models 1 and 2

> In this hybrid setting, some mobile devices have already established shared keys in the off-line phase, and other mobile devices have not established shared keys. Similar to model 2, a key registration agent R can be used for all mobile devices. In the on-line phase, the mobile devices have already established shared keys in the off-line phase can quickly mutually

authenticate each other using peer-to-peer on-line light-weight AKA described in section 3 and establish a mobile ad hoc network. Other mobile devices can join the mobile ad hoc network via key registration agent R to obtain the shared key with R first and then perform on-line light-weight AKA described in section 4. For example, an out-of-state relief worker needs to first register with agent R before joining an established mobile ad hoc network. Note that in this case, each mobile device must establish off-line pairwise authenticated public key agreement with R so that any mobile device joins later can communicate with any other mobile device in the network.

3   Peer-to-Peer On-Line Light-Weight AKA

In this section, we describe an AKA scheme between two mobile devices by utilizing on-line/off-line framework. Each mobile device in a mobile ad hoc network functions independently without exerting any trust assumption between each other.

In our scheme, before deploying the mobile devices, each pair of mobile devices first uses an authenticated public key agreement scheme such as [5, 8] to establish an authenticated shared key. If a mobile device is not capable of performing sophisticate public key computation, other devices can perform the computation on behalf of the mobile device and load the result into the mobile device. Once an authenticated shared key (K_{AB} for mobile devices A and B) is obtained, the procedure in Figure 2, known as *Peer-to-Peer On-Line Light-Weight AKA*, can be used for on-site authentication and session key agreement between a pair of mobile devices. Basically, in the on-line phase, mobile devices A and B, each chooses a nonce, n_A and n_B respectively, to challenge the other party, where h is a hash function. The shared session keys, k_A and k_B, one for each direction of communication, are derived from the shared key K_{AB} established in the off-line phase. Both parties are mutually authenticated after three rounds of message passing. Note that a shared session key selected by entity i is denoted with subscript i. For example, k_A is a shared session key that is selected by A and shared with B for the message sent from A to B. B calculates shared key k_A based on the nonce n_A sent to B and the share key K_{AB}.

The name "light-weight" comes from the fact that this AKA procedure does not involve full-blown public key crytpography computation. Instead, hash functions are used for computing shared session keys. This scheme is very suitable for many different types of applications after a shared key is obtained. For example, mobile devices in a mobile ad hoc network for disaster recovery or mobile devices in group conferencing can utilize the peer-to-peer light-weight AKA for on-site handshaking before cooperating with each other. WLAN 802.11 can utilize the light-weight AKA after a mobile device establishes a shared key with the AAA server.

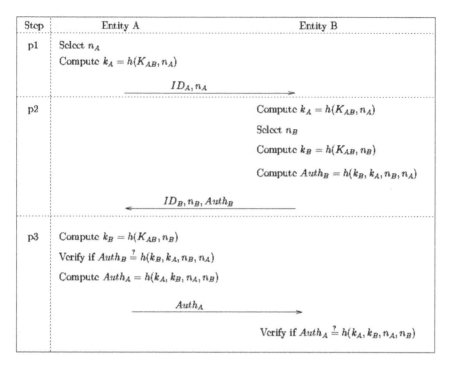

Step	Entity A	Entity B

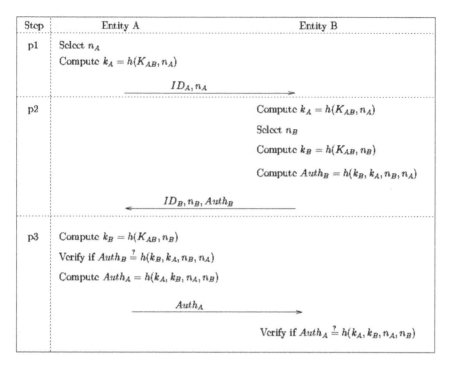

Figure 2: Peer-to-Peer On-Line Light-Weight AKA: shared key K_{AB} between mobile devices A and B has been already established during the off-line phase

4 On-Line Light-Weight AKA via a Registration Agent

In this section, we describe an AKA scheme between two mobile devices via a registration agent R. Again, each mobile device in a mobile ad hoc network functions independently without exerting any trust assumption between each other.

In the off-line phase, a mobile device first uses an authenticated public key agreement scheme such as [5, 8] to establish an authenticated shared key with registration agent R. Once an authenticated shared key is obtained, the procedure in Figure 3, known as *On-Line Light Weight AKA via a Registration Agent*, can be used for on-site authentication and session key agreement between a pair of mobile devices after deployment. Basically, in step d1, each entity individually authenticates with R by using peer-to-peer on-line light-weight AKA. In step d2, the shared session key k_A is then securely transferred to entity B, and key k_B to entity A, using h as a hash function, E as an encryption function, and D as a decryption function. Entity A can then use session key k_A for any message that is to be securely transmitted to entity B. Similarly, B uses k_B for any message to A.

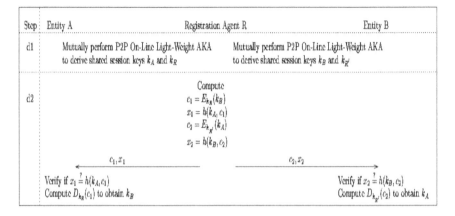

Figure 3: On-Line Light-Weight AKA via a Registration Agent: shared keys K_{AR} and K_{BR} have been already established during the off-line phase.

5 Summary and Conclusion

In this paper, with the consideration of the resource limitation in a mobile device, we change the way public key cryptography put into the mobile device and still maintain the strong security features in the services provided by the public key cryptography. In the off-line phase when timing is not critical and ample resource is available to a mobile device, the mobile device can perform sophisticate public key computations such as key agreement with other mobile devices and store the results for use in the on-line phase. In the on-line phase when timing is very critical, we propose light-weight authentication and key agreement between mobile devices.

References

1. Aboudagga, N., Refaei, T., Eltoweissy, M., DaSilva, L., Quisquater, J.-J. "Authentication Protocols for Ad Hoc Networks: Taxonomy and Research Issues." Proceedings of the first ACM International Workshop on Quality of Service and Security in Wireless and Mobile Networks. Montreal, Quebec, Canada, October 13, 2005.

2. Capkun, S., Buttyan, L., and Hubaux, J.-P. "Self-organized public key management for mobile ad hoc networks." IEEE Transactions on Mobile Computing. vol. 2. no. 1. January-March. 2003.

3. Chlamtac, I., Conti, M., and Liu, J. "Mobile ad hoc networking: imperatives and challenges." Ad Hoc Networks. vol. 1. pp. 13-64. 2003.

4. Du, W., Wang, R. Ning, P. "An efficient scheme for authenticating public keys in sensor networks." Proceedings of 6th ACM International Symposium on MObile Ad Hoc Networking and Computing (MobiHoc), 2005.

5. Harn, L., Mehta, M., and Hsin, W.-J. "Integrating Diffie-Hellman Key Exchange into the Digital Signature Algorithm (DSA)." IEEE Communications Letters. vol. 8. no. 3. March 2004.

6. Hietalahti, M. "Efficient key agreement for ad hoc networks." Master's thesis, Helsinki University of Technology. Department of Computer Science and Engineering. Espoo, Finland. May 2001.

7. Hubaux, J.-P., Buttyan, L., and Capkun, S. "The quest for security in mobile ad hoc networks." Proc. ACM Symp. Mobile Ad Hoc Networking and Computing (MobiHoc). 2001

8. Law, L., Menezes, A., Qu, M., Solinas, J., and Vanstone, S. "An Efficient Protocol for Authenticated Key Agreement." Technical report CORR 98-08. Dept of C&O, University of Waterloo, Canada. March 1998.

9. Lehane, B., Doyle, L., and O'Mahony, D. "Shared RSA key generation in a mobile ad hoc network." MILCOM. October. 2003.

10. Van der Merwe, J. Dawoud, D., McDonald, S. "Fully self-organized peer-to-peer key management for mobile ad hoc networks." Proceedings of the 4th ACM workshop on Wireless Security (WiSe). Germany. September 2, 2005.

11. Stallings, W. Cryptography and network security - principles and practice. Prentice Hall. 1998.

12. Tseng, Y.-M. "A heterogeneous-network aided public-key management scheme for mobile ad hoc networks." International Journal of Network Management. vol 17. pp. 3-15. 2007.

13. Yi, S. and Kravets, R. "MOCA: Mobile certificate authority for wireless ad hoc networks." 2nd annual PKI research workshop program (PKI 03). Gaithersburg, Maryland. April. 2003.

14. Zhou, L. and Haas, Z. "Securing ad hoc networks." IEEE Network. vol. 13. no. 6. pp. 24-30. Nov/Dec 1999.

A Recursive Method for Validating and Improving Network Security Solutions

Suleyman Kondakci

Faculty of Computer Science, Izmir University of Economics,
suleyman.kondakci@ieu.edu.tr

Abstract. We present a simple and efficient method called Validate Improve Recur (VIR) to assess and improve information security solutions. We consider validation and improvement of new security designs and evaluation of existing security implementations. The presented work will enable security administrators and evaluation facilities to improve security solutions, identify, monitor, and manage risks that may propagate to final security implementations. As a self-assessment method inexperienced users can also apply VIR to design and evaluation tasks.

Keywords: security evaluation, security design validation.

1 Introduction

Security is a matter of human activity, but while we see and react to the technological matters, we are often blind to the fact it depends fundamentally on human factors. Thus, we should distance ourselves from intuition regarding the central aspects of security risks. In regard to Turkey, majority of network owners consider the security evaluation as the least significant task in a lifecycle security planning, [1]. Others, in fact a considerable portion of the management, merely trust border protection mechanisms (firewalls) and ignore the regular risk assessment and test and evaluations. Furthermore, most managers rely on ad hoc solutions based on typical market rumors. Network administrators from several leadership companies have serious obstacles in convincing management about the severity of the security risks and threats against their systems. Due to this fact, many of the network administrators encounter problems in demonstrating the real impact of security risks. A simple approach is needed to illustrate impacts of the internal and external threats against the company assets. With this paper we intend to present a simple methodology to validate and improve security solutions including quantitative means to support the development of evaluation tools.

Some internationally recognized standards and approaches, e.g., Common Criteria (CC) and BS 7799, have published guidance and methodologies for information security evaluation and improvement. These standards and their approaches are detailed in [2-7]. For many network administrators and evaluation facilities these methodologies seem far too complicated and impractical to apply. The work presented in [8] reviews these and several other engineering standards and metrics used within IT and information security. As already known, CC and BS 7799 define standards and methodologies in information security evaluation and certification. However, they do not precisely define the methodology for risk assessment. These standards specify only that the organization should use a systematic approach to risk assessment. Hence, evaluation facilities and system administrators can benefit the VIR method presented here as a complementary risk management system. This simple and inexpensive approach is a practical means for system administrators and security professionals to validate various solutions, illustrate differences between risks found in these solutions, and propose improvements if necessary. There are a multitude of commonly encountered problems in IT environments. First of all, system administrators have difficulties in illustrating real impacts of risks to management. Impact due to lack of awareness of users (especially inexperienced) is extremely high. Furthermore, most network owners are easily manipulated by external (magazine) marketing activities. Consequently, they purchase overlapping tools and services often causing inefficient solutions. The complexities of these kinds of "ad hoc" solutions diminish the performance and quality of services. This attitude significantly affects the balance of the security solutions while causing a considerable amount of negative cash flow from the company.

The company management must make the need for balanced solutions clear. We must have a balance between the protection mechanisms and the protected environment. Unnecessarily over-equipped networks are not always the most secure networks; in fact, they may be the most expensive and inefficient ones. We consider here this aspect by validating various security solutions for the same environment in order to illustrate the most feasible solution.

Section 2 gives an overview of sources covering the methodologies applied to information security risk assessment. Section 3 presents a simple guidance that can be mainly used by the system administrators to analyze deficiencies of different solutions. Section 4 details the VIR approach and Section 5 deals with validation and improvement guided by VIR. Finally, Section 6 concludes the paper and suggests further enhancements.

2 Related Work

There are various risk analysis and management methodologies applied to information security management. Most of the methodologies are based on qualitative decision-making. Some approaches are known as probabilistic risk analysis methods, while others are quantitative. In [9] a review of information risk and security modeling is given. The paper analyzes some technical and organizational level risk and security metrics of several standards such as Common Criteria/ISO15408, NSA configuration guidelines and metrics, Information IT operational standards GMITS/ISO13335 and ITIL/ITMS, ISO7498-2, ISO17799, COSO, CobiT, and SSE-CMM/ISO21827. A fuzzy group decision-making approach in security risk assessment is discussed in [10]. An automated risk management approach is presented in [11].

The VIR method presented here is unique in its simple and practical approach to evaluating and providing guidance for recovering deficiencies on the fly during the evaluation. In particular, its simplicity can make it possible for novice users to apply it as a self-assessment tool.

Currently, most of the assessment approaches apply or adapt existing methodologies to display current impacts, but not the management of risks. Analogous to the VIR approach it is worth to mention some interesting sources, [12-14]. A quantitative risk analysis method is considered in [12]. This study proposes an integrated analysis model, containing asset, threat and vulnerability evaluations, adapted from software risk management techniques to assess the security risk quantitatively and optimize company resources. The work given in [13] proposes a risk assessment framework that combines features from fault trees and influence diagrams for assessing the 'surety' of information systems. Adaptation of business process modeling presented (BMP) in [14] is another comprehensive approach used both for risk analysis and security design for improvement. Thus, several BPM techniques have been reviewed in [14].

3   Simple Self-assessment Framework

The VIR methodology utilizes one basic questionnaire containing specific control objectives and techniques to determine current status on the security program of the assets that can be evaluated. Initially, the assessment will audit the existence of current security policies and mechanisms. If these are not available, the system administrator (or security professional) will compose a set of new security requirements, and hence obtain a new security design as a part of the assessment and improvement.

In fact, there are several unknown problems that can be difficult to identify and solve. However, a security professional can simply create various random scenarios to demonstrate different security problems and solutions. During this, a self-assessment method or tool must be presented to the security administrators and professionals. Indeed, such a method can provide invaluable functionality in information security education, and even for security test and evaluation facilities. In short, VIR is intended to provide a simulation-based method for system administrators, security professionals, and managements to determine the current status of their information security programs. The process flow of a sample simulation of a self-assessment process consists of (1) presentation of different problem scenarios with different solutions and risks, (2) illustration of impacts by validating risk propagations in the current solution, (3) demonstration of impacts found in alternative suggestions and simulations, and optionally, displaying some actual security attacks and survival scenarios.

3.1 Roadmap for Improving Solutions

The VIR approach relies on presentation of different problem scenarios and associated security solutions to diminish risks and overlapping functionalities. The scenarios consider sample solutions containing actual assets with varying importance. The tasks are separately determination of protected and unprotected overall risk, determination of risks in a custom designed/configured solution, and illustration of results using graphical charts and quantitative impact values. These are practically important to convince non-technical users, e.g., managers, without overwhelming them with theoretical substantiations.

Furthermore, the evaluator can define a new set of policies and measures to compose alternative security solutions for the company. The suggested solutions should at least provide: (1) policy, implementation, and administration procedures, (2) periodic auditing and risk assessment plan, (3) staff and user training guidance, and (4) incident response procedures and lifecycle security plan.

Every solution proposed by VIR is guided using a security design architecture containing requirements (objectives), policies, design, and security mechanism placement. The Site Security Handbook [15] is a useful source considered as system/security administrators' bible. This handbook is a guide to setting computer security policies and procedures for sites (networks) that are connected to Internet. Furthermore, X.800 Security mechanisms and services as presented in [16] describe structured standards, definitions, and architectures to facilitate security development and management tasks for product vendors and users. A practical and useful source for composing applicable security policies is the SANS security policy project found in [17]. Compared to the X.800 approach, which is rather comprehensive, the design guidance of VIR, as illustrated in Fig. 1, is compact, simple, and illustrative.

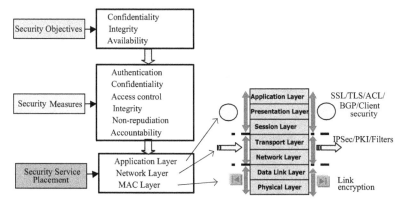

Fig. 1. The simplified roadmap for overall security design and management tasks.

Prior to evaluations, existing company objectives and policies are analyzed to obtain the status of current security implementation. If a current implementation is available, then it is evaluated. If the company cannot present any systematic security implementation, and just relies on some ad hoc safeguards, then the safeguard information is collected and evaluated. In the former case, previous and the current evaluation results are used to determine and illustrate comparable risk-impact patterns. In the latter case, the security professional will probably face the challenge of demonstrating the difference between protected and unprotected scenarios and the difference between possible NP-hard solutions.

To demonstrate various problem scenarios, commonly encountered attacks are simulated. Currently, there are several attack taxonomies readily available for use in computer simulations. A valid source [18] of security taxonomy, presented by Avizienis et al. discusses basic concepts and taxonomies of dependable and secure computing. Kuzmanovic et al. in [19] present an analysis of a class of low-rate denial of service (DoS) attacks, which are difficult for routers and counter-DoS mechanisms to detect. They also use a combination of analytical modeling, simulations, and experiments in the DoS analysis. Ellis [20] analyzes worm anatomy and models by discussing the life cycle of a worm based on a survey of contemporary worms. Such well-known attacks are easily incorporated into the attack list of VIR. For example, a virus spread scenario and a DoS attack against some critical nodes (e.g., a router, a web-server, a transaction server, or an e-mail server) can be simulated. These attacks can be launched against networks of varying protection configurations. Most importantly, these two groups of attacks (worms and DoS) typically cause the widest spectrum of harm having a remarkable potential to affect (directly or indirectly) any node residing in the network attacked. The proper realization of the attacks and the accuracy of calculated risk levels are important in order to illustrate the actual effect of simulations. The proper realization and accuracy are crucial, since the evaluator needs to convince the management by proving both the strength of the suggested solution and the effect of the harms.

4 Quantitative Assessment of Assets

We find numerous risk assessment approaches applied to different fields of science. Some theoretical and practical approaches related to information security are given in [21-24]. VIR is unique compared to these approaches.

Since the VIR approach uses quantitative risk evaluation, we need first to present definitions used in the quantitative computations. The definitions consist of basic entities such as *assets, asset values, metrics, evaluation levels, asset criticality*, and a *scoring scheme*. An *asset* denotes a network node or any IT product that is under evaluation. Every asset is assigned a weight called *asset value* to classify the weight or importance (criticality) of the asset on a scale from 0 to 5 (5 being the most important). Functionality, information classification (e.g., classified, unclassified, restricted, secret, etc), and topological location, or the degree of exposure to threats of an asset, are the key parameters for rating the assets. As seen in Fig. 2, we use *security attributes* to represent security characteristics of an asset. As a common data structure, a heirarchical tree-structure is used to store values (items) of security attributes.

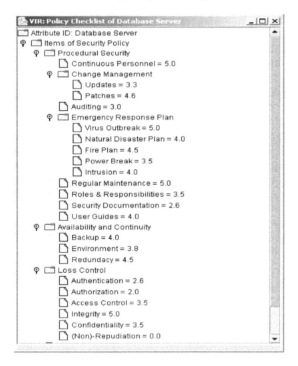

Fig. 2. The evaluation of policy of a database server with the main items *Procedural Security, Availability and Continuity*, and *Loss Control*, which can span into a tree structure.

Indeed, the data structure can store a security attribute of an asset or an entire system with several attributes. More precisely, an attribute, for example, can represent either a set of security policy items or items of an operational system under evaluation. Each attribute has an *attribute ID* describing what the attribute stands for, an *asset value, w,* describing the weight of the asset, and a list of scored *items,* representing evaluation results of the corresponding attribute items. Thus, an attribute (which is also defined as an asset) must contain at least three mandatory entries, asset (attribute) ID, asset weight (between 0 and 5), and items and their evaluation scores (btw. 0 and 5).

Fig. 2 illustrates the actual items of a database policy, but an evaluation can represent a general set of *N* items, where each item is scored according to strength or exposure to risk. For the strength evaluation, the scores represent the strength of each item, and for the risk assessment these will be chosen as appropriate to represent the risk scores. As shown above, the attributes are arranged in a tree of items. Some vulnerability items depend on other vulnerabilities, and some security strength parameters strengthen other parameters.

A typical example of an asset under evaluation can be a security mechanism, e.g., a firewall. In fact, evaluating security attributes of a firewall deals with the evaluation of the entire firewall and its associated environment. The attributes in the firewall evaluation can be remote authentication, availability (survivability under heavy traffic or DoS attacks), resistance to IP-spoofing, weakness in loss-control, authentication on a server, and maintenance cost.

Furthermore, we define a metric and a scoring scheme to quantify risk or strength of a given asset or an entire network of assets. Eq. (1) can be used to compute risks or strengths (by slightly modifying it) of a given solution. Obviously, this equation satisfies a linear definition of subranges, where the security attributes or items are treated with equal impact factors. We use a coefficient, called relative strength parameter, r, to adjust the relative strength of the items. Thus, the risk formula is

$$R^w = w\left(1 - \frac{S}{5}\right), \ S = r_j \frac{\sum\limits_{i=1}^{N} G_i}{N}, 0 < r_j < 1, (R^w, S, w) \in [0,5] \qquad (1)$$

Here, G_i denotes the risk (or strength) of the ith item, w the asset (attribute) weight, and N number of items that are evaluated for this asset, and S is the average score of an item with sub-items (or a plain item-list). The relative strength parameter r_j defines the relative strength of a group of items of an attribute. For example, for equally tempered strength factors of a policy asset we can set r_j to be constant, e.g., *1.0,* for all items. If adequate, we can also use a single relative strength parameter for each item of the asset. This can be impractical when evaluating a large network containing several assets, where each asset also has a number of items. Table 1 shows the evaluation results of main item groups of an asset representing policy-attribute of a database server with weight 4. The attribute consists of a group of items, PROCEDURAL, AVAILABILITY, and LOSSCTRL, where each group consists of further sub-items. Here, we can change rating of any group to be stronger than the others or vice versa. With equally tempered strength parameters (r = 1) for all attributes we obtain the risk factor $R^4 = 1.13$. Setting $r_2 = r_3 = 1$ and $r_1 = 0.2$ for the PROCEDURAL group we achieve higher overall risk factor as $R^4 = 1.96$. Some other test results are summarized in Table 1.

Table 1. Risk of an asset containing three main groups each having several sub-items.

Attribute	r_x	Risk (R^4)
Loss Control excluded (LOSSCTRL = 0)	$r_{3=0}, r_1 = r_{2=1}$	$R^4 = 1.87$ (M)
Availability/Continuity excluded (AVAILABILITY=0)	$r_{2=0}, r_1 = r_{3=1}$	$R^4 = 2.22$ (M)
Procedural excluded (PROCEDURAL = 0)	$r_{1=0}, r_2 = r_{3=1}$	$R^4 = 2.17$ (M)
All included	$r_1 = r_2 = r_{3=1}$	$R^4 = 1.13$ (W)

As seen, the last row depicts a weak (W) risk while others showing moderate (M) risks. This is due to the higher relative strength assigned to all groups. For instance, in the second row, AVAILABILITY group is not included in the evaluation. This has reduced the overall risk considerably.

4.1 Asset Classification and Measurement Scheme

VIR is based on simple and practical assessment criteria selected from predefined coverage metrics covering attributes of main security aspects such as confidentiality, integrity, and availability. The master measurement scale applied in VIR satisfies the required subranges to identify and classify the boundaries of four subdivided evaluation levels. Primarily, the master range is defined on an experimental scale of 0 to 5. Probabilistic models use the unit scale ranging from 0 to 1. The master scale is further divided into four evaluation levels (called subranges), *weak (W), moderate (M), severe (S)* and *very severe (V)*. These subranges (or evaluation levels) are quantified as

$$W = \{0 - 1.25\}, M = \{1.26 - 2.5\}, S = \{2.6 - 3.75\}, \text{ and } V = \{3.76 - 5\}.$$

For the strength evaluation of existing solutions or systems, *strong (S)* and *very strong (V)* are used instead of *severe (S)* and *very severe (V)*. The generic metric describing these subranges serves well as a simplified scoring scheme; however, the subranges can also be adjusted for customized needs or for different security solutions. Therefore, the boundaries for each of the subranges are designed to be flexible. This is a major requirement, because we need to apply the model to a broad range of environments representing a range of targets of evaluations with varying classifications and solutions. Computer-aided tool developers can also parameterize the scaling appropriately for the implementation of tools.

5 Validation and Improvement of Security Solutions

A security plan should pass through a design validation, the task of verifying design efficiency, to determine its strength. Validation involves several steps, such as the input data generation, defining coverage metrics, and evaluating the input data. A coverage metric defines a set of criteria that are used to determine undesired conditions in a test process. Often, security professionals prepare coverage metrics by evaluating the security requirements for the related environment. Fault detection is central to our model in which risk propagation is reduced by an iterative decision-making process. In relation to Fig. 1 (security roadmap), the evaluation process also assumes, as an initial constraint, a predefined attribute of security objectives (a set of objective items). Thus, a decision hierarchy starts with this parameter, and proceeds similarly with the assessment of policy attributes generated from the objective attributes. Further, the decision-making continues with the assessment of the security measures generated from the policy attributes. Analogous to the attributes of security objectives, predefined coverage metrics of both the policy and security measures are also used in order to complete the decision.

Please recall that, objectives describe the IT and security knowledge of the company, while policies define the protection of assets, procedural tactics, and the mechanisms used to enforce policy. Let us consider validation of a sample policy attribute shown in Table 2, which includes assessment scores and the criticality of each policy item. A policy attribute consists of a weight, its items (eventually with sub-items), and a relative strength parameter for each item. Thus, using Eq. (1) we compute the strength of each item consisting of sub-items. It may already be noted that Table 2 contains only the average scores and the criticality parameters of the items. The shaded entries of the table indicate the necessity of improvement. For example, the second and third items are not strong enough compared to their criticality parameters. Hence, an improvement must be provided so that the scores will upgrade to the *S* level, i.e., both item two and three must have values greater then or equal to *3.76*.

Table 2. The policy with assessed items and their critcality indicators (W,M, S, V).

Policy items	Score		Criticality
Administrator rights and responsibilities	4	(V)	V
User rights and responsibilities	2	(M)	S
Risk Assessment	1	(W)	S
Information Sensitivity	4		S
Password	2.3	(M)	V
Laptop Security and mobile devices	1	(W)	M
Extranet protection	1	(W)	M
Intranet protection	0	(W)	M
Anti-Virus	4.5	(S)	V
Wireless Communications	2	(M)	M
Dial-up, ISDN and Modem connections	2	(M)	M
Host Security	3	(S)	M
Audit trails and logs	1.2	(W)	M
Remote Access (authentication, authoriza tion, and confidentiality)	2	(M)	M
Acceptable Encryption (not required)	-		W
Emergency response	1	(W)	V
Connection to Internet (address and conten based filtering)	0.4	(W)	M
Connection from Internet (address an content based filtering)	3	(S)	S
Physical security	1.6	(M)	M
Backups	1	(W)	S
Change Management and Patches	2	(M)	S
Procedural policy	1	(W)	M

For larger set of data VIR uses the simple stochastic ordering method, given in [25], to compare test results against asset criticality. Stochastic ordering is also useful for allowing analysts to qualitatively compare systems. Simply, If X_1 and X_2 are random variables with distribution functions F_1 and F_2, then X_1 is said to be stochastically larger than X_2 if $F_1 \le F_2$.

The existing solution can be improved further while it is being evaluated, depending on the satisfaction of the evaluation result. If necessary, the overall assessment may be performed recursively in several consecutive steps. First, an initial solution is assessed and scored. The next step is the verification of the solution, during which the score set is mapped to one of the four subranges weak (W), moderate (M), severe (S), and very severe (V). The evaluator compares the measured score to the asset's criticality, which can be obtained by assigning a numerical importance to the asset as:

C1: *unclassified (public)*; weighed between 0 and 1.25,
C2: *classified*; weighed between 1.26 and 2.5,
C3: *restricted*; weighed between 2.6 and 3.75,
C4: *secret* weighed between 3.76 and 5.

If the criticality (importance) parameter is greater than the evaluation score, then an improvement (or a new solution) is suggested. Otherwise, upon a satisfactory result the assessment is terminated. As the Validate Improve Recur (VIR) method implies, this process can iteratively continue (recur) until a desired solution is achieved. This also implies how a network administrator can be able to make use of the self-assessment approach. VIR based security evaluation and design is crucial for critical information systems. As illustrated in the preceding example, the security designer evaluates a set of readily available security policy. The better the security policy the stronger the protection mechanisms enforced by the security policy. Therefore, the designer is motivated to improve the security policy before proceeding with the implementation of security mechanisms.

6 Conclusions

It is imperative that organizations understand the fact that the human factor is the root cause of many security incidents, which needs particular attention. Security in any system is commensurate with its risks. However, to determine appropriate and cost effective security measures is, in many cases, a complex and subjective matter. Therefore, risk analysis, like VIR method, is an essential component in enabling this to be carried out on an objective basis, and thus allowing the risk to be managed effectively by the system owner. Technical security administrators and non-technical managers have different understandings of the real impacts of security threats.

The simulation based self-assessment approach is an efficient way to evaluate, improve, and demonstrate the weaknesses in any security planning. Thus, the VIR methodology also provides a training outcome during the evaluation. In the case of weak implementations, the evaluator is guided to compose a new policy set, and this will continue recursively until a satisfactory solution is achieved. The benefit of increased security knowledge is achieved after each validation process executed by the system administrator. Obviously, this will implicitly increase the security-awareness for both the administrators and the company management if illustrated appropriately.

References

1. Kondakci, S.: Controlling Risks in Large Scale Computer Networks, Proceedings of ICSP'2003, Vol.1, No.2, International Journal of Computational Intelligence (2003) 7-10
2. Common Criteria, Common Criteria and the Common Evaluation Methodology V 3.0, http://www.commoncriteriaportal.org/, accessed (2006)
3. Consultative Objective and Bi-functional Risk Analysis: COBRA Tools, ISO/IEC 17799 Compliance and Security Risk Analysis Approach, http://www.bspsl.co.uk/17799/ (2007)
4. Common Criteria/ISO IS 15408, Version 2.1, http://csrc.nist.gov/cc/ccv20/ccv2list.htm (2007)
5. ISO/IEC, FDIS 15408-1: Information Technology - Security techniques - Evaluation criteria for IT security, http://www.gammassl.co.uk/ist33/ (2005)
6. Herrmann, D.S.: Using the Common Criteria for IT Security Evaluation ISBN:0849314046, Taylor & Frcis CRS Press (2002)
7. GAO/AIMD-00-33, Information Security Risk Assessment: Practices of Leading Organizations, United States General Accounting Office (GAO), http://irm.cit.nih.gov/itmra/gaoguid.html (2006)
8. Peltier, R.T.: Information Security Policies, Procedures, and Standards: Guidelines for Effective Information Security Management, ISBN: 0849311373, Taylor & Frcis CRS Press (2001)
9. Zivic, P.: Information Risk and Security Modeling, Proceedings of SPIE - The International Society for Optical Engineering, Vol. 5812, Elsevier Inc. (2005) 142 - 150

10. Fang, L., Kui, D., Zhiying, W.: Jun, M., Research on Fuzzy Group Decision Making in Security Risk Assessment, Lecture Notes in Computer Science, Vol. 3421, No. II, Elsevier Inc. (2005) 1114 – 1121

11. Vassilis, T., Theodore, T.: From Risk Analysis to Effective Security Management: Towards an Automated Approach, Information Management and Computer Security, Vol. 12, No. 1 Elsevier Inc. (2004) 91 – 101

12. Peter, H., Young-Gab, K., Taek, L., Moon, C-J., Yoonjung, Y., Injung., K.: A Security Risk Analysis Model for Information Systems, Lecture Notes in Artificial Intelligence, Vol. 3398, Elsevier Inc. (2005) 505-513

13. Wyss, G. D., Fletcher, S. K., Halbgewachs, R. D., Jansma, R. M., Lim, J. J., Murphy, M., Sands, P.D.: Toward a Risk-Based Approach to the Assessment of the Surety of Information Systems, American Society of Mechanical Engineers, Pressure Vessels and Piping Division, Vol. 296, Elsevier Inc. (1995) 529 – 535

14. Kokolakis, S.A, Demopoulos, A.J., Kiountouzis, E.A.: The Use of Business Process Modelling in Information Systems Security Analysis and Design, Vol. 8, IS. 3, Information Management & Computer Security (2000) 107-116.

15. Curry, D., Kirkpatrick, S., Longstaff, T., Hollingsworth, G., Carpenter, J., Fraser, B., Ostapik, F., Sturtevant, A., Long D., Duncan, J., Byrum, F.: RFC 1244 – The Site Security Handbook, IETF, http://www.ietf.org, (2006)

16. CCITT, Recommendation X.800: Security Architecture for Open Systems Interconnection for CCITT Applications, (1991)

17. The SANS Security Policy Project, http://www.sans.org/resources/policies/, accessed (2007)

18. Avizienis, A., Laprie, J-C., Randell, B., Landwehr, C.: Basic Concepts and Taxonomy of Dependable and Secure Computing, IEEE Transactions on Dependable and Secure Computing, Vol. 1 (2004) 11–33

19. Kuzmanovic, A. and Knightly, E. W.: Low-rate TCP-targeted Denial of Service Attacks and Counter Strategies, IEEE/ACM Trans. Netw. Vol. 14, No. 4 (2006) 683-696

20. Ellis, D., Worm anatomy and model. In Proceedings of the 2003 ACM Workshop on Rapid Malcode, (2003) 42-50

21. Guarro, S.B.: Risk Analysis and Risk Management Models for Information Systems Security Applications, Reliability Engineering & System Safety, Vol. 25, No. 2 (1989) 109-130

22. Bao-Chyuan, G., Chi-Chun, L., Ping, W., Jaw-Shi, H.: Evaluation of Information Security Related Risks of an Organization - The Application of the Multi-criteria Decision-making Method, Proc. of IEEE Annual International Carnahan Conference on Security Technology, Elsevier Inc. (2003) 168 – 175

23. Anderson, A.M. and Longley, D. and Tickle, A.B.: Risk Data Repository: A Novel Approach to Security Risk Modelling, IFIP Transactions A: Computer Science and Technology, No. A-37 (1993) 185 - 194

24. Tipton, H.F., Krause, M.: Information Security Management Handbook, 5th Ed., ISBN: 0849319978, Taylor & Frcis CRS Press (2003)

25. Kondakci, S: A New Assessment and Improvement Model of Risk Propagation in Information Security, Int. Journal of Information and Computer Security, Vol. 1, No. 3, (2007) 341-366

Secure Multiparty Overall Mean Computation via Oblivious Polynomial Evaluation

Mert Özarar[1] and Attila Özgit[1]

[1] Department of Computer Engineering, Middle East Technical University,
06531, Ankara, Turkey
{ozarar, ozgit}@ceng.metu.edu.tr

Abstract. The number of opportunities for cooperative computation has exponentially been increasing with growing interaction via Internet technologies. Most of the time, the communicating parties may not want to disclose their private data to the other principal while taking the advantage of collaboration, hence concentrating on the results rather than private and perhaps useless data values. To conduct such a computation while preserving the privacy of the inputs for a target case is referred to as Secure Multiparty Computation problem. In this work, the privacy preserving overall mean computation problem is analyzed. We present two protocols for two-party and multi-party case via oblivious polynomial evaluation is used as a cryptographic mechanism to assure the security of the private data. Under given security assumptions, the privacy validity of the algorithms are justified.

Keywords: Secure Multiparty Computation, Overall Mean, Oblivious Polynomial Evaluation

1 Introduction

The number of opportunities for cooperative computation has been increasing exponentially with growing interaction via Internet technologies. These computations could occur between trusted partners, between partially trusted partners, or even between competitors. Most of the time, the communicating parties may not want to disclose their private data to the other principal while taking the advantage of collaboration, hence concentrating on the results rather than private and perhaps useless data values. For example, two or more competing large organizations might jointly invest in a project that must satisfy all organizations' goals while preserving their private and valuable data [4]. For performing such computations, if none of the parties can be trusted enough to know all the inputs, privacy will become a primary concern. Hence the techniques for secure multiparty computation are quite relevant and practical to overcome the privacy gaps.

1.1 Privacy Constraints

The definition of privacy in the cryptographic community limits the information that is leaked by the distributed computation to be the information that can be learned from the designated output of the computation [15]. Although there are several variants of the definition of privacy, for the purpose of this discussion we use the definition that compares the result of the actual computation to that of an "ideal" computation. Consider a party that is involved in the actual computation of a function (e.g. a data mining algorithm). Consider also an "ideal scenario", where there is also a "trusted party" who does not deviate from the behavior that we prescribe for him, and does not attempt to cheat. In the ideal scenario all parties send their inputs to the trusted party, who then computes the function with complete input data set and sends the results to the parties of the computation.

A protocol is secure if anything that an adversary can learn in the actual world, it can also learn in the ideal world, namely from its own input and from the output it receives from the trusted party [16]. In essence, this means that the protocol run for computing the function does not leak any "unnecessary" information. Of course, there are partial leaks of information that may be considered as harmless. It is hard, however, to decide which type of leakage can be tolerated. The cryptographic community therefore aims at designing protocols that do not reveal any information except for their designated output, and in many cases such protocols can be efficiently constructed.

1.2 Secure Multiparty Computation

If multiple parties want to perform a computation based on their private inputs, but neither party is willing to disclose its own input to anybody else, then the basic problem is how to conduct such a computation while preserving the privacy of the inputs. This is referred to as Secure Multiparty Computation problem (SMC) in the literature.

In general, a secure multiparty computation problem deals with computing any probabilistic function on any input, in a distributed network where each participant holds one of the inputs, ensuring independence of the inputs, correctness of the computation, and that no more information is revealed to a participant in the computation other than that can be inferred from the participant's input and output [6]. Consider a trusted party who collects all participants' data and then performs the computation and sends the results to the participants. Without having a trusted party, some communication among the participants is certainly required for any related computation, yet we do not know how to ensure that this communication doesn't disclose anything. One solution is to allow non-determinism in the exact values sent for the intermediate communication (e.g., encrypt with a randomly chosen key) and demonstrate that a party with just its own input and the result can generate a predicted intermediate computation that is as likely as the actual values. A detailed discussion of Secure Multiparty Computation is given in [5, 10].

1.3 Adversarial Behavior

Privacy preserving algorithms are designed in order to preserve privacy even in the presence of adversarial participants that attempt to gather information about the inputs of their peers. There are, however, different levels of adversarial behavior. Cryptographic research typically considers two types of adversaries: A semi-honest adversary (also known as a *passive*, or *honest but curious* adversary) is a party that correctly follows the protocol specification, yet attempts to learn additional information by analyzing the messages received during the protocol execution [16]. On the other hand, a *malicious* adversary (*active*) may arbitrarily deviate from the protocol specification. For example, consider a step in the protocol where one of the parties is required to choose a random number and broadcast it. If the party is semi-honest then we can assume that this number is indeed random. On the other hand, if the party is malicious, then he might choose the number in a sophisticated way that enables him to gain additional information.

It is of course easier to design a solution that is secured against semi-honest adversaries, than for malicious adversaries. A common approach is therefore, first to design a secure protocol for the semi-honest case, and then transform it into a protocol that is secure against malicious adversaries. This transformation can be done by requiring each party to use zero-knowledge proofs to justify that each step that it is taking follows the specification of the protocol. More efficient transformations are often required, since this generic approach might be rather inefficient and add considerable overhead to each step of the protocol. It is remarkable that the semi-honest adversarial model is often a realistic one. This is because deviating from a specified protocol which may be buried in a complex application is a non-trivial task, and because a semi-honest adversarial behavior can model a scenario in which the parties that participate in the protocol are honest, but following the protocol execution an adversary may obtain a transcript of the protocol execution by breaking into a machine used by one of the participants.

Based on the given background, we studied the secure multiparty statistical overall mean computation problem using oblivious polynomial evaluation (OPE) technique developed by [14]. A solution for a similar problem called "weighted average problem" (WAP) has already been proposed by [11] yet the "overall mean problem" (OMP) is a little bit more complicated than WAP due to the bilinear terms in the resultant function. We believe that our contribution will demonstrate how the new paradigm can lead to practical solutions for applications involving statistical overall mean computation especially in the areas of data mining and machine learning.

The organization of the paper is as follows: Section 2 discusses the related work on SMC. Section 3 gives the OPE technique together with OMP. Then section 4 which is the heart of the paper gives the solution to privacy preserving OMP for two-party case. The extension to multiparty form is discussed in fifth section. Finally section 6 concludes this paper and proposes several future research directions.

2   Related Work

The basics of the secure multiparty computation problem is extensive since it was introduced by [17] and [18], then extended by [9] and by many others. The computation problem is first represented as a combinatorial circuit, and then the parties run a short protocol for every gate in the circuit. While this approach is appealing in its generality and simplicity, the generated protocols depend on the size of the circuit. This size depends on the size of the input domain, and on the complexity of expressing such a computation.

The usage of SMC in privacy preserving applications is presented in [13] and [14] by the introduction of "oblivious transfer". Oblivious transfer is a basic protocol that is the main building block of secure computation. It might seem strange at first, but its role in secure computation should become clear later. It was shown by [9] that oblivious transfer is sufficient for secure computation in the sense that given an implementation of oblivious transfer, and no other cryptographic primitive, one could construct any secure computation protocol.

The oblivious transfer protocol involves two parties, the sender and the receiver. The sender's input is a pair (x_0, x_1) and the receiver's input is a bit $\sigma \in \{0, 1\}$. At the end of the protocol the receiver learns x_σ (and nothing else) and the sender learns nothing. Oblivious transfer is often the most computationally intensive operation of secure protocols, and is repeated many times. Each invocation of oblivious transfer typically requires a constant number of invocations of trapdoor permutations (i.e. public-key operations, or exponentiations). It is possible to reduce the amortized overhead of oblivious transfer to one exponentiations per a logarithmic number of oblivious transfers, even for the case of malicious adversaries [16]. Oblivious polynomial evaluation is a technique based on oblivious transfer and explained in the next section.

There are a number of SMC applications distributed in broad range areas. Atallah et. al. [1] proposes preliminary work for solving computational geometry problems including scalar product, permutation, vector dominance, equality testing, point inclusion and intersection. In [4], the authors define a set of new privacy preserving cooperative scientific computation problems: privacy preserving cooperative linear system of equations problem and privacy preserving cooperative linear least-square problem. They have developed protocols to solve these problems. Du et. al. [3] also studies the problem of how to conduct the statistical analysis in a compact environment where neither of parties want to disclose their private data. The secure two-party statistical analysis problem could be solvable in a way more efficient than the general circuit evaluation approach.

OPE is applied to SMC problems especially under privacy preserving data mining concept which is introduced in [12] by designing privacy preserving ID3 decision tree algorithm for purposes. A closer paper to our one is [11] where WAP is solved by two techniques. Former is by OPE and latter is by encryption techniques based on homomorphism yet both of them are used as a tool for k-means clustering for two parties. Neural network applications are studied using OPE methods as well. Within the context of privacy-preserving data mining, [8] presented a private scalar product protocol based on standard cryptographic techniques and proved that it is secure.

3 Problem Definition, Cryptographic Tools and Criteria for Algorithm Design

In this section, the problem definition for privacy preserving two party and multiparty overall mean computation problems are defined(the corresponding algorithms are presented in sections 4 and 5 respectively). Moreover, the oblivious polynomial evaluation which is used as a cryptographic tool to construct the privacy requirements is briefly explained.

3.1   Privacy Preserving Two-Party Overall Mean Computation Problem

Suppose that Alice (party 1) has n samples $A = \{x_1, \cdots, x_n\}$, and Bob (party 2) has m-n samples $B = \{x_{n+1}, \cdots, x_m\}$. Each party wants to get the mean of their samples without revealing any private information. We are assuming that finding the mean of the union of samples from the two parties is more desirable than calculating the two samples individually.

The mean of Alice's samples is $\mu_A = (\Sigma_{i \in A} i) / n$, and Bob's is $\mu_B = (\Sigma_{i \in B} i) / (m-n)$ and overall mean of A $\cup$ B is $\mu = (\Sigma_{i \in A \cup B} i) / m$. But the computation should be done without computing the A $\cup$ B so the means are weighted with respect to their cardinalities and joined together with multiplication and then divided by the union size. Hence,

$$\mu = (\mu_A \cdot n + \mu_B \cdot (m\text{-}n)) / m \tag{1}$$

yields us the overall mean where individual means are pre-computed and proportioned by their cardinalities. The terms in the first product (μ_A, n) are only known by Alice and the terms in the second product (μ_B, n-m) are only known by Bob. Besides, the size of the union is known by neither of them. So we get the following SMC problem.

PROBLEM (Two Party OMP): Alice has n samples represented by $A = \{x_1, \cdots, x_n\}$, and Bob has m-n samples represented by $B = \{x_{n+1}, \cdots, x_m\}$, where each x_i is a real number. Alice and Bob want to compute the overall mean of A $\cup$ B without revealing any of their samples to the other principal which is according to the equation given in (1).

3.2 Privacy Preserving Multiparty Overall Mean Computation Problem

Assume that there exist k parties denoted by $P = \{P_1, \cdots, P_k\}$ each having private data sets of cardinality c_i. For party j, the mean function can be calculated as $\mu_j = (\Sigma_{\,i \in P_j}\, i)\,/\,c_j$. Without loss of generality, the overall mean for all parties is

$$\mu = (\, \Sigma_{\,i}\, \mu_i \cdot c_i\,)\,/\,(\, \Sigma_{\,i}\, c_i\,)\ ,\ 0<i<(k+1) \tag{2}$$

Our second and the main problem is given by the generalized version of the two party case where multiple parties conduct overall mean computation in the same environment without disclosing any private data represented by samples.

3.3 Oblivious Polynomial Evaluation

To design a secure protocol for computing a function f(x,y) allows two parties, a receiver who knows x and a sender who knows y, to jointly compute the value of f(x, y) in a privacy preserving way. The fact that for every computable function f(x,y) in polynomial time there exists such a (polynomially-computable) protocol is already achieved in the cryptographic research. In the OPE problem, the input of the sender is a polynomial P of degree k over some field F. The receiver can get the value P(x) for any element $x \in F$ without learning anything else about the polynomial P and without revealing to the sender any information about x. The input and output for the functionality of OPE as a two party protocol run between a receiver and a sender over a field F as follows:

- Input
    o   Receiver: an input $x \in F$.
    o   Sender: A polynomial P defined over F.
- Output
    o   Receiver: P(x).
    o   Sender: nothing.

There are various protocols to solve the OPE yet the protocol given by Naor and Pinkas [14] is preferred for OMP.

3.4 Criteria for Algorithm Design

To design a privacy preserving OMP algorithm, a number of assumptions should be made to overcome the deficiency of a trusted party. Semi-honest adversary model is chosen which is usually preferred while designing such an algorithm. The information disclosure is yielded by OPE protocol. No more breaches exist in the specific

privacy preserving one exists in the same work as well. In WAP, the target function g is defined as:

$$g : F \times F \rightarrow F \times F$$
$$g\,((x,m),(y,n)) = ((x + y)/(m + n), (x + y)/(m + n)) \tag{5}$$

The crucial difference between WAP and OMP is that the former includes linear expressions in the numerator and denominator of its function yet the latter has bilinear terms, i.e., $(x + y)$ versus $(xm + yn)$. Furthermore, the bilinear terms are coming from the individual parties so there is no obvious solution such as choosing dedicated polynomials for both numerator and denominator in the RPE problem. The construction of private-RPE $\rightarrow$ OMP is stated by placing suitable polynomials and field elements:

Reduction of private-RPE $\rightarrow$ OMP. Recall that party 1 has inputs (x,m) and party 2 has inputs (y,n), Since the reduction is from RPE, in the format of (4), party 1 needs to construct polynomials and party 2 needs to choose field elements. The polynomials for party 1 are:

$$P(w) = w + x$$
$$Q(w) = w + m \tag{6}$$
$$R(w) = wx$$
$$S(w) = wm$$

Note that all polynomials are linear and coefficients are known by party 1. On the other hand, the field elements for party 2 are:

$$\alpha = y$$
$$\beta = n \tag{7}$$
$$\gamma = -y$$
$$\psi = -n$$

The field elements y and ψ are well-defined since every field is a group and inverse with respect to the addition exists for y and n.

Let us define the $T(w_1, w_2, w_3, w_4)$ as the linear combination of P, Q, R and S polynomials:

$$T(w_1, w_2, w_3, w_4) = P(w_1)Q(w_2) + R(w_3) + S(w_4) \tag{8}$$

T is a dummy polynomial to handle the bilinear terms. Since multiplication and closed operations in a field T is also well-defined. If $(w_1, w_2, w_3, w_4) = (\alpha, \beta, \psi, \gamma)$ variable replace is done then it yields:

$$\begin{aligned}
T(w_1, w_2, w_3, w_4) &= P(w_1)Q(w_2) + R(w_3) + S(w_4)\\
&= P(\alpha)Q(\beta) + R\,(\psi) + S(\gamma)\\
&= P(y)Q(n) + R\,(-n) + S(-y)\\
&= (x+y)(m+n) + \,(-nx) + (-ym)\\
&= xm + xn + ym + yn - nx - ym\\
&= xm + yn
\end{aligned}$$

The numerator of the desired function f in (4) is constructed and the denominator is nothing but $Q(\beta)$. Using OPE, the reduction is complete by choosing suitable field elements together with constructing the polynomial, T as a combination of linear polynomials.

Two lemmas are critical for the proof that f computes the overall mean privately. The former belongs to Canetti [2] and the latter belongs to Jha et. al. [11]. Their proofs are not given here as they can be reached from original sources.

Lemma 1. (Composition Theorem for passive adversary): If g is privately reducible to f and there exists a protocol for computing f privately then there exists a protocol for computing g privately.

Lemma 2. (Private RPE): The RPE protocol privately computes RPE problem.

Theorem 1. (Two-party OMP) The protocol formed by f in (3) yields a privacy-preserving protocol for two-party OMP.

Proof: It is clear that OMP is privately reducible to RPE by choosing the numerator as T and the denominator as Q. There exists a protocol for private-RPE (inferred by Lemma 2) then by using Lemma 1, we get that given protocol is privately computes two-party OMP. It is trivial that reduction is polynomially-computable.

5   Privacy Preserving Multiparty OMP Protocol

The privacy preserving two-party OMP protocol can guide to extend the protocol for multi-party case. The protocol is presented step-by-step while explaining what is known to the both parties in crucial cases. Before starting to develop protocol, we need the following privacy-preserving protocol for cardinality summation.

Lemma 3. (Private RPE): The function h given in (9) privately computes the sum of two samples whose cardinalities are m and n respectively.

$$h : F \times F \rightarrow F \times F$$
$$h\,((x,m),(y,n)) = ((m + n), (m + n)) \tag{9}$$

Proof: The reduction to private-RPE completes the proof by joining lemma 2 with lemma 1. The unique polynomial for party 1 and the single field element of party 2 are as follows:

$$Q(w) = w+m \tag{10}$$

$$\beta = n \tag{11}$$

By computing $Q(\beta) = m+n$, party 2 gets the desired result and shares with party 1.

Multiparty OMP Protocol. The protocol for multiparty OMP can be designed from two-party OMP in the following way:

(1) Parties are ordered from 1 to k in a manner that consecutive parties are involved to two-party OMP computation. This can be done with a common share or coin-tossing into well protocol [9].

(2) Between party j and $j+1$, $0<j<k$, the two-party OMP protocol works and

$$((\mu_j, c_j), (\mu_{j+1}, c_{j+1})) \rightarrow ((\mu_j \cdot c_j + \mu_{j+1} \cdot c_{j+1})/(c_j + c_{j+1}), (\mu_j \cdot c_j + \mu_{j+1} \cdot c_{j+1})/(c_j + c_{j+1}))$$

is computed.

(3) Furthermore, party j and $j+1$ privately compute their cardinality sum via the function given in lemma 3. In other words, consecutive parties compute the partial mean and partial size of their samples.

$$((\mu_j, c_j), (\mu_{j+1}, c_{j+1})) \rightarrow (c_j + c_{j+1}, c_j + c_{j+1})$$

(4) The mean and cardinality values are updated for party $j+1$ with the new values calculated at the end of the protocol involved with the previously ordered party. (i.e. party $j+1$ gets the partial mean and partial sum of samples up to her)

(5) Apply the previous two steps for all consecutive parties. Total computation is linear in size and $k-1$ for k parties.

(6) At the end of the computation, the last party (i.e. party k) gets the overall mean together with total sample size with remaining parties.

The unimportant gap of the protocol is party j learns the size of the total previous samples yet it is not give the size of the each individual party. The only exception is for the first party, party 2 gets the size of its sample. This can be overcome by choosing the order in a circular round-robin fashion so the order of consecutive parties are preserved but only the first party changes. The probability to be the first party is 1/k which is Pareto-optimal for such a scheme.

Theorem 2. (Multiparty OMP) The protocol formed by f in (3) and h in (9) together yields a privacy-preserving protocol for multiparty OMP.

Proof: The part regarding f is done in the proof of the first theorem and the proof of lemma 3 shows us the reduction for h. Thus using lemma 1, Multiparty OMP protocol is privately computes multiparty OMP.

6   Conclusion and Future Work

In this paper, the application of the oblivious polynomial evaluation technique to secure multiparty computation problems is shown by presenting a privacy preserving algorithm for multiparty overall mean computation using oblivious polynomial evaluation as a concrete example. To form a basis, the two-party case and cardinality summation algorithms are also given as components of the main protocol.

In our future directions, we have to adapt the private-OMP to a specific case, especially in data mining or machine learning. Principal Component Analysis [7] is chosen as a test bed which suits to the problem well since the mean computations are done before forming the covariance matrix. The controlled experiments can guide us to justify that the protocol works without trouble.

Moreover, an algorithm that uses one of the homomorphic encryption schemes (HES) can be designed for the same problem. The scheme developed by Paillier [15] can be considered to be the alternatives for such a mechanism. A testing environment for the comparison of the performance analysis between OPE and HES determines the efficient algorithm under given security assumptions.

References

1. Atallah M. J. and Du W.: Secure Multi-Party Computational Geometry. In Proceedings of 7th International Workshop on Algorithms and Data Structures (WADS 2001), 165-179
2. Canetti. R.: Security and composition of multi-party cryptographic protocols. Journal of Cryptology, (2000) 13(1):143–202
3. Du W. and Atallah M. J.: Privacy-preserving cooperative statistical analysis. In Annual ComputerSecurity Applications Conference ACSAC) (2001) 102–110
4. Du W. and Atallah M. J.: Privacy-Preserving Cooperative Scientific Computations. In 14th IEEE Computer Security Foundations Workshop, (2001), 273-282
5. Du W. and Atallah M. J.: Secure Multi-Party Computation Problems and their Applications: Review and Open Problems. In New Security Paradigms Workshop (2001) 11-20
6. Du W. and Atallah M. J.: A Practical Approach to Solve Secure Multi-party Computation Problems. In New Security Paradigms Workshop (2002) 127-135
7. Duda R.O., Hart P.E. and Stork D.G.: Pattern Classification. John Wiley & Sons, (2001)
8. Goethals B., Laur S., Lipmaa H. and Mielikäinen T.: On Private Scalar Product Computation for Privacy-Preserving Data Mining. The 7th Annual International Conference in Information Security and Cryptology (ICISC 2004), volume 3506 of *Lecture Notes in Computer Science*, (2004) 104-120
9. Goldreich O., Micali S. and Wigderson A.: How to play any mental game - a completeness theorem for protocols with honest majority. In 19th Symposium on Theory of Computer Science, (1987) 218–229
10. Goldreich O.: Foundations of Cryptography: Volume 2, Basic Applications. Cambridge University Press, (2004)
11. Jha S., Kruger L. and McDaniel P.: Privacy Preserving Clustering 10th European Symposium On Research In Computer Security (ESORICS), Milan, (2005)
12. Lindell Y. and Pinkas B.: Privacy preserving data mining. In Advances in Cryptology *(Crypto 2000)*, (2000) 36–54

13. Naor M. and Pinkas B.: Oblivious transfer and polynomial evaluation. In *31st Symposium on Theory of Computer Science*, (1999) 245–254

14. Naor M. and Pinkas B.: Computationally Secure Oblivious Transfer Journal of Cryptology, Vol. 18, No. 1, 2005

15. Paillier P.: Public-key cryptosystems based on composite degree residuosity classes. In *Proceedings of Advances in Cryptology (EUROCRYPT'99)*, 1999

16. Pinkas B.: Cryptographic Techniques for Privacy-Preserving Data Mining SIGKDD Explorations, the newsletter of the ACM Special Interest Group on Knowledge Discovery and Data Mining, (2003)

17. Rabin M.: How to exchange secrets by oblivious transfer. Technical Report Tech. Memo TR-81, Aiken Computation Laboratory, 1981

18. Yao A.C.: How to generate and exchange secrets. In *27th IEEE Symposium on Foundations of Computer Science*, (1986) 162–167

Linear Approximations for 2-round Trivium

Meltem Sönmez Turan[1], Orhun Kara[2]

[1] Institute of Applied Mathematics, Middle East Technical University
Ankara, Turkey
msonmez@metu.edu.tr
[2] TUBITAK-UEKAE, Gebze, Turkey
orhun@uekae.tubitak.gov.tr

Abstract. Trivium is a hardware oriented stream cipher proposed for the eSTREAM project. No attack with complexity better than exhaustive search is available in the literature. In this paper, we model the initialization part of Trivium as an 8-round function where each round consists of 144 Trivium clocks, and analyze the security margin in terms of number of clockings. As one example, we apply Matsui's linear cryptanalysis to 2-round Trivium and give a linear approximation with bias 2^{-31}. In addition, we analyze the completeness property of the initialization function. We propose a new input to the initialization of Trivium that has better diffusion properties. Also, we conjecture that Trivium with R-clockings in initialization is secure if each register bit is affected by all the key and IV bits in R clockings.
Keywords: Linear approximations, Trivium, Stream ciphers

1 Introduction

Linear cryptanalysis, introduced by Matsui [1], is an effective known plaintext attack against block ciphers. It exploits statistical correlations between input and output bits. For a block cipher with k-bit key $(k_1, \ldots, k_k)$, n-bit plaintext $(p_1, \ldots, p_n)$ and ciphertext $(c_1, \ldots, c_n)$, the aim of the attack is to find the index sets I, J, L such that

$$\sum_{i \in I} k_i + \sum_{j \in J} p_j = \sum_{l \in L} c_l$$

holds with probability $p = 1/2 + \epsilon$, $\epsilon \neq 0$.

Some variants of linear cryptanalysis are applied to stream ciphers. The most famous example may be the correlation attacks mounted on LFSR based stream ciphers [2] [3] [4]. Linear approximations between key and keystream bits are utilized. Another example is proposed by Golić [5]. In [6], a linear approximation for $t-$functions is used to attack the TSC stream ciphers. In [7], a new method to find biased linear approximations without searching all possible linear relations individually is presented and is used to distinguish the output of the stream cipher Pomaranch.

In this paper, we give an example of linear cryptanalysis on stream ciphers. The analysis is a kind of resynchronization attack. We consider the initialization phase (key and IV loading) of a stream cipher as an iterated function. Then, we apply Matsui's linear cryptanalysis to the initialization phase by finding approximations for each iteration and combining them by piling-up lemma. As an example, we consider the initialization phase of Trivium as an 8-round function and find a linear approximation for 2-round Trivium with a bias of 2^{-31} for a subset of key and IVs.

Trivium [8], one of the focus algorithms in eSTREAM project, is a synchronous binary additive stream cipher. No attack with complexity better than exhaustive search is available in the literature. In [9], the linear sequential circuit approximations are used to evaluate the strength of Trivium against distinguishing attacks. The correlation coefficient is calculated as 2^{-72}, and the complexity to distinguish the output is $O(2^{144})$. In [10], authors try to find 288 unknown internal state bits by solving systems of equations, and this has complexity $O(2^{164})$. In [11], an attack that recovers the internal state of Trivium is presented with time complexity around $c.2^{83.5}$. Also, in terms of randomness properties, no statistical weaknesses are observed [12].

We concentrate on the initialization of Trivium which consists of 1152 clockings. Trivium is one of the fastest ciphers proposed in eSTREAM project [13]. However, initialization of Trivium with 1152-clockings may hinder the speed in platforms where resynchronization is performed very often. For instance, frequent initializations of Trivium may slow it down more than five times in a frame based encryption like GSM over-the-air privacy standard since length of each frame is 228 bits. We propose a new input to the initialization function of Trivium which provides a faster diffusion of key and IV bits into the register. Moreover, we introduce some open problems on security margins of Trivium. We conjecture that Trivium with R-clocking in initialization has security level of 2^{80} if each register bit is affected by all the key and IV bits in R-clockings.

In the following section, the framework of finding linear approximations is presented. In section 3, description of Trivium is given. In Section 4, a linear approximation for 2-round Trivium is presented. A new proposal for initialization by which it is harder to find linear approximations, is given in Section 5 and conclusion and future studies are summarized in the last section.

2   Framework

A *synchronous stream cipher* is a cipher in which the keystream is produced independent of plaintext using following equations;

$$S_0 = f_{init}(K, IV),$$
$$S_{t+1} = f(S_t),$$
$$z_t = g(S_t)$$

where S_t is the *internal state* at time t, f_{init} is the *initialization function* that use secret key K and public *initialization vector IV*, f is the *next state function* that only depends on the current state, g is the filter function that produce keystream z_i.

Stream ciphers can be modeled as a collection of Boolean functions,

$$F_i : F_2^k \times F_2^v \to F_2 , i = 1, 2, \dots$$

that generate z_i, the ith output bit of keystream using k bit key and v bit IV. Each F_i is affected from both initialization and keystream generation phases. A linear approximation to F_i's with bias $\epsilon > 2^{-k/2}$ would return one bit key information, if it is possible to resynchronize the cipher ϵ^{-2} times.

2.1   Searching for Linear Approximations

Searching for linear approximations is composed of three steps:

(i) **Selecting a subset of** z_i. In this step, the aim is to find the right hand side of the linear approximation. Selection of the subset of z_i's or equivalently the subset of F_i's is done such that $\sum z_i$ or $\sum F_i$ is affected from minimum number of internal state variables.

(ii) **Partitioning the initialization phase.** Initialization phase of some stream ciphers consists of iteratively applying the same next state function to the internal state variables. To find a linear approximation for F_i efficiently, the initialization phase is partitioned into rounds with t_i clockings so that for each round it is possible to find approximations efficiently.

As given in Figure 1, the initialization phase of the cipher can be represented as n rounds, where t_i is the number of clockings in round i. The sum of t_i's should be equal to the total number of clockings, T, in initialization.

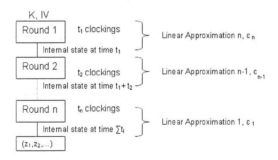

Fig. 1. Stream cipher initialization composed of rounds

In the extreme case, each round is composed of one clockings, then for each round, linear approximations with high biases can be found easily. However, as T gets large, the approximation for the whole cipher is likely to have a very small bias. On the other hand, if the number of clockings in each round is chosen to be very high, then finding linear approximations for each round becomes infeasible. This gives a trade-off between number of rounds and the selection of t_i values. Optimal selection of t_i values is an open question.

(iii) **Combining linear approximations.** The approximations found for each round are combined to find an approximation for the whole cipher. Approximations are selected so that all internal state bits are canceled out. Finally, the linear approximation based on key, IV and keystream is obtained. Bias of the approximation is found by using the piling-up lemma.

Lemma 1. *For n independent random variables $X_1, \ldots, X_n$ that take values from $\{0,1\}$, the summation given by $X = X_1 + X_2 + \ldots X_n$ has bias $\epsilon = p - 1/2$ is given by*

$$\epsilon = 2^{n-1} \prod_{j=1}^{n} \epsilon_j$$

where ϵ_i's are biases of the terms X_i's.

To attack the cipher, the found bias should be greater than $2^{-k/2}$. It is also possible to fix some of the IV bits to increase the bias, however this puts a restriction on the number of resynchronizations. In such cases, the attack requires chosen IVs. There may be other approximations valid for a subset of keys, with better biases.

3   The Stream Cipher Trivium

Trivium supports the usage of 80 bit key and 80 bit IV with internal state size of 288 bits. It is claimed to be suitable to generate up to 2^{64} bits of keystream from a pair of key and IV.

3.1   Initialization

80 bit key K $(k_1, k_2, \ldots, k_{80})$, and IV $(iv_1, iv_2, \ldots, iv_{80})$, is directly assigned to the internal state of the cipher $(s_1, s_2, \ldots, s_{288})$ and the remaining bits (except the last three) are set to zero. Then, the cipher is clocked over 4 full cycles without producing any output. The pseudo code of the initialization phase is given below.

Algorithm 3.1: LOADING KEY AND IV(K, IV)

$(s_1, s_2, \ldots s_{93}) \leftarrow (k_1, k_2, \ldots, k_{80}, 0, \ldots, 0)$;
$(s_{94}, s_{95}, \ldots s_{177}) \leftarrow (iv_1, iv_2, \ldots, iv_{80}, 0, \ldots, 0)$;
$(s_{178}, s_{179}, \ldots s_{288}) \leftarrow (0, \ldots, 0, 0, 1, 1, 1)$;
for $i \leftarrow 1$ **to** $4 \cdot 288$
  do
$\left\{\begin{array}{l} t_1 \leftarrow s_{66} + s_{91}.s_{92} + s_{93} + s_{171}; \\ t_2 \leftarrow s_{162} + s_{175}.s_{176} + s_{177} + s_{264}; \\ t_3 \leftarrow s_{243} + s_{286}.s_{287} + s_{288} + s_{69}; \\ (s_1, s_2, \ldots s_{93}) \leftarrow (t_3, s_1, \ldots, s_{92}); \\ (s_{94}, s_{95}, \ldots s_{177}) \leftarrow (t_1, s_{94}, \ldots, s_{176}); \\ (s_{178}, s_{179}, \ldots s_{288}) \leftarrow (t_2, s_{178}, \ldots, s_{287}); \end{array}\right.$
return

3.2   Keystream Generation

Keystream generation function is very similar to key and IV loading. The only difference is the filter function that generates the keystream z_i, $i = 1, 2, \ldots$. The pseudo code of keystream generation is given below.

Algorithm 3.2: KEYSTREAM GENERATION(N)

for $i \leftarrow 1$ **to** N
  do
$\left\{\begin{array}{l} t_1 \leftarrow s_{66} + s_{93}; \\ t_2 \leftarrow s_{162} + s_{177}; \\ t_3 \leftarrow s_{243} + s_{288}; \\ z_i \leftarrow t_1 + t_2 + t_3; \\ t_1 \leftarrow t_1 + s_{91}.s_{92} + s_{171}; \\ t_2 \leftarrow t_2 + s_{175}.s_{176} + s_{264}; \\ t_3 \leftarrow t_3 + s_{286}.s_{287} + s_{69}; \\ (s_1, s_2, \ldots s_{93}) \leftarrow (t_3, s_1, \ldots, s_{92}); \\ (s_{94}, s_{95}, \ldots s_{177}) \leftarrow (t_1, s_{94}, \ldots, s_{176}); \\ (s_{178}, s_{179}, \ldots s_{288}) \leftarrow (t_2, s_{178}, \ldots, s_{287}); \end{array}\right.$
return

4   Linear Approximations for 2-round Trivium

The initialization phase of Trivium can be modeled as an n-round function. As a result of making a trade-off between the number of rounds n and the number of clockings in each round, t_i is chosen to be 144 for each i. The more number of clockings in each round, the better approximations are found, however number of clockings should not be too large to prevent finding linear approximations exhaustively. For Trivium, after 144 clockings each internal state variable is affected from at most 20 previous internal state variables, therefore it is possible to obtain the exact relations and best approximations.

For 2-round Trivium, the followings

$$K = (s_0(1), s_0(2), \ldots, s_0(80)),$$
$$IV = (s_0(94), s_0(95), \ldots, s_0(173)),$$
$$z_1 = s_{288}(66) + s_{288}(93) + s_{288}(162) + s_{288}(177) + s_{288}(243) + s_{288}(288)$$

hold where $s_t(i)$ is the ith internal state bit at time t.

To check for possible trivial weaknesses of 2-round Trivium, diffusion of IV and key to internal state bits are examined using 1000 random key and IV pairs. All key and IV bits are diffused to all internal state bits, except the last seven internal state bits. However, this does not lead to any trivial weakness.

In this study, while selecting the subset of F_i's to approximate, the only restriction is the total number of internal state variables that affect the keystream bits. Each z_i is generated using the modulo 2 summation of six internal state bits. A subset of z_i's such that their summation includes less than six internal state bits after cancellations is not found after a heuristic search. Therefore, the function to be approximated is chosen to be F_1 that generates the first output bit, z_1, in terms of key and IV bits.

For 2-round Trivium, finding the equation F_1 in terms of initial state variables is not efficient, therefore an approximation is found for each round and then they are combined as given in Figure 2. The bias of the obtained approximation is found by the piling-up lemma.

The output bit z_1 is the sum of bits $s_{288}(66)$, $s_{288}(93)$, $s_{288}(162)$, $s_{288}(177)$, $s_{288}(243)$ and $s_{288}(288)$. The algebraic normal form of F_1 is found exhaustively in terms of the internal state bit of $t = 144$ as

$$\begin{aligned}
z_1 = &\, s_{144}(6) + s_{144}(16).s_{144}(117) + s_{144}(31)s_{144}(32) + s_{144}(33) + s_{144}(57) + \\
&\, s_{144}(82).s_{144}(83) + s_{144}(84) + s_{144}(96) + s_{144}(97).s_{144}(98) + s_{144}(99) + \\
&\, s_{144}(111) + s_{144}(129) + s_{144}(142).s_{144}(143) + s_{144}(144) + s_{144}(150) + \\
&\, s_{144}(162) + s_{144}(163).s_{144}(164) + s_{144}(165) + s_{144}(186) + s_{144}(192) + \\
&\, s_{144}(208).s_{144}(209) + s_{144}(210) + s_{144}(231) + s_{144}(235).s_{144}(236) + \\
&\, s_{144}(237) + s_{144}(252)
\end{aligned}$$

and its closest linear approximation is

$$\begin{aligned}
z_1 = &\, s_{144}(6) + s_{144}(33) + s_{144}(57) + s_{144}(84) + s_{144}(96) + s_{144}(99) + \\
&\, s_{144}(111) + s_{144}(129) + s_{144}(144) + s_{144}(150) + s_{144}(162) + \\
&\, s_{144}(165) + s_{144}(186) + s_{144}(192) + s_{144}(210) + s_{144}(231) + \\
&\, s_{144}(237) + s_{144}(252)
\end{aligned} \tag{1}$$

with bias $1/2 + 2^{-9}$.

Since the aim is to obtain an approximation based on key, IV and output bits, the linear approximation given above is rewritten in terms of $s_0(i)$, $i = 1, 2, \ldots, 80$ and $i = 94, \ldots, 173$ values, the remaining terms are omitted, since they are assigned to constants during initialization. Then, the equation given in Appendix A is obtained. The equation has 24 linear, 59 quadratic terms, 20 terms with degree 3. The linear approximation for the function is found as

$$
\begin{aligned}
z_1 = 1 &+ s_0(3) + s_0(6) + s_0(15) + s_0(21) + s_0(27) + s_0(30) + s_0(39) + \\
&s_0(54) + s_0(57) + s_0(67) + s_0(68) + s_0(69) + s_0(72) + s_0(96) + \\
&s_0(99) + s_0(114) + s_0(117) + s_0(123) + s_0(126) + s_0(132) + s_0(138) + \\
&s_0(144) + s_0(165) + s_0(171)
\end{aligned}
\qquad (2)
$$

with bias $2^{78} \cdot (0.25)^{59} \cdot (0.375)^{20} = 2^{-68.30}$, assuming all nonlinear terms are independent. We increase the amount of the bias by assigning zero string to certain IV and key bits.

Chosen IVs For IVs in the form $iv_{25} = iv_{26} = iv_{31} = iv_{32} = iv_{49} = iv_{50} = iv_{54} = iv_{55} = iv_{70} = iv_{71} = 0$, the bias of the equation increases to 2^{-44}. This bias is still very low and cannot be used to break 2-round Trivium. To improve bias further, also some of the key bits are fixed. Then, the bias of the second linear approximation increases to 2^{-23} for keys satisfying $k_{14} = k_{19} = k_{20} = k_{38} = k_{39} = k_{45} = k_{63} = k_{64} = k_{65} = k_{77} = 0$.

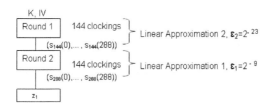

Fig. 2. Linear Approximations for 2-round Trivium

Combining two linear approximations (1) and (2), the total bias of the following approximation,

$$
\begin{aligned}
z_1 = 1 &+ k_3 + k_6 + k_{15} + k_{21} + k_{27} + k_{30} + k_{57} + k_{67} + k_{68} + k_{69} + \\
&k_{72} + iv_3 + iv_6 + iv_{21} + iv_{24} + iv_{30} + iv_{33} + iv_{39} + iv_{45} + iv_{51} + \\
&iv_{72} + iv_{78},
\end{aligned}
$$

is obtained as $2 \cdot 2^{-9} \cdot 2^{-23} = 2^{-31}$ by piling-up lemma. The upper bound on the number of resynchronizations is 2^{70}, since 10 bits of IV bits are fixed to zero. To identify a key with specified bits, we need 2^{62} chosen IV.

5   Proposal for Initialization

In this section, we propose a new method for initialization which is very similar to the original. The only difference is related to the initial assignment of state bits. Only 22 of the internal state variables are set to constants and this increase in the cost is negligible. This obviously increases the number of variables while searching for linear approximations and therefore makes it harder to find linear approximations. As a result, it may be possible to increase the efficiency of the cipher by decreasing the number of initial clockings.

The proposed initial assignment is

$$(s_1, \ldots, s_{93}) \leftarrow (iv_1, \ldots, iv_{13}, iv_{14} + k_1, \ldots, iv_{80} + k_{67}, k_{68}, \ldots k_{80}),$$
$$(s_{94}, \ldots, s_{177}) \leftarrow (iv_1 + k_1, \ldots, iv_{80} + k_{80}, 0, 0, 0, 0),$$
$$(s_{178}, \ldots, s_{288}) \leftarrow (k_1, \ldots, k_{13}, k_{14} + iv_1, \ldots, k_{80} + iv_{67}, iv_{68}, \ldots iv_{80}, 0, \ldots, 0, 1, 1, 1)$$

We propose 13 shifts while loading key bits into the first register and 13 shifts while loading IV bits into the third register. The number of shifts are chosen so that same key bits or same IV bits are not be XORed in the feedback functions of the registers during the first 80-90 clockings.

A comparison of the proposed and original method is done in terms of the completeness property. Let $G(i, j)$ be the number of key and IV bits that affect the state bit i after j clockings. The comparison of both methods is done based on $\min_i G(i, j)$ and as seen from Figure 3 the diffusion of key and IV bits are better in the proposed method. In the original method, completeness is satisfied after 525 clockings, whereas in the proposed method, 484 clockings are enough.

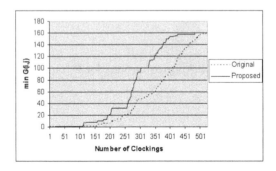

Fig. 3. Number of clockings vs. $\min_i G(i, j)$

6   Conclusion and Further Study

In this study, we mainly concentrated on the initialization of Trivium which is one of the focus ciphers of eSTREAM project. We modeled the initialization phase of Trivium as an iterated cipher with 8-rounds. For frame based applications requiring frequent resynchronizations, we question the efficiency of the initialization phase and try to attack initialization with smaller rounds. For 2-round Trivium, we obtained a linear approximation of z_1, which is valid for a subset of key and IV's. It is an open question whether there exists linear approximations for m-round Trivium where $m > 2$.

As a list of future studies, the followings can be given:

- *Use of multiple approximations.* As an extension of linear cryptanalysis, in [14], the use of multiple linear approximations is proposed. As a future work, different approximations for the same output bits can be found and combined to make better approximations.
- *Use of nonlinear approximations.* As an alternative to linear approximations, nonlinear approximations may be applied to the initialization phase to find better approximations.
- *Different modelings of initialization.* The initialization phase can be remodeled differently, using different number of clockings in each round and better approximations may be found.

The most important problem is determining the security margin of Trivium. That is, what is the minimum number of clockings for the initialization of Trivium so as to supply 80-bit security? The same question can be given for the new proposal.

Let the initialization function be complete in R-clockings. That is, each register bit is affected by all key and IV bits. Then, we conjecture that Trivium with R-clockings in initialization is secure. According to the conjecture, the original method supply 80-bit security after 525 clockings, whereas for the proposed method, 484 clockings are enough.

A F_1 for 2-round Trivium

$z_1 = 1 + s_0(3) + s_0(6) + s_0(15) + s_0(21) + s_0(27) + s_0(30) + s_0(39) + s_0(54) +$
$s_0(57) + s_0(67) + s_0(68) + s_0(69) + s_0(72) + s_0(96) + s_0(99) + s_0(114) + s_0(117) +$
$s_0(123) + s_0(126) + s_0(132) + s_0(138) + s_0(144) + s_0(165) + s_0(171) + s_0(4).s_0(5) +$
$s_0(13).s_0(14) + s_0(13).s_0(41) + s_0(13).s_0(119) + s_0(14).s_0(40) + s_0(14).s_0(118) +$
$s_0(16).s_0(17) + s_0(19).s_0(20) + s_0(19).s_0(47) + s_0(19).s_0(125) + s_0(20).s_0(46) +$
$s_0(20).s_0(124) + s_0(22).s_0(23) + s_0(25).s_0(26) + s_0(28).s_0(39) + s_0(34).s_0(35) +$
$s_0(37).s_0(38) + s_0(37).s_0(65) + s_0(37).s_0(143) + s_0(38).s_0(64) + s_0(39).s_0(40) +$
$s_0(38).s_0(142) + s_0(40).s_0(119) + s_0(41).s_0(118) + s_0(43).s_0(44) + s_0(45).s_0(46) +$
$s_0(46).s_0(125) + s_0(47).s_0(124) + s_0(49).s_0(50) + s_0(52).s_0(53) + s_0(58).s_0(59) +$
$s_0(58).s_0(164) + s_0(59).s_0(163) + s_0(61).s_0(62) + s_0(63).s_0(64) + s_0(64).s_0(65) +$
$s_0(64).s_0(143) + s_0(64).s_0(170) + s_0(65).s_0(169) + s_0(65).s_0(142) + s_0(67).s_0(68) +$

$s_0(70).s_0(71) + s_0(76).s_0(77) + s_0(79).s_0(77) + s_0(103).s_0(104) + s_0(106).s_0(107) +$
$s_0(118).s_0(119) + s_0(124).s_0(125) + s_0(127).s_0(128) + s_0(130).s_0(131) +$
$s_0(133).s_0(149) + s_0(134).s_0(148) + s_0(142).s_0(143) + s_0(147).s_0(148) +$
$s_0(151).s_0(152) + s_0(154).s_0(155) + s_0(160).s_0(161) + s_0(163).s_0(164) +$
$s_0(166).s_0(167) + s_0(13).s_0(39).s_0(40) + s_0(14).s_0(38).s_0(39) +$
$s_0(19).s_0(45).s_0(46) + s_0(20).s_0(44).s_0(45) + s_0(37).s_0(63).s_0(64) +$
$s_0(38).s_0(39).s_0(40) + s_0(38).s_0(39).s_0(41) + s_0(38).s_0(39).s_0(119) +$
$s_0(38).s_0(62).s_0(63) + s_0(39).s_0(40).s_0(118) + s_0(44).s_0(45).s_0(46) +$
$s_0(44).s_0(45).s_0(47) + s_0(44).s_0(45).s_0(125) + s_0(45).s_0(46).s_0(124) +$
$s_0(62).s_0(63).s_0(64) + s_0(62).s_0(63).s_0(65) + s_0(62).s_0(63).s_0(143) +$
$s_0(63).s_0(64).s_0(142) + s_0(133).s_0(147).s_0(148) + s_0(134).s_0(146).s_0(147)$
with bias 2^{-9}.

References

1. M. Matsui. Linear cryptanalysis method for DES cipher. In *EUROCRYPT*, pages 386–397, Secaucus, NJ, USA, 1994. Springer-Verlag New York, Inc.
2. T. Siegenthaler. Decrypting a class of stream ciphers using ciphertext only. *IEEE Trans. Computers*, 34(1):81–85, 1985.
3. W. Meier and O. Staffelbach. Fast correlation attacks on certain stream ciphers. *Journal of Cryptology*, 1(3):159–176, 1989.
4. V. V. Chepyzhov, T. Johansson, and B. Smeets. A simple algorithm for fast correlation attacks on stream ciphers. In *Fast Software Encryption*, pages 181–195, London, UK, 2001. Springer-Verlag.
5. J. Dj. Golic. Linear cryptanalysis of stream ciphers. In *Fast Software Encryption*, pages 154–169, 1994.
6. F. Muller and T. Peyrin. Linear cryptanalysis of the TSC family of stream ciphers. In *ASIACRYPT*, pages 373–394, 2005.
7. M. Hell and T. Johansson. On the Problem of Finding Linear Approximations and Cryptanalysis of Pomaranch Version 2. In *SAC*, 2006.
8. C. De Cannière and B. Preneel. Trivium - a stream cipher construction inspired by block cipher design principles. eSTREAM, ECRYPT Stream Cipher Project, Report 2005/030, 2005. http://www.ecrypt.eu.org/stream.
9. S. Khazaei, M. M. Hasanzadeh, and M. S. Kiaei. Linear Sequential Circuit Approximation of Grain and Trivium Stream Ciphers. eSTREAM, ECRYPT Stream Cipher Project, Report 2005/063, 2005.
10. H. Raddum. Cryptanalytic results on trivium. eSTREAM, ECRYPT Stream Cipher Project, Report 2006/039, 2006.
11. A. Maximov and A. Biryukov. Two trivial attacks on trivium. eSTREAM, ECRYPT Stream Cipher Project, Report 2007/003, 2007.
12. M. S. Turan, A. Doğanaksoy, and Ç. Çalık. Statistical analysis of synchronous stream ciphers. *SASC 2006: Stream Ciphers Revisited*, 2006.
13. F. K. Gürkaynak, P. Luethi, N. Bernold, R. Blattmann, V. Goode, M. Marghitola, H. Kaeslin, N. Felber, and W. Fichtner. Hardware Evaluation of estream Candidates:achterbahn, Grain, Mickey, Mosquito, Sfinks, Trivium, Vest, ZK-crypt. eSTREAM, ECRYPT Stream Cipher Project, Report 2006/015, 2006.
14. Jr. B. S. Kaliski and M. J. B. Robshaw. Linear cryptanalysis using multiple approximations. In *CRYPTO*, pages 26–39, London, UK, 1994. Springer-Verlag.

A New Data Integrity Protection Model for Free Roaming Mobile Agents

Marwa. M. Essam[1], Mohamed. A. EL-Sharkawy[1], and Mohamed. S. Abdelwahab[1]

[1]Faculty of Computer and Information Sciences, Ain-Shams University
11566, Abbasia, Cairo, Egypt

marwa.essam@gmail.com, dr_sharkawy@yahoo.com, mswahab@icicis.net

Abstract. A primary security challenge in the mobile agent paradigm is that of protecting the data collected by a free roaming agent through its itinerary. For that purpose, we propose a new data protection model. Our proposed model enables an agent to securely encrypt and encapsulate the results computed at each visited host. The key idea of the model is a sequence of interrelated encryption keys in ciphering the agent data. The construction of an encryption key at any host builds a chaining relation that links that key forwards to the host's successor and backwards to all the previously constructed keys. The model assumes the existence of a trusted server in the system to force visited hosts to execute the encapsulation algorithm faithfully. We demonstrate how the chaining relation adopted in our model prevents most of the known attacks especially the insertion of fake results and the truncation of good ones.

Keywords: Mobile agents, agent security, malicious hosts, and data integrity.

1 Introduction

A mobile agent is a software program that can be sent out from a computer into a network and roam among the computer nodes in the network. It is firstly created on some home machine, and it is dispatched to a remote host for execution. The mobile agent would execute, collect host-specific information, and generate runtime states and variables ready to migrate to another host. This process continues until the mobile agent returns to the first machine that sent him with useful information gathered through his visits to other hosts [1]. The autonomy, social ability, learning and most important mobility features offered by mobile agents have been providing it with enormous potential in many applications especially of those in the electronic commerce domain [2], [3].

Significant research and development into mobile agency have been conducted in recent years. Yet, because of some security concerns, mobile agency has failed to become a sweeping force of change [4], [5]. When an agent leaves for a new host, extreme care must be taken to prevent unauthorized modification or analysis of the agent. Agents may carry with them confidential information and logic, which shouldn't be accessible to the agent host

To protect mobile agents against malicious attack, two problems need to be solved. The first is to prevent hosts from tampering the agent code. To address this problem, computing with encrypted functions schemes is under research [6]. The second problem is to detect the tampering on the data collected by a mobile agent through its journey. This problem is especially serious for free roaming mobile agents that are free to choose their respective next hops dynamically based on the data they acquired from their past journeys.

One class of the agent data protection models tries to detect the tampering on the agent data using the concept of partial result encapsulation. That is to securely save the results of an agent's actions, at each platform visited, for subsequent verification when the agent returns to his point of origination. In this paper, we proceed to propose a new agent security model that aims to provide both data integrity and confidentiality through tampering detection. The proposed model enhances some of the existing data encapsulation schemes to improve over its limitations and to satisfy most of the security properties needed for agent data protection.

This paper is organized as follows: In section 2 the main security properties required by any model to preserve data integrity are outlined. In Section 3, a review on the previous data encapsulation models is presented. In section 4, the basic idea of the proposed model is illustrated. A security analysis of the model is presented in section 5. Finally conclusions are presented in section 6.

2 Security Properties

Given that an agent originator needs to preserve data integrity, a set of security properties have been identified by G. Karjoth in [7]. All the known data integrity schemes are analyzed against those properties. To illustrate these security properties, consider an example of a shopping agent that travels within a network of shops of size n to collect offers on a certain product. The agent journey in our example is to be decided during the agent execution (free-roaming agent). Note that we only use this shopping agent example to simplify our illustrations. However, the offer an agent collects in a host could actually represent arbitrary state information from an agent after visiting a host in any other application.

Now, assuming that an agent is at host $i<=n$, the data security properties can be illustrated as following:

- **Data Confidentiality:** Host i can't extract any of the offers added before its own.
- **Non-Repudiability:** No Host can deny submitting an offer once received by the originator of the agent.
- **Forward Privacy:** Host i can't extract the identities of the previously visited hosts.
- **Forward integrity:** Host i can't modify any of the offers added by previously visited hosts.
- **Strong Forward integrity:** If the agent revisited host i again, then none of the offers added to the agent between his two visits can be modified.

-   **Insertion Defense:** Host i can't insert any fake offers between previously added ones without being detected.
-   **Truncation Defense:** Host i can't truncate any of the previously added offers. Moreover, he can't collude with another host j to truncate in-between offers.
-   **Publicly Verifiable Forward Integrity:** Any visited host in the agent journey can check the integrity of the encapsulated offers before adding his own.

3 Related Work

G. Karjoth introduced a data encapsulation protocol in [7]. In his protocol, Karjoth suggested that an agent's computational results should be encrypted using the public key of the agent's originator. To preserve forward integrity, Karjoth suggested also that each visited host should compute some chaining value and add it with its encapsulated results. This value links a result from the currently visited host to the result generated at the previous host and the identity of the next one to visit. This chaining mechanism prevents any host from modifying the data added by any other host alone. However, nothing prevents a host from computing a new chaining value and colluding with another host later to forge the chaining mechanism and truncate in-between offers.

One-time Key based approach was suggested in [8]. In which, when the agent arrives at any host, it uses a key generation function to generate a key for a symmetric encryption of the results. Initially, the agent originator gives the agent an initial key seed for the computation of the encryption key at the first host. Then, at each subsequent host, the agent constructs an encryption key and uses that key to generate a new key seed for the next host to visit. The key generation function thus ensures that the keys form a chain that no malicious host can forge alone. However, since a malicious host may try to compute an alternative key seed to another host conspiring with it, they can simply truncate in between results.

In [9], we presented a data protection model that enhanced the One-time Key based approach. To prevent the two-colluder attack, we proposed an idea to link the key constructed at each host forwards to the host's successor and backwards to its predecessor. In our model, a host does not attach the key seed for his successor before the agent leaves him. Instead, a host computes the key seed for his successor only when the agent reaches that host and returns the next successor's identity. The key seed is then computed using the received identity. We also assumed that each agent is relayed to a trusted server after finishing its itinerary to be checked for consistency. The trusted server is responsible for reconstructing the encryption keys, authenticating the visited hosts and delivering the results safely to the agent originator. We proved that the presented model satisfied most of the data integrity properties and we also proved that the model outperforms the previous models in terms of the size and the encapsulation time overheads [10]. However, we have found that the model produces a performance problem; that is because of the execution overhead on the trusted server when the number of agents in the system increases.

Another protocol was presented by Xu in [11] In his protocol, he extended Karjoth protocol and he suggested that the chaining value computed at any host should link the current encapsulated offer backwards to the previous encapsulated offer and forwards to the next two hosts' identities. To do this in a free roaming agent scenario, when the agent migrates from a host to its successor, it carries only part of its offer. Its final offer is added and its chaining value is computed only when his successor returns back the next host's identity. Using this chaining idea, the approach defends against the insertion and the truncation attacks if no consecutive hosts colluded. However, this approach assumes the existence of a public/private key infrastructure between hosts for results encryption and signing. We find this, in addition to the computational power needed, hard to apply in an open network like the internet.

4 The Proposed Model

In the new proposed model, we try to extend the work we previously presented in [9] to reduce the load on the trusted server and in the same time satisfy the security requirements outlined in section 2. The new model follows the same data encapsulation approach adopted in the previous model. In which, the agent uses a simple key generation function to construct a key for a symmetric encryption of the results at each visited host. The construction of the key at any host builds a chaining relation that links that key forwards to the host's successor and backwards to all the previously constructed keys. In the new model, the agent is not relayed to the trusted server at the end of his journey to be checked for consistency. Instead, the agent's originator is the one that reconstructs the encryption keys and obtains the secured results. The trusted server is only used to authenticate the visited hosts for the agent originator when it requests so.

It is assumed now that the trusted server is the only party that has a permanent public/private key pair. Any host in the network may use that public key to deliver the data safely to the server. It is also assumed that, as in any internet application today, any host in the system must acquire a unique identity and a password from the server through an authenticated channel.

To illustrate the new model formulas, we use the model and cryptographic notations outlined in Table 1and Table 2 respectively. We also use the shopping agent example described in Section 2. The agent will roam the network starting from his home machine S_0 and visiting a list of servers from S_1 to S_n to collect the required offers. We describe the model as in the following subsections.

Table 1. Model Notations

Notation	Description
T	Trusted Server
S_0	Home originator of the agent ($=S_{n+1}$)
S_i $1 \leq i \leq n$	Unique identity of host i.
W	A Time-stamped value containing the agent's unique identity.
o_i	Offer from S_i with its identity included.
O_i	Encapsulated offer from host i.

Table 2. Cryptographic Notations

Notation	Description
h_i	The key seed and the chaining value for host i.
Pwd_i	A secret password of host i.
Pwd_i^-	A temporary secret password chosen by host i.
K_i	A secret key computed by host i.
r_i	A random number selected by host i.
$VerToken_i$	A token generated by host i for later host authentication.
y^-, y	A temporary private/public key pair of the agent originator.
$Enc_k(m)$	An encryption of an input m using the key k.
$H(m)$	A one-way hash value of an input m.
$X \oplus Y$	X bits are Logically XORed with Y bits.
A-› B: m	A message m is sent from host A to host B .

4.1 Agent Creation and Transmission:

To start an agent migration, the agent originator asks the trusted server for a time-stamped value including a unique agent identity for its agent. This time-stamped identity will enable the server later to uniquely identify each agent in the system.

$$T \rightarrow S_0 : \quad W \tag{1}$$

The agent originator then generates a temporary private /public key pair (y^- and y) and dispatches the agent to the first host in its itinerary along with the public key and the time-stamped identity.

$$S_0: \text{ Generate Temporary key pair } y^- \text{ and } y \tag{2}$$

$$S_0 \rightarrow S_1 : \text{Agent, W , y} \tag{3}$$

4.2 Agent at Host 1

When the agent arrives at host 1, it is executed to provide the offer o_1. To construct its encryption key, host 1 decides the next host to visit (S_2). It then sends a hash value of its identity to its predecessor (in this case, the agent originator S_0) and waits for a key seed. Making a host's predecessor generates the key seed for him will make the chained key series as we will see in the following steps.

$$S_1 \rightarrow S_0: \quad H(S_2) \tag{4}$$

When the agent originator receives the hashed identity of S_2, it computes the seed and the chaining value h_1 for host 1. The agent originator uses a hash value of its temporary private key in this computation. Since this private key is only known to the agent originator and the hash function is assumed to be collision free, host 1's computed key will be chained with host 2's identity. Thus, host 1 will be prevented later from colluding with another attacker to truncate host 2's results.

$$S_0 \rightarrow S_1: \quad h_1 = H(\ H(S_2) \oplus H(y^-)) \qquad (5)$$

Host 1 then chooses a random number r_1 and uses it with the received seed h_1 to compute its encapsulation key K_1. Using this random number is essential to maintain the confidentiality of the encryption key. Depending on the adopted encryption algorithm, the encryption key bits are selected as the first required bits of the output hash value. Thus, to use the model, a hashing function must be selected such that it produces an output greater in size than the required encryption key size.

$$S_1: \quad K_1 = H(\ h_1 \oplus H(r_1)) \qquad (6)$$

For later hosts' authentication, host 1 computes a verification token using the time-stamped agent identity, its secret password and a new temporary selected password. Host 1 will use this temporary password in the calculation of the key seed for host 2 as we will see in its encapsulation process.

$$S_1: \quad VerToken_1 = H(\ H(\ W) \oplus H(Pwd_1) \oplus H(Pwd_1^-)) \qquad (7)$$

To encapsulate its offer, host 1 uses the constructed key to encrypt the offer o_1 and produces the output: $E_{K1}(o_1)$. Host 1 also encrypts its identity, its temporary password, the chosen random number and the verification token using the originator's temporary public key. This will enable the agent originator later to reconstruct the encryption keys. Finally host 1 adds the encrypted data and sends the agent to host 2.

$$S_1 \rightarrow S_2: \quad O_1 = Enc_y(\ S_1, Pwd_1^-, r_1, VerToken_1) \text{ and } E_{K1}(o_1) \qquad (8)$$

4.3 Agent at any Consecutive Host

Following the same steps, host i where $i>1$ will encapsulate its results (Figure 1). However, host i sends its successor's identity to its predecessor host i-1 instead of the agent originator. To compute the seed h_i, as u can see in equation 9, host i-1 uses its temporary password instead of the originator's private key in the computation. It also adds its seed information h_{i-1} to the computation. Adding this seed will chain the key computed at any host i with all previously constructed keys; thus, preventing any collusion attacks.

$$S_{i-1} \rightarrow S_i: \quad h_i = H(\ H(S_{i+1}) \oplus H(Pwd_{i-1}^{-}) \oplus h_{i-1}) \tag{9}$$

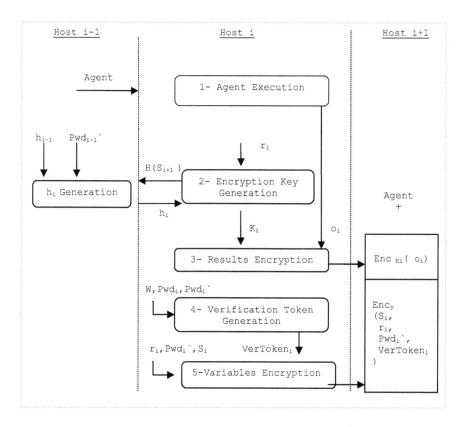

Fig. 1. A diagram denotes the encapsulation of an offer at a host i where i>1.

4.4 Returning back to the Agent Originator

When the agent reaches its originator at the end of his journey, the agent originator first decrypts using the private key y` the first part of all the offers $O_1, O_2, .. O_n$ to Obtain: $S_i$, $r_i$, $Pwd_i$` and $VerToken_i$ for $1 \le i \le n$. It then does the following:

Starting from host i= 1 to host i =n:

1. If S_i is suspected to be malicious, send S_i, Pwd_i` and $VerToken_i$ to the trusted server and wait for the server to compute the verification token again using the Pwd_i saved in its database and compare it with $VerToken_i$ to authenticate S_i.

2. If S_i is not authenticated from the server, then assume that a malicious host has pretended to be S_i and consider all offers from O_i to O_n as fake.

3. If S_i is authenticated then consider Pwd_i ` as safe and generate K_i as following:
    If $(i==1)$
    $$h_1 = H(H(S_2) \oplus H(y\text{-})) \quad \text{and} \quad K_1 = H(h_1 \oplus H(r_1))$$
    Else
    $$h_i = H(H(S_{i+1}) \oplus H(Pwd_{i-1}^{\text{—}})) \oplus h_{i-1}) \text{ and } \quad K_i = H(h_i \oplus H(r_1))$$

4. Decrypt the second part of the offer O_i using the generated key to obtain o_i. If failed to decrypt the offer, then assume that a malicious host has pretended to be the next host (S_{i+1}) and consider all offers starting from O_{i+1} to O_n as fake.

5 Security analysis of the model

Here we give an analysis of the security features in our protocol with respect to the properties outlined in Section 2.

Data Confidentiality: Since the key generation parameters for any encapsulation key are encrypted using the temporary public key of the agent originator, no one but the originator can reconstruct the keys. Furthermore, even with the fact that the key seed for any host S_i is given by its predecessor host S_{i-1}, host S_{i-1} can't figure the key K_i to reveal the offer o_i. K_i is computed using the equation: $K_i = H (h_i \oplus H(r_i))$. Then, to figure out K_i, host S_{i-1} needs to find r_i' such that $H(h_i \oplus H(r_i)) = H(h_i \oplus H(r_i'))$. This violates the assumption that the hash function H is collision-free.

Non-Repudiability: Since each host computes the verification token using its secret password, no host can deny submitting its offer given the fact that its password is shared only between him and the trusted server. Furthermore, the time-stamped agent identity included in the computation of the token will even prevent the agent originator itself later from using the tokens to submit fake offers on behalf of the visited hosts.

Forward Privacy: Although each host S_i must send the identity of host S_{i+1} to its predecessor host S_{i-1}, it is infeasible for host S_{i-1} to discover the received identity .That is because the identity is sent hashed using a one way hash function. Also, the identity of each execution environment is encrypted at each step using the originator's temporary public key.   Thus, No host will be able to extract the visited hosts' identities by examining the proposed offers.

Forward Integrity and Strong Forward Integrity: So far, we have pointed out that it is impossible for any host to figure out the keys for data encapsulation. This of course prevents the hosts from changing the added offers without leaving any detectable traces. Thus, if any malicious host x tried to manipulate an offer added by any preceding or following host S_i by replacing $E_{Ki}(Oi)$ with a dummy $E_{Ki}\cdot(Oi\grave{})$, the chained series of encryption keys will be corrupted , the agent originator will discover the forge and will mark this offer as fake.

Insertion Defense: To illustrate how our model achieves this property, consider an example of an agent that has visited shops 1,2,3,4…n and consider a malicious host x that tries to insert its offer between offer 2 and offer 3. Given the key generation function, x needs to do the following to succeed in this attack:

1) Revise the work done by both host 1 and host 2 to replace host 2's successor with x:

$S_2 \rightarrow S_1 : H(\underline{S_x})$

$S_1 \rightarrow S_2 : \underline{h_2\grave{}}= H(\ \underline{H(S_x)} \oplus H(Pwd_1\grave{}) \oplus h_1)$

S_2: Compute $K_2\grave{}$ and replace $E_{K2}(O_2)$ with $E_{K2}\grave{}(O_2)$

2) Revise the work done by host 3 to accept its successor instead of host 2:

$S_3 \rightarrow S_x : H(\underline{S_4})$

$S_x \rightarrow S_3 : \underline{h_3\grave{}}= H(\ H(S_4) \oplus H(\underline{Pwd_x}\grave{}) \oplus \underline{h_x})$

S_3: Compute $K_3\grave{}$ and replace $E_{K3}(O_3)$ with $E_{K3}\grave{}(O_3)$

3) Since h_3 has changed , host x needs to revise the work done by hosts from i=4 to n to change its computation of the key

$S_{i-1} \rightarrow S_i : \underline{hi\grave{}}= H(\ H(S_{i+1}) \oplus H(Pwd_{i-1}\grave{}) \oplus \underline{h_{i-1}\grave{}})$

$S_i : \underline{K_i\grave{}}= h_i\grave{} \oplus H(r_i)$

Given the above needed revisions, host x will never succeed in this attack unless it colluded with all hosts in the chain from the point it want the insertion to be done at.

Truncation Defense: The chaining mechanism used in the insertion defense also works for the truncation defense. But in the truncation defense our model only stands if no two consecutive hosts colluded. Suppose that as in the previous example, host x tries to truncate the offers after host 2 and adds its own. Then, to succeed, it needs to collude with both hosts 1 and 2 to replace host 3's identity with its own identity as described before.

Publicly Verifiable Forward Integrity: Because the agent originator is the only party that can reconstruct the encryption keys and verify the integrity of the encapsulated results, this property is not satisfied.

6 Conclusions and Future Work

In this paper, we proposed a new mobile agent security model. The main idea in our model is to enable the agent to encapsulate the computational results in its body, so that the agent originator can later check it for correct execution. We suggested encrypting the results at each host and using a different constructed key for the data encryption process. The constructed key at each host is to be chained forwards with the identity of the next host and backwards with the all the previously constructed keys. This chaining mechanism ensures the agent data integrity. We gave a brief security analysis on the model to show how the model satisfies most of the data integrity properties. We showed how the model achieves the insertion defense completely and the truncation defense if no two consecutive hosts colluded. Our future work will be to make an experimental study to compare our model against the other data encapsulation models and study the tradeoffs between the security features and the performance for all presented models.

[1] D. Kotz , R. S. Gray: Mobile Agents and the Future of the Internet, ACM Operating Systems Review, 33(3), 1999.

[2] C. Busch, V. Roth, R. Meister: Perspectives on Electronic Commerce with mobile agents, Proceedings of the XI Amaldi Conference On Problems of Global Security, Russian Academy of Sciences, 1998, pp.89-101

[3] V. Roth, M. Jalali, R. Hartmann ,C. Roland: An Application of Mobile Agents as Personal Assistants in Electronic Commerce", in Proceedings of the 5th Conference on the Practical Application of Intelligent Agents and MultiAgents, 2000

[4] W. Jansen: Countermeasures for Mobile Agent Security, Computer Communications,. Elsevier Press, Vol. 23, No. 17, 2000.

[5] N. Borselius: Mobile agent security, Electronics & Communication Engineering Journal , IEEE, London, UK, October 2002,

[6] T. Sander and C. Tschudin: Protecting mobile agents against malicious hosts. In Mobile Agents and Security, volume LNCS 1419, pages 44–60, 1998.

[7] G. Karjoth, N. Asokan and C. Gulcu.: Protecting the Computation Results of Free- Roaming Agents. 2nd International Workshop on Mobile Agents, Stuttgart, Germany, September 1998.

[8] J. Park, D. Lee and H. Lee: Data Protection in Mobile Agents; one-time key based approach. IEEE ISADS March 2001

[9] M. M. Essam, M.A. EL-Sharkawy and M. S. Abdelwahab: Protecting Mobile Agent from Malicious Hosts: A Data Encapsulation Approach. Proceeding of the 2nd International Conference on Intelligent Computing and Information Systems ICICIS, Cairo ,Egypt, March 2005

[10] M. M. Essam, M.A. EL-Sharkawy and M. S. Abdelwahab: A Comparative Study between Computation Results Protection Approaches for Free-Roaming Mobile Agents, Proceedings of the 5th international Conference on Informatics and Systems, INFOS2005, Cairo Univirsity, Egypt, March, 2007.

[11] D. Xu, L. Harn and M. Narasimhan: An Improved Free-Roaming Mobile Agent Security Protocol against Colluded Truncation Attack. Proceeding of the 30th IEEE Annual International Computer Software and Applications Conference volume: 2, pages 309-314, September 2006.

Secure Global Connectivity for Mobile Ad hoc Networks

K. Ramanarayana[1], Lillykutty Jacob[2]

Electronics and Communication Engineering Department,
National Institute of Technology, Calicut-673601, India
[1] E-mail: p050007ec@nitc.ac.in
[2] E-mail: lilly@nitc.ac.in

Abstract. This paper proposes a light weight solution for secure routing in integrated Mobile Ad hoc Network (MANET)-Internet. The proposed framework ensures mutual authentication of Mobile Node (MN), Foreign Agent (FA) and Home Agent (HA) to avoid various attacks on global connectivity and employs light weight hop-by-hop authentication and end-to-end integrity to protect the network from most of the potential security attacks. The framework also uses dynamic security monitoring mechanism to monitor the misbehavior of internal nodes. Security and performance analysis show that our proposed framework achieves good security while keeping the overhead and latency minimal.

1 Introduction

Mobile Ad hoc Network (MANET) has been a challenging research area for the last decade because of its versatility in routing, power constraints, security issues etc. A stand-alone MANET has limited applications because the connectivity is limited to itself. MANET user can have better utilization of network resources only when MANET is connected to the Internet. But, global connectivity adds new security threats to the existing active and passive attacks on MANET.

Many researchers proposed various solutions [1,2] to provide global connectivity to MANET. But, these proposals have not considered the security perspective of integrated network. Proposals [3,4,6,9] addressed the security threats and possible solutions for standalone ad hoc networks. These proposals have not considered the global connectivity of MANET and the related threats. Xie et al. [5] addressed the security framework for the integrated Internet-MANET. But their proposal depends heavily upon public key cryptographic algorithms, which are not desirable because of the computational overhead and latency problems. This paper proposes a framework that uses minimal public key cryptography to avoid overload on the network and uses shared key cryptography extensively to provide security.

The rest of the paper is organized as follows. Section 2 explores the related work in the area of secure   routing protocols for MANET. Section 3 presents a detailed description of the   proposed   framework. Section 4   presents   its   security and performance analysis. Finally Section 5 is about conclusions and future work.

2 Related Work

In this section we explore some of the existing secure routing protocols for MANETs.

Xie and Kumar [5] and Jiang et al. [4] use digital signature based hop-by-hop authentication in the route discovery. As Route Request (RREQ) floods in the entire network, every node in the network gets involved in the signature generation and verification process, which consumes a lot of node's resources irrespective of whether the node is included in the route or not. Moreover, public key cryptography results in long processing delay and computational overhead.

Kargl et al. [9] proposed Secure Dynamic Source Routing (SDSR) for standalone networks. According to the proposal, each node along the route appends its *Diffie-Hellman public key* and encrypted hash of calculated session key, to the Route Reply (RREP) packet, while it traverses from the destination to the source. It increases the RREP packet size enormously. A RREP packet larger than the maximum payload of 802.11 MAC frame is to be forwarded to the next hop in multiple frames. It increases delay at each node and degrades the efficiency of routing protocol. In addition to that, the online computation of session key from the *Diffie-Hellman public key* also adds delay to the route setup process.

Pirzada et al. [10] use promiscuous mode to detect the attacks such as black hole, gray hole, modification fabrication attacks, etc. But techniques using promiscuous mode fail to work when an attacker uses unidirectional antennas and also fail to detect the collaborative attacks.

3 Our Proposed Protocol: Secure Global Dynamic Source Routing Protocol (SGDSR)

3.1 Assumptions and Key Setup

We make similar assumptions as in [5,11] with SGDSR as well: i) Every Mobile Node (MN) in the MANET belongs to a certain administrative domain controlled by an agent called Home Agent (HA) and every MN shares a secret key with its HA; ii) Every node gets the digital certificate containing the [Node's Home Address, Public Key, Time of Issue, Time of Expiry] signed by a central Certificate Authority (CA) by some secure means, before entering into the ad hoc network, where CA's public key is known to HA, FA and all authorized MNs; iii) SGDSR requires pair-wise shared secret keys to be set up in each authorized node. It assumes that network has the mechanism to set up *pair wise shared secret keys* in every node which are under one administrative domain. That means, every node in the network should share a secret key with every other node in the network, and hence each node should have (n-1) shared keys in a network of 'n' nodes.

Blundo et al. [12] introduces a promising solution for establishing pair wise key set up in each node. It is a polynomial-based key pre-distribution protocol. In this proposal, the key distribution centre generates the polynomial share of node 'a', $f(ID_a,y)$, from a randomly generated   bivariate k-degree polynomial, $f(x,y)$, by substituting x=ID_a. Node 'a' can compute the shared key $f(ID_a, ID_b)$ with node 'b' by substituting y=ID_b, in its polynomial share. In this method, a node need not store all the (n-1) shared keys. It can compute the shared keys on demand with the help of its polynomial share. The memory requirement for storing polynomial is minimal.

3.2 Design Goals

The following are the design goals of SGDSR:
- To provide security against modification, fabrication, replay, and impersonation attacks on Intra-MANET routing as well as Internet-MANET routing.
- Low security overhead.
- Low route setup delay and communication overhead.

3.3 Protocol Description

SGDSR is designed to protect two communication scenarios: i) Internet-MANET communication; and ii) Intra-MANET communication.

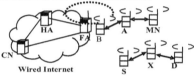

Fig. 1. Internet connectivity to MANET using Mobile IP

In Figure 1 the communication between MN and CN is Internet- MANET communication with the help of multi-hop connectivity through intermediate nodes A, B and FA. The communication between S and D through the intermediate node X is Intra-MANET communication.

3.3.1 Internet-MANET Communication
As soon as a node enters the network, it determines whether it is in its home network or foreign network by comparing the network prefix of its home address with the network address learned from agent advertisements. If they are same it comes to the conclusion that it is in the home domain else it is in the foreign network.

When a node is in the home network it can get Internet connectivity through HA. It can also take part in the Intra-MANET communication with its home address and *digital certificate* and its *polynomial share*.

When the node is in the foreign network it has to register its *Care Of Address* (CoA) with HA using Mobile IP protocol and should also obtain temporary certificate certifying its CoA and public key from CA through FA. It also needs to obtain its *polynomial share* from the FA; otherwise node is not allowed to participate in ad hoc routing. Table 1 shows the notations used in the proposed framework.

Table 1. Notations used in the framework

MN→A	Message transmission from MN to A
x,y	concatenation of two messages x and y
MN_{CoA}	MN's Care of Address
$FA_{Multicast}$	Foreign Agent's Multicast Address
FA	Foreign Agent's IP Address
ID	Route Message Unique ID
Dx	Diffie-Hellman publuc key of node x
{ }	Route Record
h_1	Hash code on (Randon nonce,Route Record)
h_n	Hash code on (h_{n-1},Route Record)
Sig_x	Signature of Node x on static part of message
$Cert_x$	Certificate of node x
FA_Req	A bit sequence indicating FA Request
FA_Rep	A bit sequence indicating FA Reply
R_Req	A bit sequence indicating Route Request
R_Rep	A bit sequence indicating Route Reply
R_Err	A bit sequence indicating Route Error
R_Report	A bit sequence indicating Route Report
K_{x-y}	Shared secret key between node x and node y
H_{x-y}	Hash code on the K_{x-y} and the specified message
H(m)	Hash on message m

3.3.2 FA Discovery and Registration Process

The salient steps involved in the secure registration of CoA with HA are depicted in Figure 2a, where I to VII show the processes at the concerned node and 1 to 6 correspond to the control message flow between different nodes. Figure 2b depicts the control messages that flow between MN, FA and HA.

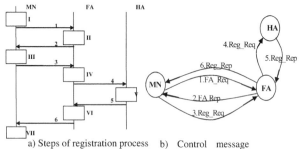

a) Steps of registration process   b)   Control   message

Fig. 2. FA discovery and registration process

Process 1: MN which enters the foreign network initiates FA discovery process by generating FA_Req message (Step 1.1 in Figure 3) with its home address as the source address and agent's multicast address 224.0.0.11 as the destination address. MN appends a hash tag h_1=hash (a random nonce, {MN}) to protect the source route from alteration. MN signs on the static message with MN's private key, i.e., Sig_{MN}= K^{-1}_{MN}[Hash (FA_Req , $MN_{HA,}$ $FA_{multicast}$, ID, D_{MN})] and broadcasts the message to its neighbors. FA_Req propagates to FA through nodes A and B.

Step 1.1 MN → A : FA_Req,MN_{HA},$FA_{Multicast}$,ID,D_{MN}, {MN_{HA}},h_1,Sig_{MN},$Cert_{MN}$

Step 1.2 A → B   : FA_Req,MN_{HA},$FA_{Multicast}$,ID,D_{MN}, {MN_{HA}A},h_2,Sig_{MN},$Cert_{MN}$

Step 1.3 B → FA : FA_Req,MN_{HA},$FA_{Multicast}$,ID,D_{MN}, {MN_{HA}A,B},h_3,Sig_{MN},$Cert_{MN}$

Step 2.1 FA → B : FA_Rep,MN$_{HA}$,FA$_{Multicast}$,ID, D$_{FA,}$ {MN$_{HA}$,A,B,FA},h$_3$,List of CoAs, H$_{FA-B,}$ Sig$_{FA}$,Cert$_{FA}$

Step 2.2 B → A : FA_Rep,MN$_{HA}$,FA$_{Multicast}$,ID, D$_{FA,}$ {MN$_{HA}$,A,B,FA},h$_2$, List of CoAs, H$_{B-A,}$ Sig$_{FA}$,Cert$_{FA}$

Step 2.3 A → MN : FA_Rep,MN$_{HA}$,FA$_{Multicast}$,ID, D$_{FA,}$ {MN$_{HA}$,A,B,FA},h$_1$, List of CoAs, Sig$_{FA}$,Cert$_{FA}$

Fig. 3. Steps involved in FA Discovery

Step1.2: Any neighbor node A of MN receives FA_Req. It checks whether it has already seen the request. It drops any duplicate and invalid FA_Req; otherwise, it makes an entry in its *route request table* and then appends its IP address to the *route record* and replaces h$_1$with h$_2$=hash (h$_1$, {MN$_{HA}$, A}). Then node A broadcasts the message to its neighbors.

Step1.3: Any neighbor B of node A receives the message and then broadcasts the FA_Req to its neighbors after doing similar process.

Process II: Upon reception of FA_Req from node B, FA validates the signature. If the signature is valid, FA computes shared session key SK$_{FA-MN}$ with the help of D$_{MN}$ in the FA_Req packet. FA initiates FA_Rep (Step 2.1). FA_Rep carries the actual IP address of FA and the list of CoAs. FA affixes H$_{FA-B}$= Hash (K$_{FA-B}$, SRM), where SRM=[MN$_{HA}$,FA$_{Multicast}$,ID,D$_{FA}$,{MN$_{HA}$, A,B, FA }, List of CoAs], and K$_{FA-B}$ is the pair wise shared key between FA and B which is calculated using FA's *polynomial share.* FA unicasts FA_Rep back to node B.

Step 2.2: Upon reception of FA_Rep, node B first computes the hash code on the buffered h$_2$ and extracted part of source route from MN to itself from the *route record* available in the FA_Rep message and checks if it is equal to h$_3$. The hash tag h$_3$ is to see that FA_Rep travels exactly in the reverse route and ensures that route is not modified. After passing the first check, node B computes the hash code of its shared key K$_{FA-B}$ and SRM in the FA_Rep and checks if it is equal to   H$_{FA-B}$. If so, node B authenticates FA. Then, it replaces H$_{FA-B}$ in the FA_Rep packet with H$_{B-A}$ and unicasts the FA_Rep to A.

Step 2.3: Finally Node A unicasts the FA_Rep to node MN if h$_2$ and H$_{B-A}$ are valid.

Process III: Node MN checks h$_1$.Also validates the signature of FA. These verifications ensure that, the learned *source route* is not a fabricated or modified one. MN calculates shared session key SK$_{MN-FA}$ using Diffie-Hellman public key D$_{FA.}$

Now both FA and MN are authenticated each other using digital certificates and can believe each other. MN chooses one CoA among the given list of CoAs and then initiates registration process with the message in Step 3 and unicasts the Reg_Req message to FA along the shortest among learned routes.

Step 3 MN → FA: Reg_Req,M$_1$,H$_{MN-FA}$

where M$_1$ =M, H$_{MN-HA}$ and M= MN$_{HA}$, MN$_{CoA}$ FA,ID, {MN$_{CoA}$,A,B,FA},List of CoAs

H$_{MN-HA}$ is the Hash(K$_{MN-HA}$,M) used for checking integrity and authentication between MN and HA, and H$_{MN-FA}$ is the Hash(SK$_{MN-FA}$,M$_1$) used for checking integrity and authentication between MN and FA.

Process IV: Upon reception of Reg_Req, FA validates H_{MN-FA}. Then FA records MN's CoA, and signs message M_1 with its private key, appends its certificate and send the message to HA as in Step 4.

Step 4 FA → HA: M_2,
   where $M_2=[M_1, D_{FA}, Sig_{FA}, Cert_{FA}]$

Process V: Upon reception of Reg_Req packet from FA, HA validates the signature of FA and H_{MN-HA}.This process ensures that: i) MN and FA authenticate each other; ii) FA has not modified the actual registration message sent by MN; and iii) message sent by FA is not altered.

After satisfying the verification results, HA calculates the session key SK_{FA-HA} with the help of Diffie-Hellman public key[14] D_{FA}.HA then registers MN's CoA and sends Reg_Rep (Step5) after signing the entire message with its private key.

Step 5 HA → FA: Reg_Rep,M, D_{FA}, Sig_{HA},$Cert_{HA}$

Process VI: Upon reception of Reg_Rep, FA validates signature of HA, then computes session key SK_{HA-FA}. FA gets temporary certificate for MN from CA with the details: [CoA, public key, time of issue, Expiry time]. Then FA sends the Reg_Rep to MN after appending the encrypted MN's *Polynomial share* as given in Step 6.

Step 6 FA → MN: Reg_Rep,M, Sig_{HA},$Cert_{HA}$, Temp-$Cert_{MN}$, $SK_{MN-FA}(MN_{Polyshare})$,$H_{FA-MN}$
   Process VII: MN validates H_{FA-MN} and the signature of HA, then records its temp-certificate and decrypted polynomial share.

The registration process ensures pair wise mutual authentication among MN, FA and HA and there by avoids any fraudulent node to impersonate or manipulate registration messages.

Once the registration process is successfully completed, HA tunnels the packets destined for MN to FA. FA sends the packets to MN through multihop communication.

3.3.3 Intra-MANET Communication

Let node S wants to communicate with node D using SGDSR protocol. Let X be an intermediate node. SGDSR permits nodes to participate in the routing protocol only after acquiring the *certificate* and *polynomial share*. If a node is in its home network   it can use its home address as its ID. If a node is in the foreign network, it has to complete the registration process first and then it can use the CoA as its ID.

Step 4.1 S → X: R_Req,S,D,ID,{S},h_1,H_{S-D}

Step 4.2 X → D: R_Req,S,D,ID,{S,X},h_2,H_{S-D}

Step 4.3 D → X: R_Rep,S,D,ID,{S,X,D},h_2,H_{D-X}

Step 4.4 X → S: R_Rep,S,D,ID,{S,X,D},h_1,H_{X-S}

Fig. 4. Steps involved in ad hoc route discovery

The source node S initiates route discovery process by generating R_Req. The sequence of steps involved is given in Figure 4.

Source S generates R_Req with its IP address as source ID, destination address as destination ID. It appends h_1=Hash (nonce,{S}), and H_{S-D} =Hash ($K_{S-D,}$ [S,D,ID]) and broadcasts to its neighbors.

A neighbor X to node S receives the R_Req and then appends its ID to the *route record* and replaces h1with h_2=hash (h_1, {S, X}) and then broadcasts the R_Req to its neighbors.

Upon reception of R_Req, node D first validates H_{S-D} to check the integrity of packet and to authenticate the sender. Upon validation, node D generates R_Rep as in step 4.3. D appends the hash code H_{D-X} =Hash ($K_{D-X,}$ [S,D,ID,{S,X,D}]) to R_Rep and then unicasts the message to node X, which in turn unicasts the R_Rep, to node S, after validating h_2 and H_{D-X}.

Finally, Node S records the source route in its route cache, after validation of h_1 and H_{X-S}. Now the node S sends the data packets using the source route.

Route Maintenance: Every node in the route keeps track of the link between itself and next hop neighbor. If the link is found broken, node generates R_Err and signs with its private key, then unicasts to the source node thorough the intermediate nodes. Upon validation of signature, source node may select an alternative route stored in its route cache. If no route is available source initiates route discovery again.

3.4 Reactive Security Mechanism

SGDSR is supported by a reactive security mechanism similar to Watch Dog [13], to mitigate the threats due to intermediate nodes. The security mechanism works as follows.

The sender node buffers the packet, transmits the packet to the next hop node and then switches itself into the promiscuous mode to over hear the retransmission by recipient node. Sender compares the buffered packet and overheard packet. From this observation sender node can find out whether the recipient node is carrying out any attacks such as black hole, modification, fabrication, impersonation, and replay attack.

If a next hop node is found guilty, the sender node informs about the misbehavior to the source node in a special packet R_Report after attaching its signature. Source node forwards the R_Report   to the FA after signing the message. It is the responsibility of FA to   inform about the misbehavior to the malicious node's HA in order to eliminate it from the network.

3.5 Optimizations

The proposed SGDSR protocol is the security extension of Dynamic Source Routing (DSR) protocol[7]. It does not allow all the optimizations possible for DSR; rather, the following optimizations are allowed:

- An intermediate node, which knows the valid route to the destination, unicasts the route request packet in the shortest known route to avoid unnecessary flooding.

- All the authorized nodes of SGDSR have a common network prefix in their IP address. Hence, host part of IP address can be used as node's ID to reduce the size of route record in the control and data packets. In this case, the *network part of address* and *subnet mask* fields should be included in the packet header in order to reconstruct any node's IP address unambiguously.

4 Security and Performance Analysis

4.1 Security Analysis: Intra-MANET Communication

The proposed SGDSR protocol is secure against most of the external attacks, because of the following three phases of defense:

1. A mobile node is permitted to participate in the routing protocol only after successful registration with its HA. This process helps:
   a. To filter out external malicious nodes from entering the network.
   b. To bind a unique IP address with the ad hoc ID of the node. IP address is not only useful to uniquely identify the node in the global communication scenario but also helps to fix accountability to the participating nodes. Any registered node found guilty can be fixed and such nodes can be eliminated from the network. This enhances trust levels among the members of the network.
2. The static part of route request messages is protected by a hash code function to detect tampering of static part by intermediate nodes. The mutable part is protected by another hash code function to restrict the route reply traversal exactly in the reverse order of learned route. This process avoids modification and fabrication attacks on the source route. End-to-end authentication in the route request phase avoids impersonation of source and destination nodes. End-to-end integrity in the route request phase avoids modification attacks by intermediate nodes. Hop-by-hop authentication in the route reply phase avoids external malicious nodes to participate in the routing protocol and thereby avoids the attacks caused by them.
3. Reactive security mechanism added into the protocol finds out the malicious operations and consequent attacks caused by the internal authenticated nodes, which can not be detected by proactive security methods. Black hole and gray hole attacks are some such attacks.

4.2 Security Analysis: Internet Connectivity

The security perspective of registration process is discussed here.

- The mutual authentication of MN, FA and HA is carried out with the help of public key and shared key cryptography techniques.

The secure registration process adopted in the protocol gives no scope for impersonation, modification, and fabrication attacks by any fraudulent node.

4.3 Performance Analysis: Computational Overhead

The computational overhead of SGDSR is very low compared to the existing protocols [5, 9] due to the following factors:

i.   SGDSR uses minimal public key cryptography in the FA discovery and registration process. Table 2 gives a comparison with [5]. SGDSR requires no sign generation/verification at intermediate nodes. SGDSR uses keyed hash function for hop-by-hop integrity and authentication, which is computationally economical, where as [5] requires four public key signature generation and verifications at each intermediate node between MN and FA to complete the registration process.

Table 2. Computational load for registration process

	Xie [5]				SGDSR			
	At MN	Int. Node	At FA	At HA	At MN	Int Node	At FA	At HA
No of Cert verifications	0	0	1	1	2	0	2	1
No. of Public key Sign gen.	1	4	3	1	1	0	2	1
No. of Public Key Sign Ver.	2	4	3	1	2	0	2	1

ii. A very important difference between [5] and SGDSR is that, proposal [5] uses public key based sign generation and verification for hop-by-hop authentication in the route request phase, which floods the entire network. Every node has to do at least one sign generation and one sign verification   irrespective of whether it belongs to the route or not. SGDSR uses light weight hash codes for this purpose, which greatly reduces the computational load as well as processing delay at each node, with out compromising security

iii. SDSR [9] appends Diffie-Hellman public key to the route reply,   for hop-by-hop authentication as well as for distribution of shared session keys among the members of the route, which increases the size of route reply packet enormously as the number of hops increases. Its adverse effects are: i) increase in communication overhead, and ii) enormous increase in processing delay due to online computation of session key and its hash value. SGDSR uses pair wise shared key pre_distribution for this purpose.

5 Conclusions and Future Work

Though there are many research proposals on 'Internet connectivity for ad hoc networks' and 'Secure routing protocols for ad hoc networks', separately, the research on the 'secure routing protocols for integrated Internet-MANET' is lacking. In this paper we proposed a secure routing protocol for global connectivity of DSR based MANET. Proposed protocol SGDSR uses hash codes extensively to minimize the computational and communication overhead. SGDSR is resistant to most common security attacks such as modification, fabrication, replay attacks and it can also detect black hole, gray hole attacks etc., with the help of reactive security mechanism.

Routing delay is another important consideration. Public key based computation intensive secure routing protocols can not work well due to the longer processing delay, especially in ad hoc networking environment. SGDSR is a carefully designed light weight protocol for secure global connectivity, and with minimal overhead and latency.

Future work includes security analysis of the proposed protocol using BAN logic and performance analysis of the protocol using OPNET simulation software. We are also working towards the design of an efficient reactive security mechanism, and the design of efficient key setup using Identity-based cryptosystem to support the protocol.

References

1. Y. Sun, E.M. Belding-Royer, and C.E. Perkins: "Internet Connectivity for Ad hoc Mobile Networks," International Journal of Wireless Information Networks special issue on Mobile Ad Hoc Networks (MANETs): Standards, Research, Applications, 9(2), April 2002
2. P. Ratanchandani and R. Kravets: "A hybrid Approach to Internet Connectivity for Mobile Ad Hoc Networks", Proceedings of IEEE WCNC,2003.
3. Panagiotis Papadimitratos and Zygmunt J. Haas: "Secure Routing for Mobile Ad Hoc Networks" In SCS Communication Networks and Distributed Systems Modeling and Simulation Conference (CNDS 2002).
4. Tingyao Jiang, Qinghua Li, Youlin Ruan: "Secure Dynamic Source Routing Protocol" Proceedings of the The Fourth International Conference on Computer and Information Technology (CIT'04) - Volume 00, (2004)Pages: 528 – 533.
5. Bin Xie, and Anup Kumar: "A Framework for Internet and Ad hoc Network Security", IEEE Symposium on Computers and Communications (ISCC-2004), June 2004
6. Kimaya Sanzgiri, Bridget Dahill, Brian N. Levine, and Elizabeth M. Belding-Royer: "A secure routing protocol for ad hoc networks",Proceedings of the 10th IEEE International Conference on Network protocols (ICNP'02)
7. D. B. Johnson, D. A. Maltz, Y.-C. Hu, and J. G. Jetcheva: "The Dynamic Source Routing Protocol for Mobile Ad Hoc Networks (DSR)", http:// www.ietf.org /internet-drafts /draftietf -manet-drIETF draft, 2004.
8. C. Perkins: "IP Mobility Support for IPv4," IETF RFC 3220, 2002.
9. F. Kargl, A. Geiß, S. Schlott, M. Weber: "Secure Dynamic Source Routing", Hawaiian International Conference on System Sciences 38, Hawaii, USA, January 2005.
10. Asad Amir Pirzada Chris McDonald, Amitava Datta: "Performance Comparison of Trust-Based Reactive Routing Protocols" IEEE Transactions on Mobile Computing, Volume 5, Issue 6, (June 2006) Pages: 695 – 710.
11. Leiyuan Li , Chigan C: "Token Routing: A Power Efficient Method for Securing AODV Routing Protocol", Proceedings of the 2006 IEEE International Conference on Networking, Sensing and Control, 2006. ICNSC '06, (April 2006) pages 29- 34.
12. C. Blundo,   A. De Santis, A. Herzberg, S. Kutten, U. Vaccaro, and M. Yung: "Perfectly-secure key distribution for dynamic conferences", In Advances in Cryptology – CRYPTO '92,LNCS 740, (1993)pages 471–486.
13. Sergio Marti, T.J. Giuli, Kevin Lai, and Mary Baker: "Mitigating Routing Misbehaviour in Mobile Ad Hoc Networks". In Proceedings of the Sixth Annual International Conference on Mobile Computing and Networking (MobiCom 2000.)
14. Whitfield Diffie and Martin Hellman: "New directions in cryptography", IEEE Transactions on information Theory,VolIT-22,no6,(Nov,1976) pp644-654.

Security Considerations
for Residential Mode on Zigbee Network

Lee-Chun Ko

Department of Internet Platform Application (X300)
Computer & Communication Research Labs
Industrial Technology Research Institute, Hsinchu, Taiwan
brentko@itri.org.tw

Abstract. An energy-efficient and low-complexity wireless sensor network protocol based on IEEE 802.15.4 standard has mainly been promoted by ZigBee alliance. This protocol has been widely developed by many companies for a variety of applications. For the security aspect of ZigBee network, it provides two security modes when initializing the network, i.e. commercial mode and residential mode. This paper briefly introduces the procedure of residential mode on ZigBee network, then we point out several security flaws when operating in this mode. We show that ZigBee network is subject to replay attacks and packets information could be revealed to the attacker. More seriously, ZigBee network is possible to reveal the encryption key when updating the encryption key. We recommend several changes to the specification that we believe will improve security for ZigBee users.

Keywords: Security, IEEE 802.15.4, Wireless Sensor Networks, ZigBee

1 Introduction

Wireless sensor networks (WSNs) attract many engineers and researchers due to its rich domain of application and research including hardware and software design. This network consists of many small sensing devices featuring low-power, low-cost and self-configuration. These sensing devices are usually deployed in large area and equipped with different type of sensors for sensing physical environment such as light, temperature and humidity etc. The sensed data are often collected and converted to digital information, and then conveyed to the back-end host by network routing mechanism for further analysis. The applications of WSNs include battle field, medical, security, disaster monitoring and automation etc. In order to achieve low-power, low-cost and large amount of deployments, battery-powered sensing devices with an 8-bit, 16-bit or 32-bit microcontroller and a small memory size is the common hardware architecture.

In WSNs packets sent over the air must be protected by cryptographic mechanism to defeat some existing attacks such as eavesdropping, packets injection and modification. Otherwise, the attacker could mount various attacks so that the network could not function normally, or more seriously, the entire network could be destroyed.

Security issues in WSNs include not only protecting packets from eavesdropping but also key distribution, key management, device authentication and secure routing etc [1, 2, 3, 5, 6, 7]. Imagine that a WSN used in burglar alarm system without using security policy, the attacker could inject packets to induce false alarm or even to suppress the real alarm. Moreover, in battle field, the attacker could sniff packets and obtain critical information from the network, or send bogus information to disturb intelligence information. Therefore, in most applications, WSNs must provide security policy so that the network is able to work securely and normally.

There are some popular protocols for WSNs that have been developed and supported by many companies [9, 10, 11]. Among these protocols, ZigBee protocol is considered the most popular in commercial market. The ZigBee protocol stack is sketched in Fig. 1. The physical (PHY) layer and medium access control (MAC) layer are defined by IEEE 802.15.4 standard [12]. ZigBee alliance defines network (NWK) layer and application (APL) layer which includes the application support sub-layer (APS), the ZigBee device objects (ZDO) and the manufacturer defined application objects. ZigBee alliance also defines security service provider (SSP) both in NWK and APL layer. NWK layer provides packets encryption and decryption by using network-wide shared key, i.e. network key. APS provides not only end-to-end packets encryption and decryption by using link key but also some security commands to ZDO for key establishment and key management.

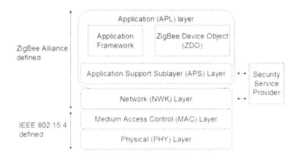

Fig. 1. ZigBee protocol stack.

ZigBee protocol provides two security modes when initializing the network, i.e. commercial mode and residential mode. In residential mode, every device uses only network keys to encrypt and decrypt packets in NWK layer, also in MAC layer if MAC security is enabled. In commercial mode, in addition to network keys, link keys are used in APS layer so as to provide end-to-end security.

In this paper the procedure of residential mode and key update on ZigBee network are introduced, and then we point out several security flaws when operating in this mode. We further describe that the ZigBee network is subject to MAC commands replay attacks even if MAC security is enabled. If MAC security is configured to access control list (ACL) mode or unsecured mode, the attacker could infer some information of network payload of packets. More seriously, ZigBee network is possible to reveal new network key when updating the current used network key.

Finally, we recommend several changes to the specification that we believe will improve security for ZigBee users.

The rest of this paper is organized as follows: Section 2 briefly introduces security mechanism used in ZigBee network. Section 3 points out several security flaws when operating in residential mode and an attack for deducing new network key. Several recommended changes for preventing these attacks are presented in Section 4. Conclusion is given in Section 5.

2   ZigBee Network

ZigBee network supports mesh and tree network topologies. As illustrated in Fig. 2, there are three different types of ZigBee device defined in ZigBee network:
Coordinator: coordinator is the most capable device and there is only one coordinator in each network. Typically, the coordinator is usually connected to the network gateway and responsible for network formation and maintenance. In security aspect, coordinator acts as a trust center for key establishment, key management and authentication with each device.
Router: Router acts as an intermediate device for forwarding packets to other devices.
End-device: End-device is the sensing element with less computing capability. End-device can only communicate with their parent such as router or coordinator.

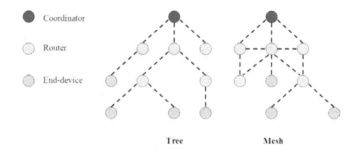

Fig. 2. The topology supported by ZigBee network.

2.1   Packet Encryption & Decryption

ZigBee uses CCM* to encrypt and decrypt outgoing and incoming packets by minor modifying CCM [4]. CCM* includes all features of CCM and additionally provides encryption-only and authentication-only capabilities. Currently, CCM* can only be used with 128-bit block cipher, such as AES-128 [13] used by ZigBee protocol. Tab. 1 shows eight different security levels defined by ZigBee protocol when executing the procedure of CCM*. Security attributes of MIC-32, MIC-64 and MIC-128 represent that the encrypted packet contains only message integrity code (MIC) but has no encryption on packet payload. The packet payload is encrypted when security attributes contain ENC.

Table 1. Security level used in CCM*.

Security level identifier	Security attributes	Data encryption	Length of MIC (bytes)
0x00	None	No	0
0x01	MIC-32	No	4
0x02	MIC-64	No	8
0x03	MIC-128	No	16
0x04	ENC	Yes	0
0x05	ENC-MIC-32	Yes	4
0x06	ENC-MIC-64	Yes	8
0x07	ENC-MIC-128	Yes	16

CCM* uses nonce to ensure that encrypting two identical plaintexts will not have two identical ciphertexts. Generally, nonce is appended to outgoing packets without encryption. The nonce used in CCM* is composed of 8-byte sender's IEEE address, 4-byte frame counter and 1-byte security control.

Once the nonce and key are available, the procedure of CCM* in NWK layer for outgoing packets can be executed. The original packet consists of NWK header and payload. The NWK payload may be generated by NWK layer itself or from APS layer. The procedure of CCM* is executed as follows: Firstly, add auxiliary (AUX) header which consists of nonce and key sequence number between the network header and the payload. Then, the NWK header, AUX header and NWK payload are included to compute MIC if security attribute contains MIC; the NWK payload and MIC are encrypted if security attribute contains ENC. Finally, The packet is then sent to the MAC layer if there is no any error occurred during executing the procedure of CCM*.

Encrypting packet payload in CCM* can be performed by dividing the packet payload into 16-byte size blocks $M_1\|...\|M_n$, and padding zero if the length of packet payload is not divisible by 16-byte. Then, computing each ciphertext block $C_1\|...\|C_n$ by

$$C_i = E(Key, A_i) \oplus M_i. \tag{1}$$

where $A_i = flags \| nonce \| counter\ i$ for $i=1, 2, ..., n,$ is XOR operation and $E(K,X)$ means AES-128 encrypting block X with key K. *flags* is 1-byte constant, 2-byte *counter* is initialized to 1 and increased by 1 after encrypting each block. The MIC is also encrypted using a similar way with *counter* set to 0.

Once each device in routing path and intended destination device receive an encrypted packet, they retrieve nonce from AUX header and use key to compute key stream. The decrypted payload and MIC are obtained by XORing key stream and the encrypted packet. Then, the MIC is verified by comparing received MIC with computed MIC whether they are the same or not. If the verification of MIC is correct, the packet is then parsed for further analysis. Otherwise, the packet is dropped and reported to the upper layer. Due to the space limitation, readers may refer to ZigBee specification for more detailed descriptions on the procedure of CCM*.

2.2   Residential Mode Authentication

When a device successfully joins to a secure network, this device must be authenticated by the coordinator before sending packets into the network. Residential mode on ZigBee network relies on maintaining network key(s) in each device for encrypting and decrypting packets. There are two different authentication procedures in residential mode, i.e. pre-configured (or pre-installed) network key and no pre-configured network key. Using no pre-configured network key is unsecured because the network key is sent to new joiner without any encryption and the network key could be possibly revealed in this transfer. We do not detail the authentication procedure of no pre-configured network key. We here focus on the authentication procedure of pre-configured network key, i.e. each sensor device is pre-configured the network key before deploying into the network. The detail description on this procedure is briefly introduced as follows.

After new joiner successfully joins to a router with pre-configured network key (by performing secure join procedure), router will send an encrypted *Update-Device* command to the coordinator to inform that a new device has joined the network. If the coordinator accepts this new joiner, it will directly send an encrypted *Transport-Key* command (illustrated in Fig. 3) with 16-byte *Key* field containing dummy key (i.e. 16 bytes 0x00) to new joiner. Then, this new joiner is considered to be authenticated. More detailed descriptions on secure join and residential mode authentication can be found in ZigBee specification.

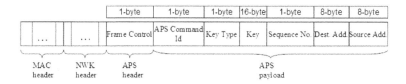

Fig. 3. The format of *Transport-Key* command.

2.3   Network Key Update

ZigBee protocol provides network key update procedure for changing the current used network key. For example, some applications may update key periodically or when certain events occurred. When deciding to update the current used network key, the coordinator firstly broadcasts an encrypted *Transport-Key* command with 16-byte *Key* field containing new network key, and then broadcasts an encrypted *Switch-Key* command to inform all devices of using this new network key. All devices use new network key for packets encryption and decryption after receiving *Switch-Key* command.

3   Security Problems

We have found several security flaws when operating in residential mode. One of security flaws induces ZigBee network to suffer from replay attacks even if MAC layer security is enabled. Another security flaw may reveal some fields of network packet payload to the attacker. More seriously, ZigBee network is also possible to reveal the new network key after performing network key update procedure described in Section 2.3 if the MAC security is configured to unsecured mode or ACL mode.

3.1   Replay Attacks

Since the residential mode does not provide end-to-end security, ZigBee defined that the MAC layer uses network keys as default keys for packets encryption and decryption. When using default keys, ZigBee specification mentioned that the MAC layer security should not use optional external frame counter (i.e. no checking incoming frame counter). Therefore, replay attacks can be mounted through this vulnerability. The NWK layer will drop replayed packets through checking the incoming frame counter of network packets. However, MAC command packets are parsed only in MAC layer. Consequently, replaying MAC command packets would be possible. IEEE 802.15.4 defined night different MAC commands such as association request, association response, disassociation notification and data request etc.

All MAC commands could be replayed. If association request and data request commands are replayed, the attacker is able to pretend a device to join to the network after the original device leaves the network. For example, a device $D1$ joins to router $R1$ by sending an association request and then a data request to $R1$. After $D1$ left the network for some reasons (ex. $R1$ forces $D1$ to leave), the attacker replays these two commands to $R1$, and $R1$ still accepts this association. This attack will cause the network existing many devices that do not actually present.

If disassociation notification command is replayed, the attacker is able to force the target device to leave the network after the target device rejoins the network. For example, a device $D1$ is notified to leave the network by receiving disassociation notification command from its parent router $R1$. Once $D1$ rejoins to $R1$, the attacker is able to force $D1$ to leave the network by replaying previous disassociation notification command sent by $R1$.

There are night different MAC commands used in the MAC layer. We may not detail the complete results of replaying each different command due to the space limitation. The readers can easily infer that replaying other commands in some circumstances will cause different types of attack, and we believe that such attacks induce serious damage in the network.

3.2   Packet Loss

If someone changes the implementation of ZigBee specification by checking the incoming frame counter for preventing replay attacks described in Section 3.1, another problem will raise as mentioned by N. Sastry and D.Wagner [8]. Their paper shows that IEEE 802.15.4 has a problem about checking incoming frame counter when using default security (i.e. default key). When using default security, each device maintains exactly one incoming frame counter for all neighbor devices. Therefore, suppose a device $D1$ receives 10 packets with frame counter 0 to 9 from another device; packets sent from other devices to $D1$ with frame counter lower than 10 will be rejected since $D1$ maintains only one largest incoming frame counter for all neighbor devices.

3.3   Revealing Network Key

IEEE 802.15.4 defines that MAC can be configured to the following modes: unsecured mode, ACL mode and secure mode. The default setting is unsecured mode. We suggest that unsecured mode and ACL mode shall never be used when operating in residential mode on ZigBee network since these two modes does not perform any cryptographic operation on packets. In addition to replaying MAC commands attack described above, the network key is possibly revealed described in this subsection.

3.3.1 Nonce Reuse

As mentioned in Section 2.1, the procedure of CCM* is similar to stream cipher; it uses nonce and key to generate key stream and produces ciphertext by XORing plaintext and key stream. Hence, it is obvious that the same nonce and key will generate the same key stream. If the attacker finds out two encrypted packets use the same nonce (this can be found by observing network AUX header), XORing these two encrypted packets will obtain the XORed result of decrypting these two packets. For example, suppose an encrypted packet $c_1 = m_1 \quad E(key, nonce)$, another encrypted packet that uses the same nonce $c_2 = m_2 \quad E(key, nonce)$, XORing these two encrypted packets will obtain

$$c_1 \oplus c_2 = m_1 \oplus E(key, nonce) \oplus m_2 \oplus E(key, nonce) = m_1 \oplus m_2 . \qquad (2)$$

Once some fields of m_1 are known, the corresponding fields of m_2 are also revealed, and vice versa.

The only one situation using the same nonce in CCM* is that the same device sends two encrypted packets using the same frame counter. Normally, this situation would not occur on ZigBee network since the frame counter is initialized to 0 and increased by 1 each time after sending one packet until sending $2^{32}-1$ packets. However, there still have some possibilities that one device reuses the frame counter, and hence reuses the nonce:

Power failure: when a device exhausts the energy or the attacker temporarily disrupts the power source.

Re-forming network: some application or maintenance requirements that need parts of devices rejoin to the network, or re-construct the entire network.

After one of situations described above occurs, the frame counter is reset to 0. The following attack shows the network key is possibly revealed by using this security vulnerability.

3.3.2 The Attack

As described in Section 2.2, in order not to reveal the network key, it is suggested to use pre-configured network key when joining the network. Using pre-configured network key causes trust center to send a *Transport-Key* command which contains dummy key to new joiner. The following attacks infer new network key when performing network key update procedure by using this *Transport-Key* commands and nonce reuse problem. The steps of the attack are described as follows:

1. The attacker collects *Transport-Key* commands sent or forwarded by target devices and constructs attack tables for each target device, these attack tables record which device sends or forwards *Transport-Key* commands using what frame counter value, and also the detailed contents of this packet.

2. The attacker temporarily disrupts the power source or by any means to induce the target devices reboot. This will cause the target devices to rejoin the network and reset the frame counter to 0.

3. Once the attacker observes that the target devices send another packet *p* using the same frame counter which is recorded in attack table, the attacker is then able to obtain some information from packet *p* by simply XORing *p* and the *Transport-Key* command that recorded in attack tables. Most fields in *Transport-Key* commands are known, such as *APS Command Id*, *Key Type*, *Key* and *Source Address*; these fields are totally 26 bytes. Therefore, the corresponding 26-byte fields of network payload of packet *p* will be revealed.

4. Another situation may reveal the new network key. If the coordinator observes that there are too many devices rejoin the network due to the step 2 or other policies that need to perform network key update procedure, the new network key is broadcast to all devices from coordinator using the a *Transport-Key* command. However, once any target device re-broadcasts the *Transport-Key* command using the same frame counter which is recorded in attack tables, the new network key is revealed by XORing two *Transport-Key* commands (one for sending dummy key and one for re-broadcasting new network key) that using the same frame counter.

One problem is how to recognize the *Transport-Key* commands. We know that the network payload consists of APS data packets or APS command packets. Unfortunately, the *Transport-Key* command is 36-byte length which is different from all other APS commands. The length of MAC header can be deduced from the contents of MAC header, the length of network header is fixed to 8 bytes, the length of AUX header is fixed to 14 bytes, the length of MIC can be deduced from security level and the length of CRC checksum is 2 bytes. Accordingly, if the attacker finds out a packet that has network payload length of 36 bytes, it has very high probability that this packet is a *Transport-Key* command. The only one exception is that APS data packets have exactly length of 36 bytes. However, 2-bit *Frame Type* in APS header indicates the packet is data packet or command packet. This can be used to further confirm that two XORed packets are both command packets. Another method can also confirm that two XORed packets are both *Transport-Key* commands by observing the XORed results of *APS Command Id* and *Key Type* fields are zero or not.

The security level can be known by sending beacon request to the coordinator or any router, and then the *Stack Profile* field contained in response packet will show which security level is current used in this network.

4 Recommendations

In Section 3 we point out several security flaws when operating in residential mode on ZigBee network. These attacks could be mounted easily so as to induce the ZigBee network to suffer from serious damage or even reveal the network key. We would like to give several changes to the specification that we believe will improve security for ZigBee users as follows:

1. MAC layer security always on: the ACL mode and unsecured mode in IEEE 802.15.4 should never be used. Using only NWK layer security is not strong enough to prevent some attacks. The cryptographic operation in MAC layer is necessary.
2. Do not use default security material: each device should maintain different incoming frame counters for all neighbor devices when operating in residential mode. This can be done by using ACL entry [12] to maintain security information about neighbor devices.
3. Modify the format of nonce in CCM*: 8-byte source address in nonce can be modified to an 8-byte random number when sending each one outgoing packet, or when the device is rebooted. This avoids nonce reuse problem even if the device is rebooted and then frame counter is reset to 0. The reason of changing source address to a random number is that each router in the routing path will decrypt and then encrypt the forwarding packets. Therefore, after the destination device receives the encrypted packets, the 8-byte source address in AUX header is the source address of previous hop router instead of the original source device. The destination device is still able to know the source address of original source device from packet herder. Hence, modifying the source address to an 8-byte random number does not affect any security requirements.
4. Make each packet length of APS command equal: from the description described in Section 3.3.2, all APS commands have different packet length. Therefore, the attacker can easily identify the purpose of each APS command. This may cause some levels of security threat. So we suggest that each APS command related to security should have equal packet length to prevent such threat.

5   Conclusion

We have briefly introduced security on ZigBee network and pointed out several security flaws when operating in residential mode. These security flaws include MAC commands replay attack and packet loss mentioned by N. Sastry and D. Wagner. We also suggest that both ACL mode and unsecured mode in MAC layer should not be used. Otherwise, the packet contents and network key will be revealed due to the nonce reuse problem. Several recommended changes to the specification for preventing these attacks are also presented.

We would like to emphasize that the next version of ZigBee specification (i.e. ZigBee 1.1), which is going to release during 2007, still have these security flaws. The system designers should note these problems when implementing the ZigBee protocol.

References

1. H. Chan, A. Perrig, and D. Song. Random Key Predistribution Schemes for Sensor Networks. In *Proceedings of the 2003 IEEE Symposium on Security and Privacy*, pages 197–213, 2003.
2. W. Du, J. Deng, Y. S. Han, and P. K. Varshney. A Pairwise Key Pre-distribution Scheme for Wireless Sensor Networks. In *Proceedings of the 10th ACM Conference on Computer and Communications Security* (*CCS'*03), pages 42–51, 2003.
3. L. Eschenauer and V. D. Gligor. A Key-management Scheme for Distributed Sensor Networks. In *Proceedings of the 9th ACM Conference on Computer and Communications security* (*CCS'*02), pages 41–47, 2002.
4. R. Housley, D. Whiting, and N. Ferguson, Counter with CBC-MAC (CCM), submitted to N.I.S.T., June 3, 2002. Available from http://csrc.nist.gov/encryption/modules/proposedmodes/ccm/ccm.pdf.
5. C. Karlof, N. Sastry, and D. Wagner. Tinysec: A Link Layer Security Architecture for Wireless Sensor Networks. In *Second ACM Conference on Embedded Networked Sensor Systems* (*SensSys'*04), pages 162–175, 2004.
6. C. Karlof and D. Wagner. Secure Routing in Wireless Sensor Networks: Attacks and Countermeasures. *Elsevier's AdHoc Networks Journal, Special Issue on Sensor Network Applications and Protocols*, pages 293–315, 2003.
7. A. Perrig, R. Szewczyk, J. D. Tygar, V. Wen, and D. E. Culler. Spins: Security Protocols for Sensor Networks. In *Proceedings of Seventh Annual ACM International Conference on Mobile Computing and Networks* (*Mobicom'*01), pages 189–199 2001.
8. N. Sastry and D. Wagner, Security Consideration for IEEE 802.15.4 Networks. In *Proceedings of the 2004 ACM Workshop on Wireless Security*, pages 32–42, 2004.
9. http://www.dustnetworks.com
10. http://www.zigbee.org
11. http://www.z-wavealliance.com
12. IEEE 802.15.4, Wireless Medium Access Control (MAC) and Physical Layer Specification for Low-Rate Wireless Personal Area Networks (LR-WPANs). New York, NY : IEEE, October 2003.
13. FIPS Pub 197, Advanced Encryption Standard (AES), Federal Information Processing Standards Publication 197, US Department of Commerce/N.I.S.T, Springfield, Virginia, November 26, 2001. Available from http://csrc.nist.gov/.

Success Rate of Reflection Attack on Some DES Variants

Esen Akkemik††‡, Orhun Kara† and Cevat Manap†

† TÜBİTAK UEKAE PK.74, Gebze, Kocaeli, TURKEY
‡METU Institute of Applied Mathematics, Ankara, TURKEY
{esena,orhun,cmanap}@uekae.tubitak.gov.tr

Abstract. In this paper, we study the success rate of the reflection attack on Feistel ciphers. We observe that the success rate of the attack depends on the number of pre-images of 0 in the round function for a given key. For the DES round function we calculate the number of the pre-images of 0 for any round key and classify the round keys with respect to this property. We use this classification to calculate the success rate of the reflection attack on a DES-variant cipher and propose an improved reflection attack with a higher sucess rate.

Key words: Reflection attack, success rate, Feistel cipher, DES, cryptanalysis, fixed point.

1   Introduction

The reflection attack[5] is a recent self-similarity analysis. The attack exploits the similarities between the round functions used in the encryption process and the round functions used in the decryption process. The reflection attack has been applied to several Feistel ciphers with very simple key schedules including GOST[7], a modified version of Magenta[4] and 2K-DES[2]. On the other hand, in [6], the reflection attack has been mounted on Blowfish, which has a very complicated key schedule.

The reflection attack tries to detect the fixed points that occur during the encryption process by looking at the plaintext-ciphertext pairs. Successfully detected fixed points are then used to recover a subkey. In [5] the round function of Feistel cipher is assumed to be a random function. Under this assumption it is proven that under certain conditions the probability that a plaintext-ciphertext that passes the detection process generates a fixed point is approximately $1/2$. We call this probability the detection rate of the reflection attack. In this paper we consider the DES round function, that is: we remove the randomness condition on the round function. We show that, in this case, the detection rate is $5/12$ for the DES round function under the same conditions. We define the success rate of an attack as the probability that the attack works for a random key. The success rate of the reflection attack depends on the probability that the round function generated by a subkey contains 0 in its image for some particular round. We focus on 2K-DES, a DES variant which uses the DES round function with an increased number of rounds, in order to determine the success rate of the reflection attack.

We show that for the DES round function the subkeys can be divided into three distinct subsets according to the number of the pre-images of 0 in the round functions generated by these subkeys. Using these subsets we show that the success rate of the attack on 2K-DES given in [5] is 56%. We propose a modified reflection attack with a success rate of 94%.

The structure of the paper is as follows: In Section 2, we give the details of the reflection attack on Feistel ciphers. In Section 3 we investigate the number of the preimages of 0 in the DES round function. We calculate the success rate of the reflection attack on 2K-DES and present an improved version of the attack in Section 4. The experimental results are given in Section 5. We conclude with Section 6.

2   The Reflection Attack on Feistel Ciphers

Let $x_1 \in GF(2)^{n/2}$ be the left half and $x_0 \in GF(2)^{n/2}$ be the right half of the plaintext. Then the Feistel structure can be stated as $x_i = \mathcal{R}_{k_{i-1}}(x_{i-1}) \oplus x_{i-2}$ where $\mathcal{R} : GF(2)^{n/2} \to GF(2)^{n/2}$ is the round function. The output of the i-th round operation is given as (x_{i+1}, x_i). In general, the swap operation is excluded in the last round and (x_{2r}, x_{2r+1}) is the ciphertext for a $2r$-round Feistel cipher.

The reflection attack separates the encryption process into four stages such that the third stage is the decryption of the first stage and the second stage has lots of fixed points. The fourth stage is used to recover the subkeys.

Consider the encryption process as the composition of 4 functions as $f_4 \circ f_3 \circ f_2 \circ f_1$. Let $f_1^{-1} = f_3$ and $f_2(x) = x$ for some $x \in GF(2)^n$. When $f_1(y) = x$ for some $y \in GF(2)^n$, we have

$$f_4(f_3(f_2(f_1(y)))) = f_4(f_3(f_2(x))) = f_4(f_3(x)) = f_4(f_1^{-1}(x)) = f_4(y),$$

and the encryption process can be approximated by f_4.

In order to apply the attack on a $2r$-round Feistel cipher, we can choose f_1 as the first $r - 1$ rounds, f_2 as the r-th round without the swap operation, f_3 as the next $r - 1$ rounds (rounds $r + 1$ to $2r - 1$) with an initial swap and without the final swap and f_4 as the last round with an initial swap. The functions f_1, f_2, f_3, f_4 are depicted in Figure 1. The condition $f_1^{-1} = f_3$ is satisfied when subkeys are related in a certain way [3], [8]. f_2 has fixed points when the round function outputs 0. Since f_4 is one round of encryption it is easy to detect the occurence of fixed points and recover the subkey.

The attack is applied to a Feistel cipher as follows: The input of f_2 is (x_r, x_{r-1}) and the output is (x_r, x_{r+1}). Assume that the output of the r-th round function is zero, i.e., $x_{r+1} = \mathcal{R}_{k_r}(x_r) \oplus x_{r-1} = x_{r-1}$. Then, we have a fixed point for f_2. The input of f_3 is (x_r, x_{r+1}) and the output is (x_{2r-1}, x_{2r}). The input of f_1 is (x_1, x_0) and the output is (x_r, x_{r-1}). Since f_3 is the inverse of f_1 we have $(x_1, x_0) = (x_{2r-1}, x_{2r})$. In f_4, we can detect the fixed points using

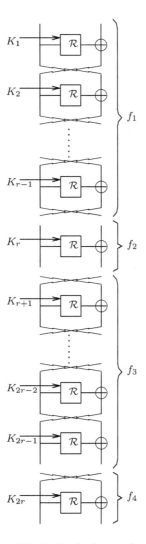

Fig. 1. f_1, f_2, f_3 and f_4

the relation $x_{2r} = x_0$. Every plaintext-ciphertext pair satisfying $x_{2r} = x_0$ is expected to have a large probability of generating a fixed point by Theorem 2 in [5]. For each plaintext-ciphertext pair satisfying $x_{2r} = x_0$, we solve the equation

$$x_{2r+1} = \mathcal{R}_{k_{2r}}(x_{2r}) \oplus x_{2r-1} = \mathcal{R}_{k_{2r}}(x_{2r}) \oplus x_1,$$

to recover the subkey k_{2r}.

3   Preimages of 0 in DES Round Function

The round function of DES has 80-bit input (48 of which is the subkey) and 32-bit output. The details of the round function is depicted in Figure 2. Let L be the input to the round function, E be the expansion operation, K be the subkey, X be the expansion of L, $C = X \oplus K$ be the input to the S-box layer and R be the output of the S-box layer. We denote the i-th bit of a bit string X by X^i. We have 16 linear relations, $X^{i_t} = X^{j_t}$ between the bits of X, since the

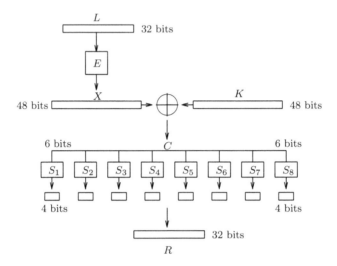

Fig. 2. DES round function

E operation makes copies of some bits of L. These relations can also be written in terms of bits of C and K.

$$X^{i_t} = X^{j_t} \iff C^{i_t} \oplus K^{i_t} = C^{j_t} \oplus K^{j_t} \iff C^{i_t} \oplus C^{j_t} = K^{i_t} \oplus K^{j_t}$$

We consider $R = 0$ and by inverting the S-box layer we have 2^{16} possible values for C. For each of these 2^{16} values we have a system of equations that must be satisfied by K. Each system can be labelled by a 16-bit vector $[v^1, v^2, \ldots, v^{16}]$ such that the corresponding system is $K^{i_t} \oplus K^{j_t} = v^t$.

We construct these systems for each C. We compute that $1/4$ of the possible systems are not produced by any of the C values, $1/2$ of the possible systems are produced by exactly one C value and the remaining $1/4$ of the possible systems are produced by two different C values. Let p be the probability that the DES round function outputs 0 for some key.

If K satisfies one of the systems that are not generated by any C, then the round function never produces 0 when used with K, since an expanded input

t	i_t	j_t	t	i_t	j_t
1	0	46	9	24	22
2	1	47	10	25	23
3	6	4	11	30	28
4	7	5	12	31	29
5	12	10	13	36	34
6	13	11	14	37	35
7	18	16	15	42	40
8	19	17	16	43	41

Table 1. The values of i_t and j_t in the relations for X.

XORed with K does not produce an S-box pre-image of 0. Hence, $p = 0$ for such subkeys. We denote the set of such subkeys by T_0.

If K satisfies one of the systems that are generated by only one C, then the round function outputs 0 for only one input value. Hence, $p = 2^{-32}$ for such subkeys. We denote the set of such subkeys by T_1.

If K satisfies one of the systems that are generated by two different C values, then the round function outputs 0 for two different input values, each value corresponding to one of the C values. Hence, $p = 2^{-31}$ for such subkeys. We denote the set of such subkeys by T_2.

4   Reflection Attacks on 2K-DES

2K-DES is one of the modified versions of DES given in [2]. 2K-DES does not have a key scheduling algorithm, instead it uses two independent 48-bit keys K_1 and K_2. K_1 is used in odd numbered rounds and K_2 is used in even numbered rounds in a total of 64 rounds. The round function of 2K-DES is the same as the round function of DES.

4.1   Previous Attack

In [5], a straighforward reflection attack is mounted on 2K-DES. The attack is applied to the encryption function to recover K_2 and applied to the decryption function to recover K_1. We only explain how to recover K_2, since the same process is used to recover K_1.

The 32nd round uses K_2 as its subkey. We assume that the round function generated by K_2 contains 0 in its image, i.e., $K_2 \in T_1 \cup T_2$. When the 32nd round function outputs 0, we get a fixed point for rounds 1 to 63. We can detect the fixed points by checking $x_0 = x_{64}$ and solve the equation $x_{65} = R_{K_2}(x_{64}) \oplus x_1$. Assuming that the 32nd round function is random, each detected plaintext-ciphertext pair has an approximately 50% chance of generating a fixed point; hence, only half of the detected plaintext-ciphertext pairs are expected to give the right K_2 value as a solution of the equation.

In Section 3, we show that the probability that the round function generated by some subkey contains 0 in its image is $3/4$. Since the attack requires both that K_1 and K_2 are in $T_1 \cup T_2$. Hence, the attack works on $3/4$ of K_1 and $3/4$ of K_2 and the success rate of the attack in [5] is $9/16$. Moreover, when $K_1, K_2 \in T_2$ the complexity of the attack decreases by 50%.

In this paper we remove the assumption that the round function is random and calculate the detection rate which is the probability that a detected plaintext-ciphertext pair generates a fixed point. We denote the event that a fixed point occurs as F and the event that plaintext-ciphertext pairs detected as D. Using Bayes' Theorem we get

$$\Pr(F|D) = \frac{\Pr(D|F) \cdot \Pr(F)}{\Pr(D)} = \frac{\Pr(D|F) \cdot \Pr(F)}{\Pr(D|F) \cdot \Pr(F) + \Pr(D|\bar{F}) \cdot \Pr(\bar{F})}.$$

We tabulate the values of $\Pr(F|D)$ and $\Pr(F)$ for keys in the sets T_0, T_1 and T_2.

Set	$\Pr(F)$	$\Pr(F|D)$ (detection rate)
T_0	0	0
T_1	2^{-32}	$\approx 1/2$
T_2	2^{-31}	$\approx 2/3$

Table 2. The values of $\Pr(F)$ and the detection rate for the DES round function

4.2   Improved Attack

We give an improved version of the reflection attack on 2K-DES by using the three sets of subkeys given above.

We modify the attack to include the cases when one of the subkeys is in T_0 and get a higher success rate than the sucess rate of the previous attack. The attack is depicted in Figure 3.

Without loss of generality assume that $K_1 \in T_0$ and $K_2 \in T_1 \cup T_2$. We recover K_2 by the reflection attack and focus on recovering K_1. Since the 34-th round function uses K_2, we can have 0 as the output of the 34-th round and get a fixed point for the rounds 4 to 64, i.e., $x_4 = x_{64}$ and $x_3 = x_{65}$. Consider the first 3 rounds of 2K-DES. For these rounds we have the following equations:

$$\mathcal{R}_{K_1}(x_1) = x_0 \oplus x_2 \tag{1}$$
$$\mathcal{R}_{K_2}(x_2) = x_1 \oplus x_3 \tag{2}$$
$$\mathcal{R}_{K_1}(x_3) = x_2 \oplus x_4 \tag{3}$$

In Eq. (2), K_2 and the output of the round function are known. From the output we can determine the 2^{16} possible values of C. Since we know K_2, we have

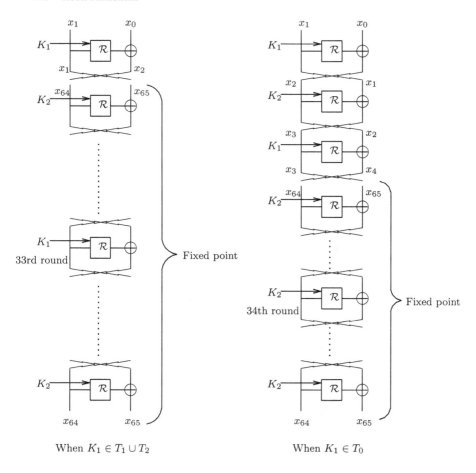

Fig. 3. The Improved Attack on 2K-DES

2^{16} possible values for X. The probability that a 48-bit vector is an expanded version of a 32-bit vector is $2^{32}/2^{48} = 2^{-16}$ and we expect to have only $2^{16} \cdot 2^{-16} = 1$ possible value for x_2 in Eq. (2). We put x_2 in Eq. (1) and Eq. (3) and determine the possible C values. Since we know x_1 and x_3, we know the expanded versions of these values, and by XORing them with C, we have two sets with 2^{16} possible values for K_1. We know that the right subkey is in both sets and the probability that at least one random subkey in one set is also in the other set is

$$1 - (1 - (2^{16} - 1)/2^{48})^{2^{16}-1} = 0.000015 \approx 2^{-16}.$$

Hence, we expect to have only the right subkey in the intersection of these two sets.

When the output of 34-th round function is not 0, the number of possible values for K_1 has a binomial distribution with parameters $N = 2^{16}$, $P = 2^{-32}$; hence, the expected size of the intersection of two sets is $N \cdot P = 2^{-16}$.

With 2^{34} known plaintexts, we expect the right subkey to appear around $2^{34} \cdot 2^{-32} = 4$ times and $2^{34} \cdot 2^{-16} = 2^{18}$ random subkeys to appear around once.

The improved attack fails only when both K_1 and K_2 are in T_0; hence, the success rate of the improved attack is

$$1 - \frac{1}{4} \cdot \frac{1}{4} = \frac{15}{16}.$$

5   Experimental Results

We verify our results on the sizes of the sets T_0, T_1 and T_2. For K_1 (and also for K_2) we expect $N/4$ of N keys to be in the set T_0, $N/2$ to be in T_1 and $N/4$ to be in T_2. Hence, for M random plaintexts, we expect $N/4$ keys to produce no fixed points, $N/2$ keys to produce $M \cdot 2^{-32}$ fixed points and $N/4$ keys to produce $M \cdot 2^{-31}$ fixed points. In our experiments, we choose $M = 2^{36}$ and $N = 48$. Hence, we expect 12 of the keys to produce no fixed points, 24 of the keys to produce 16 fixed points and 12 of the keys to produce 32 fixed points.

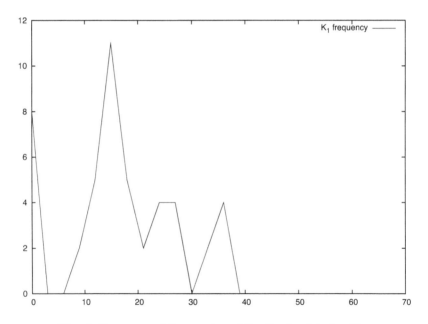

Fig. 4. Frequency of the number of the fixed points for K_1

For a random key we encrypt 2^{36} plaintexts with 2K-DES and obtain the corresponding ciphertexts. We determine the plaintexts where the plaintext is the same as the output of the $63rd$ round and call these plaintexts the fixed points for K_2. We also determine the plaintexts where the ciphertext is the same as the output of the $1st$ round and call these plaintexts the fixed points for K_1.

We repeat this process for 48 random keys and record the number of the fixed points for K_1 and K_2. In Figure 4 and Figure 5 we plot the number of the keys that generate fixed points with respect to the number of fixed points generated. The results show peaks around $32, 16$ and 0, which is in accordance with our expectations.

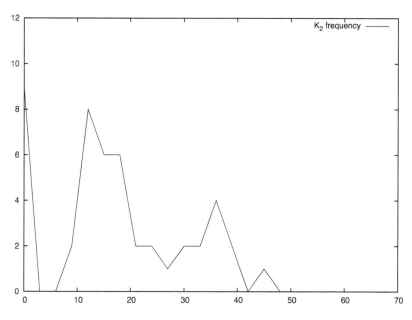

Fig. 5. Frequency of the number of the fixed points for K_2

6   Conclusion

We show that the success rate of the reflection attack depends on the structure of the round function and the subkeys used. In order to show this dependence we consider the DES round function. We classify the subkeys used in the DES round function into three disjoint sets depending on the number of the pre-images of 0. Utilising these sets we show that the success rate of the reflection attack on 2K-DES in [5] is $9/16$. We propose an improved version of the reflection attack on 2K-DES which has a higher success rate of $15/16$. We also show that the detection rate of the reflection attack is $5/12$ when the DES round function is used. We state an open problem: determine the effect of pre-images of 0 on the security of DES.

References

1. E.Biham, *New Types of Cryptanalytic Attacks Using Related Keys,* J. of Cryptology, Vol.7, pp.229-246, 1994.
2. A. Biryukov, D. Wagner, *Slide Attacks,* Proceedings of FSE'99, LNCS 1636, pp.245-259, Springer Verlag, 1999.
3. D. Coppersmith, *The Real Reason for Rivest's Phenomenon,* Proceedings of Crypto'85, LNCS 218, pp.535, Springer Verlag, 1986.
4. M.J. Jacobson, Jr., K. Huber, *The Magenta Block Cipher Algorithm,* First AES Conference, 1998.
5. O. Kara, *Reflection Attack on Product Ciphers.* Available at `http://eprint.iacr.org/2007/043`, 2007.
6. O. Kara, C. Manap, *A New Class of Weak Keys for Blowfish,* accepted at FSE'2007.
7. J. Kelsey, B. Schneier, D. Wagner, *Key Schedule Cryptanalysis of IDEA, G-DES, GOST, Safer, and Triple DES,* Proceedings of Crypto'96, LNCS 1109, pp.237-251, Springer Verlag, 1996.
8. J. H. Moore, G. J. Simmons, *Cycle Structures of the DES with Weak and Semi-Weak Keys,* Proceedings of Crypto'86, LNCS 263, pp.9-32, Springer Verlag, 1987.
9. B. Schneier, *Description of a New Variable-Length Key, 64-bit Block Cipher (Blowfish),* Proceedings of FSE'94, LNCS 809, pp.191-204, Springer Verlag, 1994.
10. *Data Encryption Standard,* U.S. Dept. of Commerce, National Bureau of Standards, FIPS Pub. 46, 1977.

Concurrency Issues in Rule-Based Network Intrusion Detection Systems

Mustafa Atakan[1] and Cevat Şener[2]

[1] Middle East Technical University, Ankara 06530, TR
mustafa.atakan@gmail.com
[2] Middle East Technical University, Ankara, 06530, TR
sener@ceng.metu.edu.tr

Abstract. As the bandwidth of present networks gets larger than the past, the demand of Network Intrusion Detection Systems (NIDS) that function in real time becomes the major requirement for high-speed networks. If these systems are not fast enough to process all network traffic passing, some malicious security violations may take role using this drawback. In order to make that kind of applications schedulable, some concurrency mechanism is introduced to the general flowchart of their algorithm. The principal aim is to fully utilize each resource of the platform and overlap the independent parts of the applications. In the sense of this context, a generic multithreaded infrastructure is designed and proposed. The concurrency metrics of the new system is analyzed and compared with the original ones.

1 INTRODUCTION: RBNIDS and Performance Problem

Rule Based Network Intrusion Detection System (RBNIDS) is one of the approaches in IDS world. It uses expert systems to analyze network trail data for intrusion attempts of pending or completed security violations [1]. One of the main goals of such systems is to be responsive as in preemptory manner, not very late as in reactionary manner where it processes traffic logs instead of instant network packets [2]. However, since the preemptory model is resource-intensive it may present a problem: When it is not as fast as the speed of the packets accumulating in the network buffer of the kernel, the IDS software begins to miss newly incoming packets (where the actual compromising sequence may be hidden) after the buffer is full.

There are many solutions to overcome this drawback such as increasing the hardware resources of the system or choosing just header over content analysis. However, those are not the focus of this paper. The purpose here is to try full utilization of computer resources by introducing a concurrency paradigm to a typical RBNIDS algorithm. By categorizing the independent parts of a RBNIDS, one can fairly deploy those parts to the resources in order to get maximum speedup [3].

2   IMPLEMENTATION: Producer-Consumer Model for Concurrency in Snort

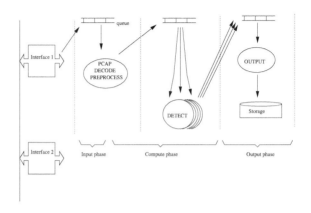

Fig. 1. Multithreaded overview of producer-consumer model

3   Introducing Concurrency to a Preemptory RBNIDS

A typical RBNIDS has three phases. Input, detect, and output. Suppose that in a sequential model, each of input and output phases takes 5% of the execution time and detect phase takes the remaining 90%. For the sake of simplicity, both input and output phases are supposed to be non-overlapping and they are inherently sequential. Then, the Amdahl's Law[3] says that the best speedup that new concurrent program can achieve is ten where input and output phases will still consume 10%.

In order to obtain maximum speedup for a sequential RBNIDS, various techniques can be proposed: For instances, the detection algorithm can be re-coded using a fast finite state automata. However these kinds of methods work well up to some point since independent parts of the algorithm aren't categorized and resources of the platform may not be fully utilized by these parts. So, in order to achieve this full utilization, a multithreaded model should be applied to the algorithms of RBNIDSs.

---

[3]

$$Speedup = \frac{T}{T_p} \qquad Speedup_{overall} = \frac{1}{P_{seq} + \left(\frac{P_{par}}{Speedup_{part}} \right)} \qquad (1)$$

4   CASE STUDY: Snort

Snort is a libpcap-based [5] packet sniffer and logger that can be used as a lightweight NIDS. Since it is highly capable, single threaded, has a programmable detection engine [6], very popular all around the world and most importantly open source [4], it has chosen as a case study tool.

The new multithreaded model works in an assembly line. The output of the first phase (PCAP+DECODE+PREPROCESS) is actually the input of the next phase (DETECT) in the pipeline. As seen from Fig. 1, these phases are deployed to three different thread groups. Since the nature of expertise of each thread group is different and they are using different system resources, the chance of occurrence of a bottleneck for all of them at the same time is very rare. While one group of threads in the pipeline is blocked (doing I/O for example), the other groups of threads are possibly doing their tasks.

The first thread group (PCAP+DECODE+PREPROCESS) does fetching, decoding and preprocessing of each packet repeatedly as a single thread. The time required to process a single packet has to elapse in order to get the next packet. Since it doesn't have to deal with detection and outputting, an efficiency increase is theoretically expected. Each processed packet is inserted into a FIFO queue (whose integrity is maintained by mutexes) where it waits to be handled for detection. The detection threads dequeue the packets, process and queue the action records into another queue that the output module will use. When there is no packet in the queue, they wait on a condition variable to be signaled; hence stealing no CPU cycles. The output module, which is created as another thread, dequeues the action records from the second queue and does its task. When there is no output record in the queue, the behavior of the output thread is the same as the detection threads.

The multithreaded snort is developed on Linux (debian 3.0) operating system. The running kernel is 2.4.23 and the thread library is LinuxThreads (version 0.9) where each thread is a separate process. The kernel scheduler itself schedules the threads just like Unix processes [7], which supplies the main strength resulting in simpler, more robust thread library, especially with respect to blocking system calls [8]. Although there are several user-level thread libraries available for Linux, they are deficient in functionality, performance, and/or robustness [9].

5   TEST RESULTS and COMPARISON

5.1   Test Media

The new application has four threads, one for input, two of them for detect and last one is for output. The original snort (version 1.9.0) and the new multithreaded snort are tested on a two-processor PC[4].

Twenty test files are obtained from METU (Middle East Technical University) bridge using tcpdump and each contains 500000 packets. They are replayed over a cross cable with *tcpreplay* tool with a maximum speed of 45 Mbps.

---

[4] Pentium-III/1133 Mhz, 926MB memory

5.2   Testing with Seven Different Configuration Files

In order to understand their effect on detect phase, seven different configuration files are composed for the tests. First two files contain a few simple rules. The third is the METU original snort configuration file in which there are 4905 rule entries. The forth, fifth, sixth and seventh configuration files contain the third one plus 50, 100, 500, 1000 randomly generated content matching rules respectively.

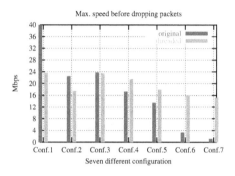

Fig. 2. Maximum speed of packet sent before dropping begins

In the first tests, the packet files are replayed while both snort versions run exclusively. An average value for twenty different test files run is calculated and used as a result for each configuration file. The aim of this test is to determine maximum packet speed that both versions of snort can handle.

As seen from the figure 2, performance of the multithreaded snort is worse than the original one in first two configurations. In other words, the original snort is capable of handling more packets than the multithreaded one before packet dropping begins. The reason is, as seen from the content of configuration files, the duty of detection phase is very small. Since there exists some overhead caused by schemes such as mutexes and context switching, multithreaded snort is obviously beaten. However, as the content of the configuration file gets larger, the new snort starts to show better performance. The more additional rules are added into configuration, the more speedup is gained. The performance difference between each version gets larger when duty on the detection phase increases. Through Conf.3 to Conf.7, the dramatic increase in handling capability of multithreaded design is easily seen.

As a second test, when the files are replayed from the local disk with the same configuration files, the execution time of the multithreaded snort decreases. Adding more rules brings out slight convergence to *perfect speedup*.

The same tests are also performed on Linux kernels 2.4 and 2.6 as a third test. Kernel 2.6 runs multithreaded snort more efficiently since it handles extreme loads more smoothly without breakdown and scheduling storms [10]. It shows its power, which is mainly caused by O(1) scheduler, when there are high loads.

6   CONCLUSION and FUTURE WORK

In this study, a typical RBNIDS algorithm is analyzed with respect to concurrency paradigm and a model is planned and implemented to obtain speedup. To prove that the model works as expected, a famous NIDS software, namely snort, is chosen as a case tool. It is the premier work that snort execution paths are deeply identified, documented, and a structural multithreaded schema is designed for its algorithm. The proposal and the results from the tests tell us that there are independent paths in an ordinary sequential RBNIDS and by a re-implementation, a pretty nice speedup is easily obtained even if the execution platform is limited on resources. The main goal of the design is multiplexing the detect phase of snort in order to make it more responsive the pending packets. The consequences of the study are good focal points for such systems.

As a future work, the algorithm of output module can be changed in such a way that it outputs over a fast media (i.e. a different interface) since the original one introduces performance impact when disk speed is too slow. Instead of a one thread and local disk access, the output phase can be modeled with parallel threads which are responsible for inserting records into a remote database.

Yet another, the proposed infrastructure can be redesigned to run as "distributed processes" in a cluster systems rather than SMP machines. Since clusters are very cheap to buy and scale up, any performance improvements on these systems are well appreciated.

References

[1] T. F. Lunt, A Survey of Intrusion Detection Techniques, Computers and Security, ch. 12, pp. 405-418, June 1993

[2] Maximum Linux Security, http://mandrake.petra.ac.id:8888/info/max/BkPg155x247.htm 1999

[3] Andrew S. Tanenbaum, Modern Operating Systems, Prectice Hall, 2 ed, 2001

[4] Martin Roesch, Lightweight Intrusion Detection for Networks, http://www.snort.org/docs/lisapaper.txt, 2003

[5] Van Jacobson and Craig Leres and Steven McCanne, LIBPCAP, Lawrence Berkeley National Laboratory, http://www-nrg.ee.lbl.gov, 1994

[6] Steven McCanne and Van Jacobson, The BSD Packet Filter: A New Architecture for User-level Packet Capture, USENIX Technical Conference Proceedings, 1993

[7] Xavier Leroy, Linuxthreads - POSIX 1003.1c kernel threads for Linux, http://pauillac.inria.fr/ˇxleroy/linuxthreads/README, 1996

[8] Xavier Leroy, The LinuxThreads library, http://pauillac.inria.fr/ˇxleroy/linuxthreads, 2003

[9] Xavier Leroy, Internals of LinuxThreads, http://pauillac.inria.fr/ˇxleroy/linuxthreads/faq.html, 2003

[10] Jeremy Andrews, Linux: New Scheduler Proposal, http://kerneltrap.org/node/view/341, 2002

The Use of the Google Search Engine for Accessing Private Information on the World Wide Web

Ş. Ahmet Gürel, Erhan Basri, and Yıltan Bitirim

Department of Computer Engineering, Eastern Mediterranean University,
Famagusta, TRNC
{ahmet.gurel, erhan.basri, yiltan.bitirim}@emu.edu.tr

Abstract. This paper discusses how the search engine Google can be used to access private information on the World Wide Web. Google hacking, the use of search operators of the Google search engine enables hackers to perform complicated queries to retrieve information located at misconfigured web-servers. In this paper, private information that might be sensitive is classified and Google hacking techniques that can be used to retrieve this information is investigated. It is found out that Google hacking enables access to private information even if they are not published publicly and kept unlinked.

Keywords: privacy, private information, Google hacking, Google search operators, World Wide Web

1 Introduction

Nowadays, Internet is used as a source of information. Various types of information can be accessed through the internet. Some examples are documents, government information, advertising, and commerical applications. Even though the Internet provides advantages on information access, such as scientific articles referencing web pages and online goverment reports [1], private information such as contact information can also be accessed on the Internet. However, the access of private information can create privacy risks for Internet users. These risks occur because security flaws of operating systems or web applications or configuration mistakes on servers connected to Internet can result in unwanted access of the private information of the users to the public. In 2002, 9.9 million users suffered from personal information theft in USA [2]. Furthermore, there are 800.000 personal databases under threat today according to research of University of California at Los Angeles (UCLA) [3].

Since there are a huge number of documents on the web, there is a need to index and filter these documents according to the information necessities of the Internet users [4]. Therefore, different search engines (such as Google, Yahoo, and MSN) are developed to index the web documents and help users to retrieve most relevant information on the web [5].

The search engine Google is the most popular search tool and is one of the most visited web sites [1]. Google, when first established in 1998, had indexed 25 million of web pages, which increased to 8 billion at June 2005, with 17 million of images and 6600 printed catalogs. Nowadays, Google has 25 billion of web pages and 1.3 billion of images indexed in its database. Google services encounter more than 200 million of queries every day. In the year 2006, Google had up to 54% of the world search engine markets, which provided Google a net income of 1.47 billion dollars [6]. Today, search engine tool is not the only service that Google provides. Google also has advertisement services, e-mailing services, some mobile services, and web applications and software tools which are used in the business world [7].

However, success of Google as a search engine has resulted in certain security risks for web sites, and Internet users. The term "Google Hack" refers to using Google to retrieve certain information by hackers [12]. Since Google provides an extensive list of search options through basic and advanced search operators, an experienced hacker can make use of these search operators to retrieve sensitive information such as what type of server is being used, administrator passwords, and server side programming language [8]. Although research has been done for Google hacking, there is no known study that investigates the effect of Google hacking techniques on private information of users or companies. In this article, Google hacking is investigated for retrieving private information over the Internet and Google Hacking techniques for this purpose are revealed.

This paper is organized as follows. In Section 2, the architecture of Google search engine is briefly discussed. In Section 3, query formulation for the Google search engine is presented. In Section for 4, the methodology for experiments to obtain private information and a classification of private information are outlined. In Section 5, experiments that are performed to obtain private information using search queries are presented. And finally, the affect of Google Hacking on private information is demonstrated with concluding remarks in the last section.

2 Architecture of Google

The architecture of Google search engine consists of different components, as shown in the figure 1.

Google services use more than one Server Farm (collection of servers), working with Linux as operating system, and using C, C++ and Python programming languages for server-side coding [9, 10]. In 2006, it is estimated that Google has more than 450,000 servers around the world [6].

Google collects web pages by using a program called Web Crawler (Web Spider or Web Robot), which surfs the Internet and expands the web pages list. A web page can also be added directly by using Google's site. Web crawler program used by Google is called "Googlebot". List of URLs (Uniform Resource Locator) is sent to the Googlebot by a URL Server, with assigned unique ID tags, called "docID"s, for each URL. Then those web pages are sent to the "Store Server". Web pages are compressed by "Store Server" and then stored into a "Repository". For indexing, "Indexer" and "Sorter" work together. Compressed documents in the "Repository" are taken by "Indexer". "Indexer" uncompresses those documents, parses them and converts them into a set of word

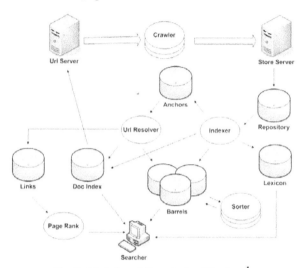

Fig. 1. High Level Google Architecture[1]

occurrences called hits. "Indexer" classifies these set of words and sends them into the storages called "Barrels". "Indexer" also defines the keywords in the web pages. The component called "Doc Index" is responsible for storing the documents' IDs, locations in the "Repositories", statuses, contents and URL title. "Lexicon" component stores a huge collection of words and pointers. "PageRank" is an algorithm used by Google to rank the pages according to their relevancy. The algorithm evaluates the pages by taking hits in the pages and the positions of these hits, (like hits in the title, hits in the body part, etc.), into account [9].

3 Query Formulation

While using search engines, search queries are essential for obtaining the most relevant results. For queries, Google provides basic and advanced search operators. To obtain results faster and more efficiently, search query formulation is done by using one or more specific operators. For general information, Google basic search operators can be used, but for something more specific than general, Google's advanced search operators can be used.

---

[1] The figure is compiled from the study of Brin and Page [9].

3.1 Basic Search Operators

Basic search query, contains basic search operators. These operators are '+','-' ,'~','.',' " " ','|' and wildcards. Google's basic queries are not case sensitive, and they can include up to 10 terms. Google eliminates stopwords (such as "**the**", "**of**", "**a**", "**i**", "**to**", "**and**") from the query [11].

To bring the pages which include a specific word, plus operator (+) can be used. In opposite of the plus operator, minus operator (-) is used to bring the pages which do not include a specific word. Google search queries can also include wildcard operators. The asterisk operator (*) is used as a wildcard that will be replaced by zero or more characters, while the dot operator (.) is replaced by one character that separates two words. To bring the pages which include words relevant to a specific word, tilda operator (~) is used. To bring the pages which include a specific phrase, quotation operators (" ") are used, by surrounding the phrase. Finally, Google search queries also support a *logical or* (|) operator , to bring the pages which satisfy one of the conditions separated by the operator.

3.2 Advanced Search Operators

Google's queries can consist of special terms called "Advanced Operators", which are used to create advanced filters to obtain more accurate results. To create even advanced filters, most of these advanced operators can be used in combination with special characters, boolean operators, and other advanced operators. An advanced operator syntax is strictly in this form: *operator:search-term*. According to the syntax, there should be no space before or after the colon (:), operators should be written by using lower case characters, and if the *search-term* includes more than one word, it can be quoted by using double quotation marks [11, 12].

To create a filter according to the titles of the pages, the advanced operator *intitle* can be used. For example, the query *intitle:EMU* brings the pages with *EMU* in their web page titles. But, instead of using the *intitle* operator twice, for bringing the pages which include two specific words in their titles, the operator *allintitle* can be used. For example, the query *intitle:EMU intitle:TRNC* brings the pages which include both EMU and TRNC in their titles, which can also be expressed by the query *allintitle:EMU TRNC*.

To create a filter according to the URLs of the pages, *inurl* and *site* operators can be used. The operator *inurl* is used to bring the pages which include a specific word in their URLs. For example, the query *inurl:emu* brings the pages with *emu* in their web addresses, like http://www.**emu**.edu.tr and http://cmpe.**emu**.edu.tr. Morever, the operator *site* can be used to bring the pages which are located on a specific URL. For example, the query *site:cmpe.emu.edu.tr* filters the pages which are located on the web address http://cmpe.emu.edu.tr. Pages which include links to a specific URL can also be listed by using the operator *link*. For example, the query *link:www.emu.edu.tr* lists the pages which link to the web address http://www.emu.edu.tr

Pages, which include a specific word in their body parts, are searched by use of the operator *intext*. For example, although the query *intext:academic* lists the pages which includes the word *academic* in their body parts. For pages which include more than one specific words in their body parts, instead of using the operator *intext* for each specific word, the operator *allintext* can be used. For example, although the query *intext:EMU intext:academic* brings the pages which include both of the words *EMU* and *academic* in their body parts, the query *allintext:EMU academic* can be used for the same purpose.

For listing the documents through the Internet according to their file types, the operator *filetype* can be used. However, this operator must be used with other advanced operators or basic search queries. For example, the following query lists the Microsoft Excel files (with extension .xls) which include the term "TRNC": *filetype:xls TRNC*. As an another example, the following query lists all the Microsoft Word files in http://www.emu.edu.tr: *filetype:doc site:www.emu.edu.tr*.

To bring the pages which include numbers in a specific range, the advanced operator *numrange* can be used. The range is specified in this format: *rangebegin-rangeend*. For example, the following query filters the pages which include the numbers between 995 and 1005: *numrange: 995-1005*.

To retrieve the image of a web page from the cache memory of Google servers, the operator *cache* can be used. The following example brings the last cached home page of the Eastern Mediterranean University's (EMU) web site, from the Google cache: *cache:www.emu.edu.tr*

For retrieving the definitions of a specific word from the Internet, the operator *define* can be used. For example, the following query brings the definitions of the word "university": *define:university*.

4 Experiments

The search engine Google can also be used to search for unguarded data which should be hidden, like private information or technical information on a server computer. Google hacking is a term which is used to express the usage of Google search engine as an aggressive hack tool [13]. Long has published the Google search queries that can be used to reveal the security flaws of a server computer in his study [11].

Our research, however, deals more about the security of personal and organizational information and how Google hacking enables access to private information. Personal information includes ID numbers, contact information, chat logs, e-mail accounts, passwords, credit card numbers, etc. and organizational information includes employee list, salary list, financial reports, private communication logs of a company, etc. To analyze the effect of Google hacking for accessing private information, first of all, private and organizational information are categorized in groups: (1) IDs, (2) contact information, (3) confidential documents, (4) personal passwords, and (5) private communication data. These groups are discussed by giving some sample hack scenarios below.

IDs are unique numbers like citizen identity numbers, social security numbers, taxpayer numbers, driver license numbers, etc. When revealed, IDs can be used to perform personal operations in the name of the victim, without his or her consent. For example, when a hacker obtains a person's social security number, the hacker can get the victim's personal details. By using those details, the hacker can withdraw credits from a bank and leaving the victim in debt [15].

Contact information includes phone numbers, e-mail addresses, home addresses, etc. When a hacker obtains contact information of a person, the hacker can use that information for unsolicited advertisements. If an e-mail address is obtained, it can be used for spam mails without consent of the user.

Confidential information is a kind of information that is used internally by an organization. Confidential information can include an organization's secrets, strategies, long-term or short-term plans, financial status, employee list, salaries, etc. When the strategies of an organization are obtained, other organizations can illegally benefit from the secrets they contain. When the obtained information is about the salary list, another organization can start a mass employee theft by offering higher salaries.

Personal passwords can be described as passwords for e-mail accounts, instant messaging accounts, portal accounts, server administration accounts, etc. When a hacker obtains a password, that password can be used to harm a person's account, or an organization, or a server disk. If a hacker obtains a personal password like an e-mail password, the hacker can read private information or send malicious e-mails. If the password is the administrator password of a portal, then the hacker can access all private information of portal members. If the password belongs to an online-banking account, then the hacker can perform illegal operations on the victim's bank account.

Private communication data include chat logs, chat contact lists, e-mail address books, etc. These kinds of information can reveal the private daily life or business secrets of a victim. If the communication data include passwords, the hacker who obtained the data can use those passwords to access the victim's personal accounts.

5 Experimental Result Samples

Based on the classification provided in Section 4, this section presents some examples of successful queries to retrieve private information using Google search engine. The type of query formulated is presented with advance search operators that might be used for the purpose. Furthermore, screen shots of the results of the queries are also presented. Sensitive information in these pictures are blurred out on purpose to provide confidentiality.

Microsoft Excel files (with extension .xls), Microsoft Word Documents (with extension .doc), and text files (with extension .txt), can be used to store ID information. If so, the following queries can be used to search for ID information: *filetype:xls "TC kimlik no"* (as shown in figure 2), *filetype:xls "ssk sicil no"*, *filetype:txt "my social security number is"*.

Contact information can be found again in popular document types like ".xls", ".doc", ".txt". As shown in figure 3, the following query displays the Excel documents which includes the word "gsm no": *filetype:xls "gsm no"*.

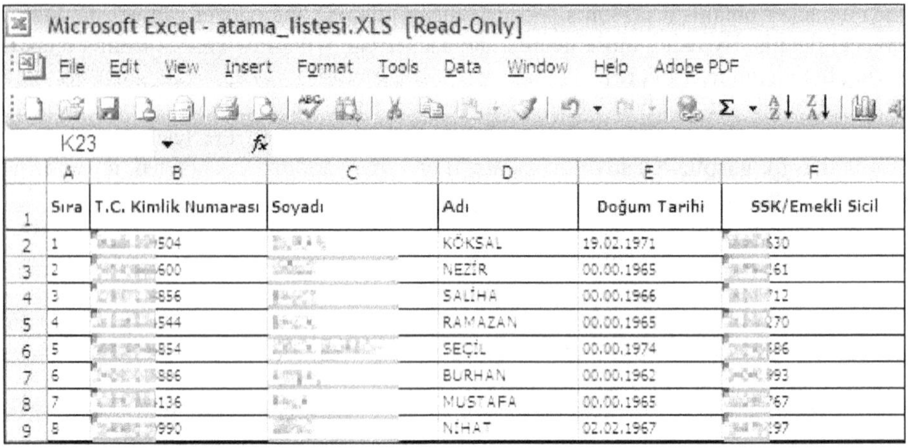

Microsoft Excel - atama_listesi.XLS [Read-Only]

	A	B	C	D	E	F
1	Sıra	T.C. Kimlik Numarası	Soyadı	Adı	Doğum Tarihi	SSK/Emekli Sicil
2	1	504		KÖKSAL	19.02.1971	630
3	2	600		NEZİR	00.00.1965	61
4	3	856		SALİHA	00.00.1966	12
5	4	544		RAMAZAN	00.00.1965	270
6	5	854		SEÇİL	00.00.1974	586
7	6	886		BURHAN	00.00.1962	993
8	7	136		MUSTAFA	00.00.1965	767
9	8	990		NİHAT	02.02.1967	197

Fig 2. Result of the query *filetype: xls "TC kimlik no" ("T.C. Kimlik no": Republic of Turkey ID No, "Soyadı": Surname, "Adı": Name, "Doğum Tarihi": Date of Birth, "SSK/Emekli Sicil no": Health System Number*)

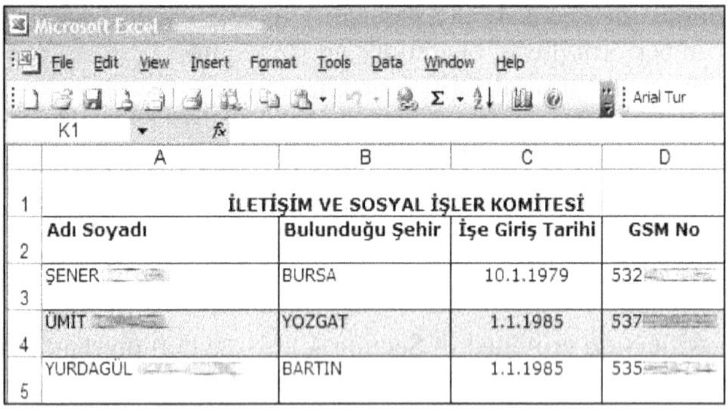

Microsoft Excel

	A	B	C	D
1	İLETİŞİM VE SOSYAL İŞLER KOMİTESİ			
2	Adı Soyadı	Bulunduğu Şehir	İşe Giriş Tarihi	GSM No
3	ŞENER	BURSA	10.1.1979	532
4	ÜMİT	YOZGAT	1.1.1985	537
5	YURDAGÜL	BARTIN	1.1.1985	535

Fig. 3. Result of the query *filetype:xls "gsm no" ("Adı Soyadı": Name Surname, "Bulunduğu Şehir": City, "İşe Giriş Tarihi": Date of Employement, "GSM No": Mobile Phone number)*

Furthermore, Outlook Express E-mail folder files (with extension .dbx), Outlook Personal Folder files (with extension .pst), MSN Contact list files (with extension .ctt), contact list files of the instant messaging program Trillan (with the file name mystuff.xml), contact list files of the instant messaging program Aim (with the file name buddylist.blt) can be used to store contact information. Moreover, tabular file formats like Comma separated values (with extension .csv) and Microsoft Excel files (with extension .xls) can store contact information. An example of how this type of contact information can be revealed is shown in figure 4.

```
<?xml version="1.0"?>
<messenger>
    <service name=".NET Messenger Service">
        <contactlist>
            <contact>    _  @hotmail.com</contact>
            <contact>     @nqrd.dk</contact>
            <contact>     @hotmail.com</contact>
            <contact>    _     @hotmail.com</contact>
            <contact>     @hotmail.com</contact>
            <contact>     @groenbjerg.net</contact>
            <contact>     @hotmail.com</contact>
```

Fig. 4. Result of the query *filetype:ctt "msn"*

Confidential information can also be found in document types like Microsoft Excel files (with extension .xls) Microsoft PowerPoint files (with extension .ppt) and Adobe Acrobat files (with extension .pdf). For example, the following queries can be used to search for internal confidential presentations: *filetype:ppt confidential "for internal use only"*. The result of the query *filetype:ppt confidental "for internal use only"* is shown in figure 5.

Fig. 5. Result of the query *filetype:ppt confidental "for internal use only"*

Server log files contain information such as passwords usernames, and access times for a web site, and they can be searched to obtain these types of information by a hacker. For example, HKEY_CURRENT_USER nodes of Windows Registry files (with extension .reg) can be checked to find passwords, as seen in figure 6.

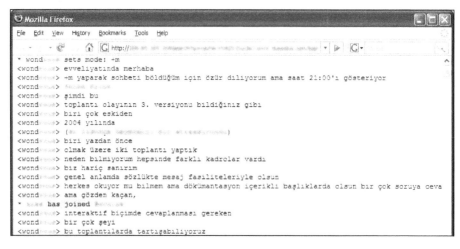

Fig. 6. Result of the query *filetype:log inurl:"password.log"*

Chat log files and Outlook Personal Folder files (with extension .pst) can include private communication information. Chat log files store the logs of dialogs between two peers and Outlook Personal Folder files store the inbox of an e-mail user, including the contents of sent and received e-mails. An example of how chat log files can be obtained is shown in figure 7.

Fig. 7. Result of the query *"has joined" filetype:txt*

6 Conclusion

Private documents in a server disk can be accessed by Google search, if those documents are configured as accessible documents. Moreover, Google can access and index file and directory structures of a server disk, if the server is configured to allow directory listing, and this makes it easier to hack into that server by using Google search [14].

With successful query formulation, a Google hacker can obtain private information. These exploits are results of both misconfigured web-servers, and user knowingly publishing their private information on the Internet. For example, contact information kept on a web site can be retrieved by simply searching document files include the word "contact". Furthermore, the ID of a person can be retrieved to allow identity theft [11].

Hackers can also use Google for stealth purposes. To stay unidentified, hackers can use Google's cache memory by working on the memory image of the target site, thus by not even visiting the target site and leaving any signature [14, 16]. Furthermore, when private information has been hacked, it might not be possible to trace this activity.

As it can be seen in the sample experiments private information of a person or a company can be revealed with search techniques using the Google search engine. Therefore, private information such as ID numbers, contacts etc. can only be protected through eliminating misconfigurations at servers and through control of what information is published. Google hacking enables hacks to private information even though they are not published publicly and kept unlinked. Since some of hacking techniques utilizes misconfigured web servers, it is important that administrators apply patches frequently.

There are some tools to prevent Google hack, by locating the security flaws in a web page. These tools use Google Hack Database (GHDB) [17] or their own database to locate these security flaws. Some known the tools are Gooscan [18], Sitedigger [19], Goolink [20] and Athena [21]. Although these tools can be used for scanning security flaws, they are not specialized to scan the security of private information. Therefore, future work on this topic includes the development of a tool to detect security flaws specifically for private information that will be available online for public use.

References

1. J. Bar-Ilan. Comparing rankings of search results on the web. Information Processing and Management: an International Journal, 41(6):1511-1519, December 2005.
2. Federal Trade Commission. Identity theft survey report, pp. 6 -7, September 2003. http://www.consumer.gov/idtheft/pdf/synovate_report.pdf
3. US University notifies 800.000 of database hack. Computer Fraud & Security, pp. 3, January 2007.
4. R. B.Yates and B. Riberio-Neto. Modern information retrieval. Addison Wesley, 1999.
5. Y. Shang and L. Li. Precision evaluation of search engines. World Wide Web: Internet and Web Information Systems, 5(2):159-173, 2002.
6. http://en.wikipedia.org/wiki/Google (visited on November 8, 2006)

7. F. Mantar. Google desktop. December 2005.
http://ftp.cs.hacettepe.edu.tr/pub/dersler/BIL4XX/BIL447_YML/belgeler/GoogleDeskt
op_20221815.pdf (visited on January 19, 2007)

8. E.I. Tatlı. Google reveals cryptographic secrets, July 2006.
http://th.informatik.uni-mannheim.de/people/tatli/pub/ghack_crypto.pdf (visited on
January 6, 2007)

9. S. Brin and L. Page, The anatomy of a large-scale hypertextual web search engine,
Computer Networks and ISDN Systems, 30(1):107-117, 1998.

10. http://www.python.org/about/quotes/   (visited on January 27, 2007)

11. J. Long, Google hacking for penetration testers. Syngress Publishing Inc. Rockland,
MA, 2005.

12. T. Calishain and R. Dornfest. Google hacks, O'Reilly, 2003.

13. E. I. Tatlı. Google ile güvenlik açıkları tarama, 2006.
http://th.informatik.uni- mannheim.de/people/tatli/resources/pdf/googlehacking.pdf
(visited on February 13, 2007)

14. S. A. Mathieson. Google – Swiss army knife for hackers? InfoSecurity Today,
November/December 2005.

15. http://www.ssa.gov/pubs/10064.html (visited on December 9, 2006)

16. E. Rabinovitch, Project your users against the latest web-based threat: malicious code
on caching servers, IEEE Communications Magazine, 45(7):20-22, March 2007.

17. http://johnny.ihackstuff.com/ (visited on April 17, 2007)

18. Gooscan, http://johnny.ihackstuff.com/downloads/task,doc_details/gid,28 (visited on
April 17,2007)

19. Sitedigger,
http://www.foundstone.com/index.htm?subnav=resources/navigation.htm&subcontent=/
resources/proddesc/sitedigger.htm (visited on April 17, 2007)

20. Goolink, www.ghacks.net/2005/11/28/goolink-scanner-beta-preview-2/ (visited on
April 17, 2007)

21. Athena, http://www.buyukada.co.uk/projects/athena (visited on April 17, 2007)

DARIS: A Probabilistic Model for Dependency Analysis of Risks in Information Security

Suleyman Kondakci

Faculty of Computer Science, Izmir University of Economics,
suleyman.kondakci@ieu.edu.tr

Abstract. This paper presents an abstract concept of security planning processes using a simple probabilistic model to express conditional risk factors. This analytical work emphasizes relationships related to major security planning phases. The work discusses the chain of logical events that describe dependence in risk propagation. This unique work can provide a guidance for security planning and risk management applicable to various engineering fields. Especially, for security education, we need to provide approaches that are theoretically sound as well as practical and realistic. The risk analysis method provided by DARIS can also provide a theoretical basis in education of information security.

1 Introduction

The process of lifecycle security planning involves at least four phases; as the initial phase (i) requirements analysis, (ii) design, (iii) implementation, and (iv) test and evaluation. Risk assessment during the initial phase is an essential requirement needed to reduce cumulative risks that may propagate through the subsequent phases of a security planning process. In this work we present a probabilistic risk propagation model called Dependency Analysis of Risks in Information Security (DARIS). The root of the risks that is likely to encounter in the first phase is due to the misspecification of the *security objectives*. Security objectives represent the fundamental knowledge on security aspects and the specification of assets and IT environment. Estimation of risk levels by use of numerical values, probabilistic means, and associated algorithms/models is termed as *quantitative risk assessment*.

In the following, Section 2 presents related work. Section 3 deals with the methodology and context of this paper. Section 4 describes the probabilistic dependence analysis of risks in a security planning. Section 5 concludes the paper.

2 Related Work

Statistical and probabilistic models for risk assessment have been widely applied in diverse fields. Among many, three such areas are discussed in [1-3]. We do not find methodologies similar to the one presented here. Especially, it was difficult to find comparable theoretical methods in IT security risk analyses dealing with weakness propagation. A limited number of practical approaches motivated only by business needs are found in the literature. It is, however, difficult to address abstract approaches directed towards teaching risk management in academia. We agree with the statement given in [4] that teaching information security is a daunting task. Although, there are several valuable textbooks used in computer security education, e.g., [5], it is difficult to find a complete source for a through risk analysis task. Our approach is a novel source for theoretical risk assessment in security education.

3 Assessment Process and Metrics

There are several common approaches to security planning presented by different sources. For example, information security process is described by [6] has the following flow:

Assessment→Policy→Implementation→Training→Audit→Assessment.

In similar notation, DARIS has the iterative flow of steps shown in Fig 1.

Fig. 1. The DARIS security design and assessment and improvement phases.
The improvement process starts with the assessment of objectives (O) and proceeds similarly with the assessment of policy (P) generated from the objectives. Further, the decision-making continues with the assessment of the Implementation generated from the policy. Each phase of assessment produces a risk value corresponding to *weak (W), moderate (M), severe (S),* or *very severe (V)*. If an undesired risk level (e.g., W) is found then an improvement is applied and the assessment is iterated. The improvement is done via necessary modification on objectives, policy, or implementation while performing the chain of assess-improve-iterate tasks. Finally, an overall test and evaluation is executed in order to determine the status of the entire solution. A simple metric is used to quantify and classify the assessment results in order to determine the degree (W, M, S, and V) of risk. We use four sub ranges and a master scale of scores varying between 0 and 5. Thus, we can use the following scheme for risk sub ranging:
$$W = \{0 - 1.25\}, M = \{1.26 - 2.5\}, S = \{2.6 - 3.75\}, \text{ and } V = \{3.76 - 5\}.$$

4 The Conditional Model of Decision Making

The DARIS framework relies on a probabilistic decision hierarchy based on the conditional probability theory. This probabilistic model can also be used to generate necessary probability distributions that can be incorporated into some mapping functions to quantify risk levels. An event-oriented decision making model, which is interoperable with DARIS is presented in [7]. Recently, due to increasing terror activities, homeland security has been under focus for creating effective decision-making mechanisms. A common approach to risk analysis for homeland security decision-making is given in [8]. Dependence in a security assessment process can be illustrated by a Venn diagram shown in Fig. 2 as a collection of sets and their relationships.

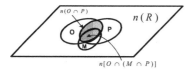

Fig. 2. The Venn diagram representation of intersection of O, P, and M.

We assume that dominating threats are rooted from nonexistent or weak security objectives. Thus, we define three data sets, **O** for the security objectives, **P** for the policies, and **M** for the countermeasures. The set **R** represents the entire space of all possible security attributes for a given environment.

Fig. 2 depicts an environment where the assessment of a security planning is conducted. In order to combine with an empirical study, this probabilistic model can be simulated to support a wide range of predefined scenarios. Thereafter, appropriate input attributes can be selected to match a variety of security solutions. For example, in the initial phase, it may be required to observe the effect of weak security objectives against moderate or severe security objectives. To do so, randomly generated items of the security objectives with various weights and scores are produced. This, in turn, produces different item scores to reflect the scoring subranges to match weak, moderate, severe, or very severe. The items are evaluated and classified according to the subranges they are supposed to fall in. The scoring scheme is detailed in [7].

Here, we introduce the theorems of the risk dependency in the security planning (or solutions). If necessary, related material and statistical theories and practices can be found in [9-10]. A simple relational hierarchy of dependence, as a roadmap to decision making, can justify the overall joint relationship in a security process.

Theorem 1. If an organization is not aware of security aspects and has not yet defined its security objectives (O), then the organization cannot claim to have defined an appropriate security policy (P), and hence no efficient countermeasures (M) can be designed via inappropriate security policy.

In order to calculate the probability of O and P events, it is convenient to use conditional probabilities as an intermediate framework. Thus, for two dependent events (O and P, objective and policy) the compound probability is given by

$$\Pr(O \cap P) = \Pr(P \mid O)\Pr(O) = \frac{\Pr(P \cap O)}{\Pr(O)}\Pr(O) \tag{1}$$

Where $\Pr(P \mid O)$ denotes the probability that event P occurs knowing that event O has occurred. Existence of valid security objectives is a prerequisite that must be fulfilled in order to define successful security policies. Empirically, the objective events should be determined/collected from a random experiment with. $\Pr(O) > 0$ Obviously, as illustrated in Fig. 2, denotes the conditional relative frequency of P and O, where $n(O \cap P)$ is the total number of successfully defined policy outcomes given O objectives, and n(R) is the total number of outcomes (sample space). Namely, $n(O \cap P)$ depicts the number of outcomes leading to the occurrence of both O and P from n(R) outcomes of desired events.

Theorem 2. The probability of an event representing the countermeasure or the practical safeguard (M) is dependent on the existence of the event P, which occurs for the security policy. Hence, the compound probability for the security measure and policy is given by

$$Pr(M \cap P) = Pr(P) Pr(M \mid P) = Pr(P) \frac{Pr(M \cap P)}{Pr(P)} \qquad (2)$$

Similar to the $Pr(P \mid O)$ conjecture, the number of outcomes leading to the occurrence of both P and M out of np(N) outcomes can be computed. Thus, the conditional relative frequency of P and M. Here $n(P \cap M)$ depicts the number of outcomes leading to the occurrence of both P and M, and np(N) is the total number of organizations with successfully defined security policies, i.e., $n(O \cap P)$ .Proof of Theorem 1 and 2. Since the security policy (P) is derived from objectives (O) and security measures (M) are derived from the security policy, it is obvious that measures are also dependent on the objectives. Hence, the proof of Theorem 1 and 2 can be derived in a combined form given below. Thus, the probability Pr(P) assumes (or depends on) the occurrence of O, and Pr(M) assumes the occurrence of P. Starting with equations (1) and (2), the conditional probability $Pr(P \mid O)$ is calculated by

$$Pr(P \mid O) = \frac{Pr(P \cap O)}{Pr(O)} \qquad Pr(M \mid P) \text{ is } Pr(M \mid P) = \frac{Pr(M \cap P)}{Pr(P)} \qquad (3)$$

The definitions of dependence between O, P, and M, and Eq. (3) imply that, probability Pr(P) is dependent on probability Pr(O), and probability Pr(M) is dependent on probability Pr(P), respectively. The overall dependent probability Pr(P) can be calculated by using the total probability formula, i.e.,

$$Pr(P) = \sum_k Pr(P \mid O_k) Pr(O_k). \qquad (4)$$

We can also use the Baye's Theorem to prove this equation. To proceed, we need to consider the non-empty sample space Ω having k mutually exclusive events of security objectives.

$$\Omega = \bigcup_k O_k, (O_k \cap O_j) = \phi, \forall (k \neq j). \qquad (5)$$

Since one of the events $\Omega = O_1, O_2, \cdots, O_N$ must occur for a successful policy P, then

$$P = P \cap \Omega = P \cap \left(\bigcup_k O_k \right) = \bigcup_k (P \cap O_k). \qquad (6)$$

Hence,

$$Pr(P) = Pr\left(\bigcup_k (P \cap O_k) \right) = \sum_k Pr(P \cap O_k), \qquad (7)$$

$$Pr(P) = \sum_k \frac{Pr(P \cap O_k)}{Pr(O_k)} Pr(O_k) = \sum_k Pr(P \mid O) Pr(O_k).$$

Similarly, for Pr(M), the overall dependent probability for the security measure is obtained by

$$Pr(M) = Pr\left(\bigcup_k (M \cap P_k) \right) = \sum_k Pr(M \cap P_k), \qquad (8)$$

$$Pr(M) = \sum_k \frac{Pr(M \cap P_k)}{Pr(P_k)} Pr(P_k) = \sum_k Pr(M \mid P_k) Pr(P_k).$$

5 Conclusions

Specification of security requirements, design of security policies and countermeasures are all conditionally dependent, which can be described by a conditional probability model. We have shown that there is a correlation between knowledge and policy design, and in turn, between policy and security implementation. Proper knowledge (objective) is necessary for preparing practically sound policies, and strength of policies reflects onto final implementations. If no security specifications have been setup, then we cannot expect the environment can determine its security policy effectively, via which necessary countermeasures can be enforced. It was hereby shown that, in a lifecycle security, overall efficiency of security measures is strongly associated with the degree of security-awareness. We believe that DARIS framework can be a useful theoretical ground in training of information security.

References

1. Kumamoto, H., Henley, E. J.: Probabilistic Risk Assessment and Management for Engineers and Scientists 2nd Ed., ISBN: 0-7803-6017-6, IEEE Press (1996)
2. Bedford, T.: Probabilistic Risk Analysis, Foundations and Methods, ISBN: 0-52177320-2, Cambridge University Press (2003)
3. Bruske, S. Z.: Wright, R. E., Geaslen,W. D., Potential uses of probabilistic risk assessment techniques for space station development, NASA STI, USA (1985)
4. Gutierrez, F.: Stingray: A Hands-on Approach to Learning Information Security, Proceedings of SIGITE'06, ACM (2006) 53-58
5. Stallings, W.: Cryptography and Computer Security: Principles and practices, 4th ed., Pearson Education, Inc. ISBN: 0-13-187316-4, Upper Saddle River, NJ 07457 USA (2007)
6. Maiwald, E.: Network Security A Beginner's Guide 2nd ed., Emeryville, CA:McGraw-Hill/Osborne (2003)
7. Kondakci, S.: A New Assessment and Improvement Model of Risk Propagation in Information Security, Int. Journal of Information and Computer Security, Vol. 1, No. 3, (2007) 341-366
8. Sims, J. R., Balkey, K. R., Ayyub, B. M., Feigel, R. E.: A Common Approach to Risk Analysis for Homeland Security Decision-Making, Engineering Technology Management, Elsevier Inc. (2003) 181 - 186
9. Ghahramani, S.: Fundamentals of Probability With Stochastic Processes, 3rd ed. Pearson Education Inc., ISBN: 0-13-129849-6 (2005)
10. Everitt, B.: Chance Rules: An Informal Guide to Probability. Risk and Statistics, Copernicus (1999)

Improved Threat Modeling Process for Grids

Mohammad Othman Nassar

Arab Academy for Banking and Financial Sciences
Amman-Jordan
moanassar@yahoo.com

Abstract. Although nowadays organizations are now more aware of security issues than ever, and many powerful technologies are available for securing computer systems, but the protection techniques applied in organizational computer systems may often be mismatched against real threats. Threat modeling can help us to know, and deal with those attacks that are relevant and real. to get a comprehensive threat model we must use a suitable threat model process, although numerous works have been published on threat modeling including threat modeling processes, I believe there is a lack of integrated, systematic approach toward threat modeling for grids that can be considered as a complete solution to this problem. This paper propose a new step called "the mode of use" as a part of threat modeling process; this step is added to help in creating a comprehensive threat model for grids.

Keywords: information systems security, grids, threat modeling.

1 Introduction

The general goal of the Grid computing paradigm [1] is to provide resources, including processor cycles, data sources, special equipment, and even people, as easily as electricity provided. Protecting against Internet attacks is not an easy mission. Creating formally secure systems is not feasible with current techniques, so the available solution is to be a ware of threats as much as possible to protect our systems; we can complete this task using threat modeling. Threat modeling is a systematic technique for identifying potential threats to computer systems and software. Someone may ask why we do not just skip threat modeling, and simply adapt the system's security requirements from "industry's best practices" or standards. It is easy to answer this question if we know that those practices and standards usually need some customization for the target system. Threat modeling can help us to know and deal with those attacks that are relevant. Given finite budgets in time and money, threat modeling allows designers and implementers to marshal their resources for practical protection against real threats [3]. Although organizations are now more aware of security issues than ever, and many powerful technologies are available for securing computer systems, but the protection techniques applied in organizational computer systems may often mismatched against real threats [3]. This problem must motivate the security designers to implement a comprehensive threat modeling on the target system instead of using measures recommended in the security standards. Today some grids are used in a different ways than the past [1], This will arise many new security challenges to grids; these challenges must be considered as a part of threat modeling process.

2 Related Work

Many researchers address the problem of threat modeling, in[7] the authors define a threat model for grid computing at a very high level, and without focusing on specific security challenges of the grids that comes from the new "mode of use" for grids , in [6] only three specific attacks on clusters are presented, in [5] the unique properties of cluster security are addressed, but limited threat model is presented, in [8] a threat model is presented without focusing on the specific threats that the grid may face. in [1] only limited set of threats that came from the new "mode of use " for grids are presented, in[2] a threat model is presented for group communication in general, this model dose not state the specific requirements for grids.

3 Threat Modeling Process

To create a quality threat model, we need systematic and detailed process [2]; it is not enough to brainstorm the actions or intentions of potential attackers to a system. The proposed threat modeling process (which includes a new concept called the "mode of use", this concept is adapted from [1]) consists of the following four high-level steps:

3.1 Characterizing the System

The main idea for this stage is about understanding the system components and their interconnections, to create a system model that describes the main characteristics of the system. The security designer needs to build a system model that reveals the essential characteristics of the system. Depending on the type of the system, it may be modeled using one of a number of different approaches for modeling [8]. As an example, an application is modeled by a Data Flow Diagram, and a networked system by a Network Model.

3.2 Identifying Assets and Access Points

Identification must be made of the system's assets and entry points through which an attacker will seek access to these assets. An asset is any element of a system which provides critical functionality, examples on assets include: Message availability, Processes, Cryptographic key content, Network bandwidth, and Message integrity. Entry points to a system are what an attacker must use to acquire access to assets. Examples on entry points include: Past, current, or future grid members, Hardware ports, file system read/write, Communication channels, and Open sockets.

3.3 Identifying the "Mode of Use" For the Grid

Today some grids are used in a different ways than the past, one new use of grids is explained in [1]; the authors introduce a new usage for grids called on demand paradigm, where the computational peak loads or free resources are outsourced to organizations offering computational power or the required resources. this will arise many new security challenges to grids, such challenges comes from the fact that we need to be able to dynamically rent resources with only a minimal administrative overhead. To understand why we add the "mode of use" to threat modeling process lets consider the following example, Most Grids run code from third party partners and for practical reasons, it is infeasible to audit all of this code. This is a major change compared to traditional grids running a controlled base of known source code [5]. Recently, this has led to attacks on third party components of Grids and clusters as documented in the 2005 SANS Top-20 list which added a new cross-platform threat category [11]. So using the ordinary grid modeling process without considering the "mode of use" will not allow us to deal with and consider such new attacks. So considering the "mode of use" as a part of threat modeling process will make the threat modeling process capable to address any new trends in using grids.

3.4 Identifying the Threats

The goal of this step is to identify threats to the system using the information gathered so far. A threat is the adversary's goal, or what an adversary might try to do to a system [10]. The best method for threat enumeration is to step through each of the system's assets and access pointes, reviewing a list of attack goals for each asset, this must be done while keeping in mined the mode of use for the grid. The output of threat identification process is a threat profile for a system, describing all the potential attacks, each of which needs to be mitigated or accepted.

4 CIAA Model

We choose to use the well-known security aspects of Confidentiality, Integrity, Availability, and Authentication (CIAA) as the basis for our threat modeling of grid systems, this method is dynamically extensible to new threats, Also CIAA is valid over different time periods and different systems [3]. Thus, threats could be classified by these properties as shown in the following context.

Confidentiality Attacks: Confidentiality attacks are passive attempts to read information without proper authorization [2]. Brief discussion for some attacks is provided:

Session Eavesdropping Attacks: Because of the dispersal of the processes throughout a typically large network Traffic between member processes in grid system it is virtually impossible to hide.

Session Traffic Analysis: even when encryption is done, the attacker can observe traffic flow on the network and make corresponding deductions based on when, where, message type, and volume patterns.

Integrity Attacks: Integrity attacks attempt to actively modify information without proper authorization. Brief discussion for some attacks is provided:

Message Injection Attacks: An attacker may inject messages into the system which will be viewed as legitimate system traffic by grid members.

Traffic Modification Attacks: An attacker may intercept packet data, change or delete some bits, then forward data as if no changes occurred.

Traffic Deletion Attacks: an attacker may delete data on the communication channels.

Traffic Mirror-Rerouting Attacks: An attacker re-directs traffic elsewhere without affecting the original stream's destination.

Availability Attacks: Availability attacks attempt to make grid communication services unavailable for a period of time. Brief discussion for some attacks is provided:

Denial-of-Service (DoS): we have different types that affect grid, two of them will be presented: first, Masquerading Sender DoS Attacks: here the attacker can flood traffic to all members of specific group after gaining access to the system by joining that group through an authentication attack. Second, Masquerading Receiver DoS Attacks: After gaining access to the system by joining a group through an authentication attack, the attacker may join the group as many receiver processes this will increase the overhead of the system because it must expand to handle traffic to these masquerading new members which consumes bandwidth and processing resources.

Network Infrastructure Attacks: the underlying network infrastructure such as (routers, switches, hubs, etc) may face an attack.

Authentication Attacks: these attacks occur when an attacker masquerades as a legitimate group member identity. Brief discussion for some attacks is provided:

Group Access Control Attacks: an attacker may join a group by stealing identity credentials such as (password, certificate, and keys) and masquerading as a legitimate group member.

Group Membership Information Attacks. An attacker may modify or destroy that data if he gets access to group membership information.

Traitor Attacks. A traitor in this context is a group member who shares content or keys with unauthorized parties such as attackers.

Non-Repudiation Attacks. Non-repudiation is about preventing the sender of a message from denying that he sends it.

5 Conclusions and Recommendations

The existing threat modeling process dose not address the new security requirements for grids that come from the new "mode of use" for grids, thus this paper spot light on this need as a part of threat modeling process, but the need for deeper analysis for those threats that came from the new modes of use for grids still needed. creating comprehensive threat model needs continuous search for new threats to be added to the threat list, it is recommended that each system will have complete review for the threat modeling process, because the unique requirements that any system may have.

References

[1] Smith, M., Friese, T., Engel, M., Freisleben, B., Koenig, G.A., and Yurcik, W.: Security Issues in On-Demand Grid and Cluster Computing. 2nd International Workshop on Cluster Security (Cluster-Sec) held in conjunction with the 6th IEEE International Symposium on Cluster Computing and the Grid (CCGrid), Singapore, May(2006).

[2] Hester, j., Yurcik, W., and Campbell, R. H.: An Implementation-Independent Threat Model for Group Communications. SPIE Security and Defense Conference / Program on Data Mining, Intrusion Detection, Information Assurance, and Data Networks Security (OR21), Orlando FL USA, April (2006).

[3] Myagmar, S., and Yurcik, W.: Why Johnny Can Hack: The Mismatch Between Vulnerabilities and Security Protection Standards. IEEE International Symposium on Secure Software Engineering (ISSSE), McLean, March (2006).

[4] Mogilevsky, D., Lee, A. J., and Yurcik, W.: Defining a Comprehensive Threat Model for High Performance Computational Clusters. ACM Computing Research Repository (CoRR) Technical Report cs.CR/0510046, October (2005).

[5] Pourzandi, M., Gordon, D., Yurcik, W., and Koenig, G. A.: Clusters and Security: Toward Distributed Security for Distributed Systems. 1st International Workshop on Cluster Security (Cluster-Sec) held in conjunction with the 5th IEEE International Symposium on Cluster Computing and the Grid (CCGrid), May (2005).

[6] Torres, M., Vaughn, R. B., Florzez, G., Liu, Z., and Bridge, S. M.: attacking a high performance computer clusters. Proceedings of the 15th annual Canadian information technology security symposium, May (2003).

[7] Naqvi, S., Riguidel, M.: Threat model for grid security services. European grid computing conference, (2005).

[8] Myagmar, S., Lee, A. J., and Yurcik, W.: Threat Modeling as a Basis for Security Requirements. Symposium on Requirements Engineering for Information Security (SREIS), in conjunction with 13th IEEE International Requirements Engineering Conference (RE), Paris, France, August (2005).

[9] Sheyner, O., Haines, J., Jha, S., Lippmann, R., and Wing, J.: Automated Generation and Analysis of Attack Graphs, In Proceeding of IEEE Symposium on Security and Privacy, April (2002).

[10] Swiderski, F., and Snyde, W.: Threat Modeling. Microsoft Press, (2004).

[11] SANS Institute, "The Twenty Most Critical Internet Security Vulnerabilities," November (2005), http://www.sans.org/top20/.

Achieving Private SVD-based Recommendations on Inconsistently Masked Data

Ibrahim Yakut and Huseyin Polat

Department of Computer Engineering, Anadolu University, Eskisehir, 26470, Turkey
Phone: +90 222 321 3550, Fax: +90 222 323 9501
E-mail: {iyakut, polath}@anadolu.edu.tr

Abstract. Users' concerns about private data might be different and they want various privacy levels. Therefore, they might decide to mask their data differently to achieve required privacy levels. Providing collaborative filtering (CF) services on inconsistently masked data is challenging. In this paper, we discuss how to achieve singular value decomposition (SVD)-based CF services on inconsistently masked data.

1 Introduction

Collaborative filtering (CF) is a recent technique, which provides recommendations. The goal in CF is to predict how well an active user (a) will like the target item (q) that he/she did not buy before based on the preferences of a community of users [4]. Privacy is a major concern for computer users who contribute their private information for CF. Many CF schemes are vulnerable and can be mined for preferences of users [1]. The privacy risks introduced by CF systems are severe and many like unsolicited marketing, price discrimination, being subject to government surveillance, and so on [2]. Privacy protection is vital for CF because users can provide more truthful and reliable data for CF purposes.

Randomized perturbation techniques (RPT) are used for data disguising to achieve privacy-preserving collaborative filtering (PPCF). Polat and Du [5] employ the RPT to protect users' privacy while producing accurate predictions, where users perturb their ratings in the same way. However, privacy concerns might vary from a user to another. Data sensitivity and the value of the information may also differ among different users. Moreover, there are various factors in data disclosure like sharing of users' data with others, type of data, and the purpose for which data is collected. Due to these reasons, users might decide to disguise the private data differently using various parameters.

2 Related work

Canny proposes models for PPCF allowing users to control their data, where homomorphic encryption is applied [1]. Data is masked in the same way and users actively participate in the CF process. Unlike his schemes, we do not use cryptographic techniques, users disguise their data differently, and they do not actively participate in the CF process. Polat and Du [5] employ the RPT for PPCF. In their schemes, data disguising is done by each user using the same method. We explore how to produce accurate referrals on incompatibly masked data. In [6], providing predictions on inconsistently disguised data using memory-based algorithms is discussed. However, we show how to achieve CF services on inconsistently masked data using model- or SVD-based algorithms.

Sarwar et al. [7] propose an SVD-based CF algorithm. The sparse user-item matrix (A) is filled using the average ratings for items. The filled matrix is normalized (A_{norm}) and A_{norm} is factored into U, S, and V. The matrix S_k is obtained by retaining only the largest k singular values. Then, $U_k\sqrt{S_k}$ and $\sqrt{S_k}V_k^T$ are computed. The scalar product of the uth row of $U_k\sqrt{S_k}$ and the qth column of $\sqrt{S_k}V_k^T$ is calculated and the result is de-normalized to find the prediction for any user u on q, where $\overline{v}_u$ is mean rating for user u:

$$p_{uq} = \overline{v}_u + \left[U_k\sqrt{S_k}(u) \cdot \sqrt{S_k}V_k^T(q) \right]. \tag{1}$$

3   SVD-based PPCF on Inconsistently Perturbed Data

Users might decide to disguise their data differently, as follows:

1. Some users who have no privacy concerns divulge their data, while many users, who are worried about disclosing the private data, perturb their data.

2. We can further classify the users who perturb their data into sub-groups.

a. Perturbing Data. Users can employ either uniform or Gaussian distribution with mean (μ) being 0 to generate random data for data masking.

b. Level of Perturbation. To achieve required privacy and accuracy, users are able to select the standard deviation (σ) values of random data differently. Users choose the σ values uniformly randomly over the range $(0, \gamma)$.

c. Number of Cells. Users might decide to mask different numbers of cells. Users might perturb different numbers of ratings. Some users mask all of their ratings, while others decide to perturb some randomly selected ratings. Users may want to conceal rated items because it might be more damaging revealing rated items. To hide rated items, users might fill all unrated items' cells with random numbers. Rather than doing this, they randomly select some unrated items' cells to fill with noise data. Given a ratings vector, users might decide to randomly choose some of the ratings vector's cells to mask. These chosen cells consist of ratings and empty cells of unrated items.

Users send inconsistently perturbed data to the server, which does not know the type of perturbing data, the σ values, and the amount of masked data due to inconsistent data masking. We modify the SVD-based CF algorithms to achieve required privacy and accuracy. We factor sparse matrix and employ user mean votes for normalization. We normalize ratings by converting them into z-scores and predictions can be computed, as follows, where σ_u is the σ of user u's ratings and we call this algorithm as the modified algorithm 1 (MA1):

$$p_{uq} = \overline{v}_u + \sigma_u \times \left[U_k\sqrt{S_k}(u) \cdot \sqrt{S_k}V_k^T(q) \right] = \overline{v}_u + \sigma_u \times P, \tag{2}$$

We normalize votes using bias-from-mean approach and predictions can be computed, as follows, where we call this algorithm as the modified algorithm 2 (MA2):

$$p_{uq} = \overline{v}_u + \left[U_k \sqrt{S_k}(u) \cdot \sqrt{S_k} V_k^T(q) \right] = \overline{v}_u + P. \tag{3}$$

In [5], it is shown how to estimate SVD of A' when users mask private data in the same way and disguise all ratings and empty cells. It becomes a challenge to estimate the SVD of A' when users disguise their data differently. The estimations are based on the scalar product and sum computations. Moreover, to get rid of the contribution of the random numbers in the diagonal entries of $A'^T A'$, we need the average standard deviation (σ_r) values of random numbers. It is still possible to estimate the scalar product and sum computations on inconsistently masked data, considering different scenarios, explained previously:

First, suppose that some users mask their ratings, while others not. In this case, we can estimate the sum and the scalar product of such vectors because random numbers are drawn from some distributions with μ being 0. Second, suppose that some users employ uniform, while others use Gaussian perturbing data for data masking. Since random numbers are drawn from distributions with μ being 0, again, it is possible to estimate the sum and scalar product of such vectors. Third, users employ different σ values. The sum and the scalar product can be estimated from data disguised by users who employed various σ values due to μ being 0. Finally, users might mask various numbers of cells. Due to the same reasons, we are able to estimate the sum and scalar product when users disguise different numbers of cells. In such cases, we are able to get rid of the contributions of random numbers in the diagonal entries of $A'^T A'$. In sum, it is still possible to provide SVD-based CF services on inconsistently masked data.

Our schemes do not introduce extra storage and communication costs, while they cause extra computation costs, which are negligible. Privacy introduced due to uniform or Gaussian perturbing data for consistently data masking is analyzed in [5] and can be similarly analyzed. Since users employ inconsistent data masking, privacy improves. The server tries to figure out the true ratings and rated items. Due to randomly inserted noise data, it will not be able to learn rated items. Since users disguise normalized ratings, it becomes difficult for the server to obtain original votes without knowing the mean ratings and standard deviations of the ratings, which are only known by users. Due to inconsistent data perturbation, the server will not be able to learn how and how much data is masked. To obtain true ratings, the server should know the type and the parameters of the perturbing data and the amount of disguised data. However, it does not know such information due to inconsistent data masking. Therefore, it can be said that the inconsistent data perturbation improves privacy.

4   Experiments

We performed trials using Jester [3] and MovieLens public (MLP) data collected by the GroupLens Research Project (www.cs.umn.edu/research/Grouplens). We employed the *Mean Absolute Error* (MAE) and the *Average Relative Error* (ARE) as evaluation criteria [1, 4, 8]. We generate random values based on various data disguising ways. We then add those noise data to the private values, and find variably perturbed user-item matrix, A'. We divide data into training and testing sets. Although we employ all users' data in MLP, we randomly select 1,000 users for training from Jester. We randomly select 10% of the ratings as test data, where we set k at 10 for both data sets. For test items, we withhold their ratings and try to predict their values using our proposed schemes. We compare the predictions that we found based on disguised data with the withheld ratings. We run data disguising 100 times, find overall MAEs and AREs, and display final values. We performed the following experiments using the MA1:

Experiment 1. We performed experiments to show how accuracy changes with varying numbers of users who disguise their data, where they perturb their ratings only. We defined x_u as the percentage of the users who disguise their private data and conducted experiments while varying x_u from 0 to 100. We employed Gaussian distribution employing predetermined σ being 2. We showed our results in Table 1. Accuracy worsens with increasing x_u values because increasing numbers of users send masked data.

Table 1. Accuracy With Varying x_u Values

x_u (%)	0	30	60	100
MLP	0.7723	0.8043	0.8193	0.8322
Jester	3.4192	3.6174	3.7836	3.9847

Experiment 2. We conducted experiments using both data sets, where we employed uniform or Gaussian distributions with σ being 2 to generate the random numbers. We defined x_g as the percentage of users who perturb their data using Gaussian, while the remaining users use uniform perturbing data. We ran experiments while varying x_g from 0 to 100. We only showed MAEs for Jester in Table 2. As seen from Table 2, accuracy improves with increasing numbers of users employing uniform perturbing data. However, the accuracy loss due to using Gaussian perturbing data is small.

Table 2. Accuracy With Varying x_g Values

x_g (%)	0	30	60	100
Jester	3.9021	3.9381	3.9517	3.9847

Experiment 3. We show how accuracy changes with varying levels of perturbation. We performed experiments for γ being 1, 2, 3, and 4. As expected, the results on uniformly randomly selected σ values are better than the ones when users employ predetermined values due to smaller randomness. We computed the overall MAEs and showed our outcomes in Table 3. As seen from Table 3, our results become better with decreasing levels of perturbation due to less randomness. Although accuracy worsens with increasing σ values, the results are still promising when γ is 4, where the ARE is 8.14 % for MLP.

Experiment 4. We performed experiments to show how varying amounts of perturbed cells affect accuracy. We defined x_c as the percentage of the cells to be perturbed. We used Gaussian distribution with σ being 2 to generate random numbers. We varied x_c from 0 to 100. We computed the overall MAEs, which are similar for both data sets. When x_c is 0, the MAE is 0.7723 for MLP, while it is 0.8633 when x_c is 100. For MLP, we lost 2% accuracy when we changed x_c from 60 to 30. Although with increasing x_c values, we disguise more cells and that makes accuracy worse, the ARE is only 5.69 % when x_c is 30 for MLP.

Table 3. Accuracy vs. Level of Perturbation

γ	1	2	3	4
MLP	0.7798	0.7984	0.8283	0.8408
Jester	3.4679	3.7422	3.8751	4.1254

5   Conclusions and Future Work

We showed how to achieve SVD-based CF tasks on differently perturbed data. We elucidated various ways of data disguising. We modified and/or simplified the SVD-based CF algorithms. We performed experiments to evaluate the overall performance of our schemes. We will study how to extend our schemes to other algorithms. We will explore how to increase accuracy when some aggregate data is disclosed. We will deeply evaluate schemes proposed for the MA2.

References

1. J. Canny. Collaborative filtering with privacy via factor analysis. In *Proceedings of the 25th ACM SIGIR'02 Conference*, pages 238–245, Finland, August 2002.
2. L. F. Cranor. 'I didn't buy it for myself' privacy and E-commerce personalization. In *Proceedings of the 2003 ACM Workshop on Privacy in the Electronic Society*, pages 111–117, Washington, DC, USA, October 2003.
3. D. Gupta, M. Digiovanni, H. Narita, and K. Goldberg. Jester 2.0: A new linear-time collaborative filtering algorithm applied to jokes. In *Proceedings of the Workshop on Recommender Systems: Algorithms and Evaluation, 22nd ACM SIGIR'99 Conference*, Berkeley, CA, USA, August 1999.
4. J. L. Herlocker, J. A. Konstan, A. Borchers, and J. T. Riedl. An algorithmic framework for performing collaborative filtering. In *Proceedings of the 22nd ACM SIGIR'99 Conference*, Berkeley, CA, USA, August 1999.
5. H. Polat and W. Du. Privacy-preserving collaborative filtering. *International Journal of Electronic Commerce*, 9(4):9–36, 2005.
6. H. Polat and W. Du. Effects of inconsistently masked data using RPT on CF with privacy. In *Proceedings of the 22nd ACM Symposium on Applied Computing, Special Track on E-commerce Technologies*, Seoul, Korea, March 2007.
7. B. M. Sarwar, G. Karypis, J. A. Konstan, and J. T. Riedl. Application of dimensionality reduction in recommender system–A case study. In *Proceedings of the ACM WebKDD Web Mining for E-commerce Workshop*, Boston, MA, Aug. 2000.
8. U. Shardanand and P. Maes. Social information filtering: Algorithms for automating "word of mouth". In *Proceedings of the ACM Conference on Human Factors in Computing Systems*, pages 210–217, Denver, CO, USA, May 1995.

Micro-Architectural Side-Channel Attacks & Branch Prediction Attack

Çetin Kaya KOÇ

Istanbul Commerce U., Turkey, & Oregon State U., USA

Abstract. We give an overview of side-channel attacks on commodity processors, particularly for computers running as servers. These attacks, named as micro-architectural attacks, exploit the cache and branch prediction behavior of the processor. The branch prediction attacks have been shown to be quite successful, and require that software and hardware architects develop countermeasures against such attacks.

Web Services Security: Protocols, Implementations, and Proofs

Karthik bhargawan

Microsoft Research Cambridge, UK

Abstract. This talk will overview recent work in verifying security properties for protocols being standardized as part of the XML Web Services framework. I will introduce the WS-Security and its associated standards and discuss how these protocols differ from traditional cryptographic protocols. Through examples, we will see how to write and generate models and implementations of these protocols. I will discuss flaws in some protocols and their implementations, and for others we shall see how to achieve formal proofs of correctness under the Dolev-Yao threat model. All our papers and some of our verification tools are available online at http://Securing.WS.

The Performance Results of ECDSA Implementation on Different Coordinate Systems

Serap Atay, Ph.D.

Izmir Institute of Technology, College of Engineering, Department of Computer Engineering, Urla, Izmir, Turkey
serapatay@iyte.edu.tr

Abstract. Elliptic Curve Cryptography has a high computational cost due to arithmetic operations of point addition and point doubling. But the cost can be reduced if different coordinate systems utilized. This paper shows that the performance of an elliptic curve digital signature algorithm (ECDSA) can be significantly increased by using different coordinate systems.

Key words: ECDSA, Elliptic Curve Arithmetic, Affine coordinate system, Projective coordinate system, computational costs of EC arithmetic.

1 Introduction

Elliptic curves are proposed for the asymmetrical cryptography by Koblitz & Miller in 1986 separately. As shown in Table 1, ECC can guarantee the same level of security with shorter key lengths. The National Institute of Standards and Technology – NIST recommended 2048 and 3072 bits key length for RSA and 224 or 283 bits key length for ECDSA until 2008 [2]. Therefore; Elliptic Curve Cryptography (ECC) has the increasing implementation area for the mobile devices and smartcards which all have the limitations in terms of processing power, power, memory, and of communications bandwidth.

Table 1. Nist recommended key lengths for equivalence security levels [2].

RSA and Diffie-Hellmann Key Length (bits)	ECC Key Length (bits)
1024	160
2048	224
3072	256
7680	384
15360	521

However, the software implementation of ECC has already speed problem due to heavy computational requirements of arithmetic operations. The computational costs can be reduced greatly if these arithmetic operations are done in standard Projective or Jacobian planes instead of Affine coordinate system [1].

The digital signature is a very critical tool for e-government, e-commerce and financial services. It can be used easily and rapidly from any where and any time. The digitally signing can guarantee the integrity, authentication and non-repudiation of any document in a virtual world.

U.S. Government Federal Information Processing Standard (FIPS) has defined the Digital Signature Standard (DSS) in 1991. In DSS the security is based on the computational intractability of DLP in sub-groups of Z_p^*. Elliptic Curve Digital Signature Algorithm-ECDSA is the elliptic curve analogue of the DSA. And, the security of ECDSA is based on the computational intractability of the ECDLP. ECDSA has been accepted;

- as an ISO standard "ISO 14888-3" in 1999.
- as an ANSI standard "ANSI X9.62" in 1999.
- as an IEEE standard "IEEE P 1363" in 2000.
- as a FIPS standard "FIPS 186-2" in 2000.

This paper is made up of four sections including the introduction. While the section 2 provides the fundamental background information on ECC and EC arithmetic operations on different coordinate systems, section 3 covers the ECDSA implementation on different coordinate systems with the operational performance. The last section concludes the findings.

2 Elliptic Curves in Asymmetrical Cryptography

ECC schemes are public-key mechanisms that provide the same functionality as RSA. All elements of asymmetrical cryptography use the finite groups of abstract algebra and any finite group can be created by the modulus operation with a prime number p (mod p) of the positive integer numbers Z^+. The group can be shown with F_p.

Elliptic curves are third degree polynomials, and are called as Weierstrass equation as in the form of $y^2 + a_1xy + a_3y = x^3 + a_2x^2 + a_6$ where the coefficients a_i are in the finite field. The field is defined by modulus operation of any prime number which is $p > 3$, the field is shown by F_p and is called as a prime field. Elliptic curves are then defined as $E : y^2 = x^3 + ax + b$, $a,b \in F_p$ and shown as $E(F_p)$ [3].

E is an elliptic curve over prime field $E : y^2 = x^3 + ax + b$, and all pairs of (x, y), $x, y \in F_p$ which satisfy the equation E and define a new group over the F_p. This is suggested first time by Jacobi in 1835 [4]. When a line is drawn through the two point on the curve or a tangent line drawn at a point on the curve which intersects the curve at another point then the operations of both the point addition and point doubling are obtained which is in turn called as the "chord and tangent" rule. The idea here is of using of the lines through known points to produce new points on the elliptic curve.

The $(x,0)$ point defines a parallel line to y axis and intersects the elliptic curve at infinity. Therefore the $(x,0)$ point is called as "the point at infinity" and shown with O [4].

All protocols of elliptic curve cryptosystems use the solution (x, y) pairs of elliptic curve polynomial over the field F_p. The count of the solution points is defined by Hasse theorem as $\# E(F_p) = p + 1 - t$ and $|t| \le 2\sqrt{p}$, the t is called as trace [5].

The security of ECC has not been proven yet such as RSA or DLP. The security level of ECC is determined by the key length, application protocol, domain parameters and of the complexity of elliptic curve discrete logarithm problem.

2.1 Elliptic Curve Discrete Logarithm Problem

When the point generation is started from a base point P on the curve, the order of the created subgroup of $<P>$ should be n. To generate a public/private key pair from this base point on the curve; a random integer number d should be selected and point addition or doubling operations should be done iteratively d times until a new point on the curve is reached such as $Q \in <P>$ and $Q = d.P$, $d \in [0, n-1]$.

When the only known parameters are Q and P then the difficulty to find the value of d is called as "discrete logarithm problem - DLP" and shown as $d = \log_p Q$. The (Q, d) pair defines public and private keys respectively [3].

2.2 Point Addition and Point Doubling

In order to obtain the points of P_1 and P_2 the aforementioned chord and tangent rule is executed and, $P_1 = (x_1, y_1)$ and $P_2 = (x_2, y_2)$ are gained respectively as the points on the curve.

- If $x_1 \neq x_2$, $P_1 \neq P_2$ and λ defines the slope of the line through these two points then $\lambda = (y_1 - y_2)/(x_1 - x_2)$ and $x_3 = \lambda^2 - x_1 - x_2$, $y_3 = \lambda(x_1 - x_3) - y_1$. This operation is called as point addition.

- If $P_1 = P_2$, λ defines the slope of tangent line through P_1 then $\lambda = (3x_1^2 + a)/(2y_1)$, $x_3 = \lambda^2 - 2x_1$ and $y_3 = \lambda(x_1 - x_3) - y_1$. This operation is called as point duplication [3].

2.3 Domain Parameters

Domain parameters provide the necessary information of which they should be known by each party of whom uses the cryptographic application. ECC schemes use $D = (p, FR, a, b, P, n, h)$ as domain parameters [3]:

- FR represents the elements of the finite field F_p.

- The order of the elements of the F_p is p.

- The elliptic curve E is a polynomial which is defined over F_p and its' constants are $a, b \in F_p$.

- P is a base point over the elliptic curve; $P = (x_P, y_P) \in E(F_p)$.

- If point doubling and addition is implemented iteratively from the base point, it creates a subgroup $<P> \in F_p$. And, the order of this subgroup is n.

- The order of the elliptic curve over F_p is $\#E(F_p)$ and $h = \#E(\mathbb{F}_q)/n$ is called as a cofactor. h should be small integer value and n should be prime number to avoid the known attacks of ECDLP.

2.4 EC Arithmetic on Different Coordinate Systems

The high computational cost of EC arithmetic stems from the point doubling and point addition operations and moreover all ECC schemes use these operations very frequently. Therefore the cost reducing is important and is a hot topic. One of the ways to solve this problem is to create a new scalar multiplication method which needs less addition and less doubling operations. Still another way is to directly reduce the cost of point addition and point doubling.

Affine coordinate system is a two dimensional Euclidean plane and each point is represented by (x, y) pair. The analysis of the steps of point doubling and point addition operations of elliptic curves shows that the modular inversion has the highest arithmetical cost [1]. If these arithmetic operations are done on a standard Projective, Jacobian, Modified Jacobian or on Chudnovsky coordinate systems, the modular inversion operations can be eliminated and hence the total cost of point operation is reduced. The projective coordinate system replaces $x = X/Z$, $y = Y/Z$ and the Jacobian coordinate system replaces $x = X/Z^2$ and $y = Y/Z^3$, they use three dimensional space and a point is represented with (x, y, z) triple. The Modified Jacobian and Chudnovsky are the variants of Jacobian coordinate system. They are designed to improve the performance specifically for point addition and/or point duplication. Therefore Modified Jacobian represents a point with (X, Y, Z, aZ^4) to improve the speed of point doubling and Chudnovsky represents a point with (X, Y, Z, Z^2, Z^3) to improve the speed of point addition. While the best performance values of point addition are observed in Chudnovsky, the Modified Jacobian is better for point doubling [1].

Affine coordinate system uses minimum bandwidth due to representing of point (x, y) pair, but other coordinate systems use greater than two values and therefore need greater bandwidth [1].

3 ECDSA Implementation on Different Coordinate Systems

ECDSA scheme was implemented by using NIST curves [6] and the performance values of signature generation and verification for Affine, Projective, Jacobian, Modified Jacobian and of Chudnovsky coordinate systems were obtained.

A Pentium 4, 1.7GHz processor with a 512MB main memory plus a 256KB cache, running on Windows XP operating system utilized as the hardware and, ANSI C compiler with GMP software library [7] were used as the implementation domain. All operations are repeated 100 times and the speed-up values measured in milliseconds.

As a result, the best performance values are obtained in mixed coordinate system as shown in Table 2 & 3 and their graphics are depicted as Fig. 1 & 2.

Table 2. Performance values for signature generation on different coordinate systems.

Coordinate Systems	Signature Generation (msec)				
	192 bits	224 bits	256 bits	384 bits	521 Bits
Affine	34,08	50,56	67,26	164,02	344,34
Projective	20,22	29,26	40,78	111,64	248,46
Jacobian	10,83	15,32	20,27	53,66	121,85
Chudnovsky	11,45	15,35	20,44	49,73	116,43
Modified	13,89	16,82	21,71	54,11	128,65
Mixed	9,71	13,17	17,43	44,67	105,75

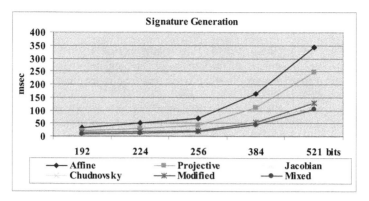

Fig. 1. The signature generation performance on different coordinate systems.

Table 3. Performance values for signature verification on different coordinate systems.

Coordinate	Signature Verification (msec)				
Systems	192 bits	224 bits	256 bits	384 bits	521 Bits
Affine	68,39	102,78	137,6	327,96	688,39
Projective	40,10	59,36	83,7	222,46	496,45
Jacobian	21,97	31,7	41,7	107,43	244,25
Chudnovsky	23,73	31,49	42,82	99,28	232,07
Modified	24,95	35,53	44,18	107,73	255,05
Mixed	19,28	26,35	34,84	44,67	209,66

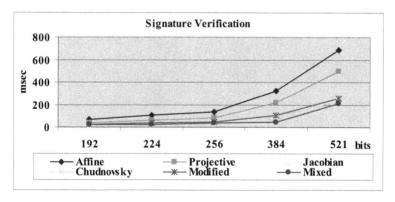

Fig. 2. The signature verification performance on different coordinate systems.

4 Conclusion

The elliptic curve cryptography use shorter key lengths for the same security level than the other members of asymmetrical cryptography such as RSA. The computational speed problem is valid for all members of asymmetrical cryptography due to intensity of mathematical operations.

In this paper we have shown that the operational speed of signature generation and verification can be increased if implemented on mixed coordinate system. Hence, the ECDSA can become a preferred solution with low computational costs.

References

[1] H. Cohen., A. Miyaji, T. Ono. "Efficient elliptic curve exponentiation using mixed coordinates", 1998, In Advances in Cryptology – Asiacrypt'98 (Beijing), vol. 1514 of Lecture notes in Computer Sciences, s. 51-65.
[2] W. T. Polk, D. F. Dodson, W. E. Burr, "Cryptographic Algorithms and Key Sizes for Personal Identity Verification", April 2005 http://csrc.nist.gov/publications/nistpubs/800-78/sp800-78-final.pdf, Information Technology Laboratory National Institute of Standards and Technology, MD, 20899-8930, s. 6.
[3] D. Hankerson, A. Menezes, S. Vanstone, "Guide to Elliptic Curve Cryptography", 2004.
[4] Dale Husemöller, "Elliptic Curves", Graduate Texts in Mathematics, 1987.
[5] N.Koblitz, "A Course in Number Theory and Cryptography", 1994.
[6] http://www.nist.org
[7] http://www.swox.com/gmp/, GMP software development library.

Remote Security Evaluation Agent for the RSEP Protocol

Suleyman Kondakci

Faculty of Computer Science, Izmir University of Economics
suleyman.kondakci@ieu.edu.tr

Abstract. This paper presents a unique remote security evaluation agent together with data structure and risk assessment methodology of the security evaluation protocol RSEP. RSEP was designed for test and evaluation of information systems over the Internet and open networks. Application of this novel protocol can be used by security and risk assessment facilities to risk management. IT owners can apply RSEP to ensure efficient proactive lifecycle protection.

Introduction

This paper deals with the risk assessment process of the remote security evaluation protocol RSEP, [1]. A secure software agent called Evaluation Agent (EA) using this protocol performs remote security evaluation. The approach presented here also relates to some of the internationally recognized security evaluation standards. The quantitative assessment discussed here can provide a complementary function in risk management for these standards. The standards, named as Common Criteria (CC) [2] and BS 7799 [3] and [4], define standards and methodologies in information security evaluation and certification, but do not, however, define the methodology for risk assessment. The standards also suggest that the organization should use a systematic approach to risk assessment. Therefore, we believe that both the CC and BS 7799 certification facilities and others can apply the risk assessment function of RSEP in their certification programs if desired.

In a comprehensive information security research project, we developed a novel security evaluation framework including the protocol named Profile Based Remote Security Evaluation Protocol (RSEP). The framework provides the methodology and specifications of operations for "seamless", inexpensive, and simple implementation of evaluation tools and techniques for remote and local risk management programs. Unusually, it avoids annoying, ad hoc technical tests, procedural questionnaires and complicated inspection procedures that are encountered in traditional IT evaluation methodologies. RSEP is also efficient in improving security awareness by providing means for early detection and tracking of risks, threats, and vulnerabilities for proactive risk management operations. Hence, the RSEP scheme is intended to enable IT owners to implement their own proactive lifecycle security policies. This is especially required for critical information systems.

Results of pilot tests that we have performed have shown improvements in the security of the sample test environments. It has shown increased user-awareness, improved management of loss-control (e.g., confidentiality, integrity, access control, and authentication), lowered virus outbreaks, and better change management. The change management issues mainly focused on disaster recovery, patches, and regular backup operations. The main concept of the framework is detailed in [1]. Most of the literature referred in [1] will naturally be relevant here too. The RSEP framework is based on a proactive approach to Information Assurance (IA). It can also enable involvement of the IT owner/security administrator actively in the test and evaluation process. That is, either an RSEP-compliant evaluator or the local staff can conduct physical security tests. However, it is preferably recommended that the local staff perform both the maintenance and the tests. They will then save the test results into a formatted security document called Security Profile (SP). The RSEP evaluator will then evaluate the contents of the SP. The structure and interpretation of the contents of SPs are presented in [1] and recaptured in Section 3 in this paper. Company SPs contain security attributes composed for each asset or a specific group of assets residing in a given IT environment. The strength of each asset is tested, evaluated, and graded according to its classification (or criticality). For example, a router (either a hardware or software unit) is defined as an asset. The router has its own SP, which contains all required test attributes for its security evaluation. For example, one of these attributes is its strength against source routing. Routers are sensitive to routing misbehaviors that are often activated externally by malicious act. A discussion of routing misbehavior in mobile ad hoc networks is given in [5]. Another feature of the router is the filtering capability of spyware and spam messages. Following the evaluation, these properties (attributes) are scored and the test results are saved into the SP. The remote RSEP evaluator will then download the SP and run the algorithm presented here to evaluate the SP. In turn, the asset's *assurance level* will be evaluated and the contents of the SP will be updated with a risk score and a short report about eventual vulnerabilities or improvements. Thereafter, the evaluator will upload the updated SP back to the target location. As mentioned, the remote evaluator can also execute the active tests while evaluating the system.

We use the Common Criteria (CC) term *target of evaluation* (TOE) for the assets and systems that are under evaluation. The TOE is defined by the CC as the IT product or system and its associated guidance documentation that is the subject of an evaluation. Details on the CC methodology are given in [2]. However, we do not consider or involve any of the CC methodologies in the RSEP framework. Following this brief introduction to RSEP, an example of the data structure for a particular SP is presented. We will also discuss the communication security and the quantitative risk assessment approach using the evaluation agent (EA) of the RSEP framework.

In the following, Section 2 briefly introduces features and the communication architecture of the RSEP protocol and its communication with evaluation agents. Further, it introduces the RSEP method and its benefits. Section 3 discusses the remote evaluation process, the data structure, and the algorithm used to compute the quantitative risks and strengths. A practical example completes this section, and Section 4 concludes the paper.

2. RSEP Framework and Evaluation Agents

Proactive and seamless security evaluation and risk management is one of the key application areas of RSEP. A major function of this protocol is evaluation and risk management in general, and specifically the security of protection systems. It provides models and means to remotely monitor protection mechanisms in a timely manner. Thus, in this manner, dynamic threat patterns can be captured and analyzed in order to mitigate current and future security risks encountered in the protection mechanisms. The RSEP framework provides inherent proactive lifecycle security. As a primary step in achieving this, this framework forces the staff of TOEs to continuously track threats and vulnerabilities and recover from incidents. We use the term *evaluation agent* (EA) to represent the evaluator tool or a principal. An EA is a principal (a software agent or code) that remotely collects and processes the evaluation information under the control of a RSEP evaluator. The sequence of an evaluation process is as follows:

1. Connection establishment and authentication between EAs and TOEs
2. SP delivery to the EA (after a valid authentication)
3. Evaluation and update of the SP contents
4. Return of the SP to the TOE

Communication of Agents and SP-services

The specifications of the RSEP architecture require secure data exchange between EAs and TOEs adhering to the standards given in [6]. Interactions begin with a connection establishment (a hello session) phase, which is a fundamental requirement for mutual identification and authentication of the evaluating agent, EA, and the SP-service. As described in [1], the SP-service defines functions by which security profiles and IDs stored and exchanged securely in a disk file. The SP-service waits for requests from EAs and delivers the required documents (i.e., SP and the ID of the TOE) to the EAs. SP-services can be designed to function either as distributed (decentralized) or centralized. The distributed model is designed as a stand-alone application that runs as a server process (daemon) on the evaluated node (i.e., TOE). The initial hello message from the EA requests a valid digital ID from the SP-service for the TOE under evaluation, in the process, the EA must also present a valid digital ID. The SP-service associated with the TOE should verify the validity of the digital ID sent from the EA. On receipt of a digital ID of from EA, the SP-service verifies the EA. On successful verification, the SP-service packs and sends a hello message containing the TOE's certificate comprised of its digital ID and the associated SP. By receiving a valid SP and TOE-certificate, the EA terminates the hello session and begins the evaluation of the received SP. This process could either be executed as a batch or real-time processes depending on the current resource requirements and the priority of the evaluation request.

Part of the SP exchange process is illustrated in Fig. 1. As already noticed, the security of the data exchange between the EA and TOE is ensured by the combination of two basic services: authentication and confidentiality, respectively. An initial certificate (e.g., digital ID) exchange included in the hello session is required in order to protect against masquerading and "man in the middle" type of attacks. Interactions are securely performed by a dedicated encryption scheme shown in Fig 1, which is implemented in the secure exchange protocol applying the elliptic curve cryptography, presented in [7]. Cryptographic techniques based on the elliptic curve algorithm are described in the standards proposal of ISO/IEC [8]. The interactions assume that the communicating parties, EA and TOEs, have already exchanged required public keys. A fast key exchange technique with elliptic curve systems is given in [9].

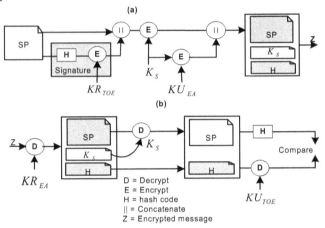

Fig. 1. The secure SP exchange between EA and SP-service

Fig. 1 shows a modified public-key scheme that is used for the SP exchange of the RSEP protocol. The modification is actually achieved by encrypting the hash code of the encrypted SP coming from the TOE with a secret-key. Besides, the secret-key, also called temporary session key (K_s) is also encrypted with the public key of the EA (KU_{EA}) and then concatenated with the encrypted hash code. As shown in Fig. 1 (a), the SP's hash code is encrypted with the TOE's private key KR_{TOE}. The generalized scenario shown in Fig. 1 (a) illustrates the transfer of an SP of a given TOE from its SP-service. A SHA-256 based digital signature with a 256-bit hash code is generated to ensure the message authentication, see [10] for the SHA algorithm. The hash code is encrypted with the public-key cryptographic algorithm using the private key of the TOE (KR_{TOE}) and the result is concatenated to the SP. The result is then encrypted with the temporary session key, K_s, which is a random key generated by utilizing the BBS algorithm, [11]. Further, the temporary session key K_s and the SP are encrypted using receiver's (the evaluation agent EA) public key, KU_{EA}. The receiver, EA, uses its own private key, KR_{EA}, to decrypt the message and obtain both the SP and the session key. This is illustrated in Fig. 1 (b). The EA uses the decrypted session key to decrypt the SP. Then it uses the sender's public key, KU_{TOE}, to decrypt and recover the hash code. The EA generates a new hash code from the SP just received, and compares it with the decrypted hash code. If the two hash codes match then the SP is considered authentic, however, if no match is found, the receiver replies with an error code, and terminates the authentication session. Note that, as obvious, with this scheme, confidentiality of communication is also ensured.

The Evaluation Process

The evaluation process assumes a valid contract between an evaluation facility and the TOE owner. After issuing the contract, a cryptographic public key exchange operation is performed either locally or remotely over the Internet. The key exchange covers a secure digital certificate exchange operation between the evaluation facility and the TOE owner, see [12] for an overview of the key management issue. The evaluation agent, EA, starts the evaluation process by requesting a valid SP from a remote SP-service. The SP-service also serves the TOE under test. Upon the arrival of the requested SP, the agent program parses the SP and extracts the required security attributes to evaluate the associated TOE. Recall that each TOE is associated with an SP whose contents are also needed to compute the overall evaluation, and store the result. After parsing and computing the overall risk for the TOE, a number of fields in the SP are updated and returned to the SP-service. In turn, the SP-service encrypts and saves the received SP. The majority of the fields that are updated by the EA include: evaluation date, evaluator ID, risk level, resulting assurance level (AL), and a final report containing a list of vulnerabilities and eventual recommendations to the TOE owner. It should be noted that, the evaluator, if required, might conduct remote attacks and record a list of additional current exploits and vulnerabilities. However, this issue requires legislation in tact like reformed hacking. The following code snippet denotes a part of a sample SP designed for a database server (under evaluation) of which the tree-structured XML infoset, [13], is shown in Fig. 2.

```
<xs:element name="ProceduralSecurity">
    <xs:element name="ChangeManagement">
            <xs:element name="Updates" type="xs:decimal" fixed="3.3" />
            <xs:element name="Patches" type="xs:decimal" fixed="4.6" />
    </xs:element>
  <xs:element name="DisasterRecovery">
            <xs:element name="FirePlan" type="xs:decimal" fixed="4.5" />
            <xs:element name="PowerBreak" type="xs:decimal" fixed="3.5" />
            <xs:element name="VirusOutbreak" type="xs:decimal" fixed="5.0" />
            <xs:element name="Intrusion" type="xs:decimal" fixed="4.0" />
            <xs:element name="NaturalDisaterPlan" type="xs:decimal" fixed="4.0" />
    </xs:element>
</xs:element>
```

As mentioned earlier, each SP represents two sets of data, the ID and the attribute parts. Thus, this code shows the attributes (partly) part of the SP. The attribute items are evaluated according to their strength levels, and scored, and the scores are saved in the corresponding fields of the SP. For the overall assessment, this sequence of operations is necessary for each attribute found in a given IT environment.

The Quantitative Risk Assessment

Obviously each SP, in general, is a tree of a finite set of XML infoset (or elements). As shown in Fig. 2, each parent node stores the value for a given test attribute, e.g., Procedural Security. Thus, each of the parent nodes may span across several child nodes to identify the sub-attributes or items of a given attribute.

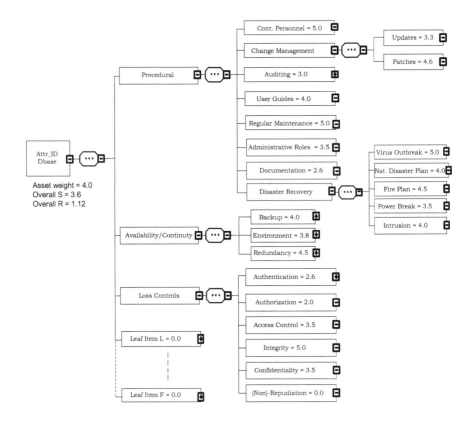

Fig. 2. . The tree representation of attributes of a sample SP of a database server.

That is, each attribute may be comprised of a finite set of sub-attributes or items in a recursive manner. This hierarchical spanning is, naturally, in the form of a tree, which may grow to several levels to hold the necessary subitems. A scalar weight between 0 and 5 depicts the asset value, which is a close conformance to the criticality level of the asset. Asset weight of 0 defines the minimum significance while 5 depicting the maximum significance. Note that, although we find these weights adequate for the sample calculation, one can freely apply other metrics, for example, 0-100 scale can also be used to score the strengths and asset weights.

The root node is the asset ID and the remaining nodes are a hierarchical collection of subtrees. A node without children is called a leaf (or external) node; other nodes are called internal nodes. In the sample evaluation of the SP shown in Fig. 2 we have *F* leaf nodes and **3** main branches (Procedural, Availability/Continuity, and Loss Controls), at the same level as *F*, each consisting of further subtrees. It is assumed that each child node can have other leaf nodes and/or subtrees of varying number of elements. The overall risk value, R, is always computed and saved into the corresponding field of the SP.

Each item in the SP is locally evaluated and scored between 0 and 5. For example, the above code shows a part of the procedural security attribute tagged as DisasterRecovery whose child elements were evaluated and scored accordingly. The average score from the child nodes (internal nodes) of the DisasterRecovery is computed and incorporated into its parent node for overall risk computation algorithm. A generalized form of the recursive risk calculation algorithm is given in Eq. (1) and (2).

$$e_{(L-1)(g),h} = \frac{\sum_{i=1}^{k_{L(j)}} e_{L(j),i}}{k_{L(j)}}; \ where \begin{cases} L = 0,1,2,...,n \\ j = 1,2,3,...,m_L \\ 1 \leq g \leq m_{L-1} \end{cases} \tag{1}$$

and

$$S = e_{0(1),1} = \frac{\sum_{j=1}^{F+m_{(L=1)}} e_{1(j),1}}{F + m_{(L=1)}}; S \in [0,5] \tag{2}$$

Where, $e_{(L-1)(g),h}$ represents the average quality (or strength) of item h computed from the items of level L for parent node g that stays at level $(L-1)$, and S represents the asset-dependent overall strength factor. In other words, the parameter $e_{(L-1)(g),h}$ now represents the average value computed for node h and transferred to offspring g at level $L-1$. Thus, the variables defined are

- $L(j)$ = The jth offspring at level L
- $k_{L(j)}$ = The number of attribute items of jth offspring at level L
- m_L = The number of offsprings at level L

Note that, variable $e_{L(j),i}$ represents the score of the ith item belonging to offspring j of level L. The example shown in Fig. 3 consists of three levels, three offsprings, and 0 F leaf nodes under the root node. The variable $k_{L(j)}$ depicts the number of items that can be of different sizes for each of the offsprings. For example, as shown in the XML infoset-tree, Fig. 2, let us assume that DisasterRecovery has 5 items, ChangeManagement 2 items, but the Procedural in total 8 items, and so on.

To compute the overall risk parameter, we begin with the leaf nodes of the lowest-level offsprings and calculate the average score of each level in a recursive manner and a bottom-up fashion. In the literature it is called *postorder traversal* of a tree. For example, the sub-result, $e_{(L-1)(g),h}$, is computed from level L and then fed into level $L-1$. This computation will continue recursively until each node in the entire tree had been traversed until the 0th level, $L = 0$, is reached. The root node stays at level 0. Each offspring is computed separately by Eq. (1), and the overall result, S, is calculated by incorporating the results, $S1, S2, ... SN$, from all offsprings and leaf items under the root node by Eq. (2). Considering the overall risk parameter of a single asset with weight w we have

$$R^w = w\left(1 - r_j \frac{S}{5}\right); \ (R^w, w, S) \in [0,5], 0 < r_j < 1. \tag{3}$$

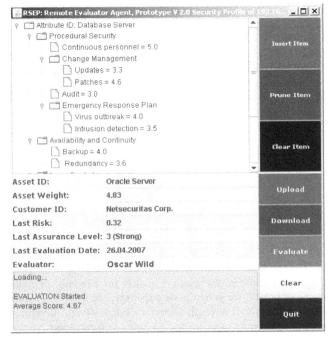

Fig. 3. The remote evaluation of attributes of a sample SP representing a database server.

Where, R^w gives the value of the risk parameter and w depicts the asset weight of the asset evaluated. A relative strength parameter, $r_j = (0,...,1)$, is used as a local factor, e.g., strength of an attribute can be adjusted with this coefficient to express its strength relative to other attributes. Simply, for a set of n assets, the total risk parameter becomes

$$TR = \sum_{a=1}^{n} (R^w)_a; 0 \le TR \le 5n \qquad (4)$$

Please Note that, the calculation does not require the recursive calculation for plain attribute structures that only contain leaf (or child) nodes under the root and no other sub-attributes. It simply applies Eq. (5). Where the sum (dividend) denotes the sum of the scores of the leaf nodes given under a plain structure, w denotes the asset weight, and N denotes the number of leaf items (nodes) belonging to the single attribute.

$$R^w = w\left(1 - r_j \frac{\sum_{i=1}^{N} G_i}{5N}\right); (R^w, G_i, w) \in [0,5], 0 < r_j < 1. \qquad (5)$$

4. Summary and Conclusions

The remote security evaluation protocol RSEP defines a new concept of remote evaluation methodology while promoting several areas of application. One of these might be the analysis of risks and vulnerabilities in information systems. It also defines a protocol-independent communication architecture and data structure to store test properties for TOEs. The test properties (stored in SPs) denote the evaluation information needed during the evaluation and for providing proactive lifecycle security. The RSEP data structure makes use of XML infoset. This interoperable data structure enables also the design of the test properties in a hierarchical fashion. The hierarchical ordering of security attributes are necessary due to interdependency of security items that are defined for assets or systems. For the security professional the quantitative risk evaluation has been simplified by issuing a simple recursive algorithm on the evaluation data stored in SPs. As mentioned, SPs are stored in XML infoset (XML tree), which is required for its interoperability features and minimal resource usage, and especially need for delivering highly effective solutions in mobile communication devices and applications. The information stored in SPs represent the overall security picture of a given asset. Testing the security of the systems and verifying the updated SPs of the tested systems efficiently achieve the lifecycle security. This feature is especially important for the secure operation of the protection systems and for the critical systems. The tests can be conducted over the Internet and open networks securely. In these tests the roles of the remote evaluator and evaluating agents (principal codes or tools) are important. The RSEP policy delegates responsibilities between the local staff (network owner/system administrator) and the evaluators, who determine the assurance level and risk factors for the evaluated systems. On the other hand, the local staff should ensure the continuous protection and some of the fundamental physical test activities. Through its seamless test and evaluation approach, this remote risk assessment methodology brings a new dimension into the rapidly growing ubiquitous computing world. In addition, the quantitative risk assessment is considerably simplified for the self-assessment to the extent, which can be conducted by network owners who may even be inexperienced personnel.

The risk assessment approach of RSEP can easily be applied to any quantitative risk assessment operation. Though RSEP is very simple for determining quantitative results, it uses an informal method to compute quantitative risk values. The quantitative risk assessment methods are much harder to realize than those of the qualitative methods. This problem can likely be reduced by a formalized quantitative risk assessment standard, which is a shortcoming of RSEP to be defined further.

References

1. Kondakci, S.: A Remote IT Security Evaluation Scheme: A Proactive Approach to Risk Management, Proceedings of IEEE International Workshop on Information Assurance (2006) 93-102
2. CC Evaluation Document V3.0, http://www.commoncriteriaportal.org/ (2007)

3. ISO/IEC FDIS 15408-1: Information Technology - Security Techniques - Evaluation Criteria for IT Security, Part 1, http://www.gammassl.co.uk/ist33/27N4241.pdf (2007)
4. The ISO 17799 Toolkit: http://www.iso17799-made-easy.com/ (2007)
5. Marti, S., Giuli, T., Lai, K., Baker, M.: Mitigating Routing Misbehavior in Mobile Ad hoc Networks. In Proceedings of MOBICOM 2000, pages (2000) 255–265
6. ISO/IEC 9798-3, Information Technology–Security Techniques – Entity Authentication Mechanisms – Part 3: Entity authentication (2000)
7. Aydos, M., Yanik, T., Koç, Ç. K.: High-speed Implementation of an ECC-based Wireless Authentication Protocol on an ARM Microprocessor. IEE Proceedings - Communications, 148(5) (2001) 273-279
8. ISO/IEC 15946, Information Technology – Security Techniques – Cryptographic Techniques Based on Elliptic Curves, Committee Draft (CD) (1999)
9. Schroeppel, R., Orman, H., O'Malley, S., Spatscheck, O.: Fast Key Exchange With Elliptic Curve Systems, Advances in Cryptology – Crypto '95, Lecture Notes in Computer Science, 963 (1995), Springer-Verlag, 43-56.
10. FIPS 180-1, Secure Hash Standard, Secure Hash Algorithm, SHA-256, http://csrc.nist.gov/publications/fips/fips180-2/fips180-2withchangenotice.pdf (2006)
11. Blum, L., Blum, M., Shub, M.: A Simple Unpredictable Pseudo-random Number Generator, SIAM Journal on Computing, No. 2 (1986)
12. Fumy, S., Landrock, P.: Principles of Key Management, IEEE Journal on Selected Areas in Communications, Vol. 2, No. 5 (1993)
13. XML 1.1: Extensible Markup Language: W3C Recommendations 4th February 2004, http://www.w3.org/TR/xml11/ (2007)

Test Case Generation for Firewall Implementation Testing using Software Testing Techniques

Tugkan Tuglular

Department of Computer Engineering, Izmir Institute of Technology, Gulbahce Koyu, Urla, Izmir, Turkey
tugkantuglular@iyte.edu.tr

Abstract. The firewall implementation testing approach checks actions performed by the firewall with respect to corresponding firewall rules. This type of firewall testing can be implemented by developing test cases from firewall rule sequence, generating test packets using those test cases and injecting those test packets into the firewall. Although this method has been already defined in the academic world, an approach to generate test cases does not exist in the literature. In this work, a test case generation approach is developed using software testing techniques.

1 Introduction

Firewall tests have to be performed to verify that the firewall works as specified. In this work, a test case generation approach is developed, which defines test cases based on the firewall rule sequence. Using these test cases and possibly a real traffic database, test packets can be prepared to be injected to check if the firewall implementation is erroneous, i.e. the rules do not correspond to the actions of the firewall. Although injection based firewall testing is accepted as an inefficient way of testing firewall implementations in the literature [1], there has been no alternative method developed yet. Most of the academic work focuses on testing of firewall rules where firewall implementation is assumed error-free. Even if firewall implementation is error-free, a firewall can be hacked and programmed to behave differently from the intended security policy. In that case, real time injection based testing is one of the ways to reveal the security breach.

There are three general approaches to firewall testing; penetration testing, testing of the firewall implementation, and testing of the firewall rules [2]. Penetration testing is performed to check the firewall for potential breaches of security that can be exploited. The firewall implementation testing approach evaluates the correspondence of firewall rules with respect to the actions the firewall performs (e.g. if a rule indicates to block a packet but the firewall forwards the packet, that means a firewall implementation error exists) [2]. Testing of the firewall rules verifies whether the security policy is correctly enforced by a sequence of firewall rules or not.

The firewall implementation testing is achieved through defining test cases, deriving test packets from these test cases and sending or injecting the packets to the firewall to analyze its behavior [2]. *Heidi* states that there are two strategies to inject packets; injecting bogus packets pretending to be originated from all the hosts outside the private network and injecting bogus packets pretending to be originated from some of the hosts outside the private network, where the first strategy is extremely inefficient, and the second strategy cannot cover all possible host IP addresses, which make the testing incomplete [1]. This is very similar to software testing, where software cannot be tested for all possible values and execution paths [3]. However, a limited number of significant values and paths can be selected and exercised. This is the main idea behind the test case generation approach developed in this work.

The rest of this paper is organized as follows: Section 2 describes the firewall decision model. Section 3 explains developed test case generation approach. Section 4 presents a test case generation example. Section 5 explains future work and concludes this paper.

2   Background and Terminology

"A firewall is a network element that controls the traversal of packets across the boundaries of a secured network based on a specific security policy. A firewall security policy is a list of ordered filtering rules that define the actions performed on matching packets. A rule is composed of filtering fields (also called network fields) such as protocol type, source IP address, destination IP address, source port and destination port, and a filter action field. Each network field could be a single value or range of values. Filtering actions are either to accept, which passes the packet into or from the secure network, or to deny, which causes the packet to be discarded. The packet is accepted or denied by a specific rule if the packet header information matches all the network fields of this rule. Otherwise, the next following rule is used to test the matching with this packet again. Similarly, this process is repeated until a matching rule is found or the default policy action is performed" [4]. In this paper, a "deny" default policy action is assumed.

The common format of packet filtering rules, represented as follows, is used throughout the paper:

(<order>,<protocol>,<src_ip>,<src_port>,<dst_ip>,<dst_port>,<action>) [4]

The 5-tuple (<protocol>,<src_ip>,<src_port>,<dst_ip>,<dst_port>) is considered as the predicate of a rule. This rule predicate is critical for the test case generation approach presented in this paper.

The <order> of the rule determines its position relative to other filtering rules. The <protocol> specifies the transport protocol of the packet and may have one of these values: IP, ICMP, TCP or UDP. The <src_ip> and <dst_ip> prescribe the IP addresses of the source and destination of the packet respectively. The IP address can be a host (e.g., 193.140.248.11), or a network address range (e.g., 193.140.248.*). The <src_port> and <dst_port> fields specify the port address of the source and destination of the packet respectively. The port can be either a single specific port number, or any port number indicated by "any" [4].

As an example, the following security policy is to block all TCP traffic coming from the network 193.140.248.* except SMTP:

    1: tcp, 193.140.248.*, any, *.*.*.*, 25, accept
    2: tcp, 193.140.248.*, any, *.*.*.*, any, deny

"Some firewall implementations allow the usage of non-wildcard ranges in specifying source and destination addresses or ports. However, it is always possible to split a filtering rule with a multi-value port field into several rules each with a single-value port field" [4]. In this paper, only wildcard ranges is used for IP address and ports.

3 Test Case Generation Approach

Pressman states that test cases which (*i*) guarantee that all independent paths within a module have been exercised at least once, (*ii*) exercise all logical decisions on their true and false sides, (*iii*) execute all loops at their boundaries and within their operational bounds, and (*iv*) exercise internal data structures to ensure their validity, can be defined using software testing methods [3]. Same principles are applied in the test case generation approach developed in this work.

To be able to define test cases that fit to the model stated by *Pressman*, first sequence of firewall rules is converted to a firewall policy tree developed by *Al-Shaer and Hamed*, where each node in the tree represents a field of the filtering rule, and each branch at this node represents a possible value of the associated field. The root node of a policy tree represents the protocol field, and the leaf nodes represent the action field, intermediate nodes represent other 5-tuple filter fields in order. Every tree path starting at the root and ending at a leaf represents a rule in the policy and vice versa. Rules that have the same field value at a specific node, will share the same branch representing that value [4]. A policy tree example is shown in Figure 1.

This tree enables us (*i*) to traverse all independent paths, (*ii*) to exercise ACCEPT and DENY decisions on true and false sides of each leaf, (*iii*) to execute each path for the regions immediately adjacent to its boundaries, and (*iv*) to ensure that each rule is represented by a valid data structure. Items (*i*) and (*iv*) are trivial because of the tree structure. Item (*ii*) is achieved by the Equivalence Class Partitioning (ECP) approach. The Boundary Value Analysis (BVA) approach is used to fulfill item (*iii*). The algorithm presented in this paper is built on the model stated by *Pressman* and the approaches of Equivalence Class Partitioning and Boundary Value Analysis.

The Equivalence Class Partitioning approach divides the input domain of a software to be tested into the finite number of partitions or equivalence classes. A test case design by ECP has two steps [6]:

    (1) identifying the equivalence classes,
    (2) defining the test cases.

The equivalence classes are identified by taking each input condition – in this paper each field of the rule predicate – and partitioning it into two or more groups or classes. Two types of equivalence classes exist: valid equivalence classes represent valid inputs to the program, and invalid equivalence classes represent all other possible states of the condition [6]. In this paper, the term "valid inputs to the program" means valid values for the rule predicate and the term "invalid equivalence classes represent all other possible states of the condition" means invalid values for the rule predicate.

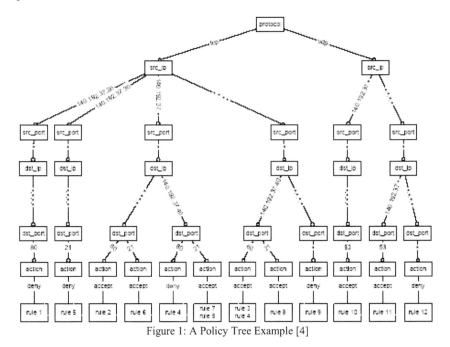

Figure 1: A Policy Tree Example [4]

Throughout the paper, valid equivalence class is named as ECv and invalid equivalence class as ECinv. Since all the fields in the rule predicate are either positive integers or a range of positive integers, it is safe to assume that there will be two invalid equivalence classes for each valid equivalence class; one is small in values (ECinv-) and the other large in values (ECinv+). For instance, assume that for a rule predicate, <src_port> field value is 80. That means ECv is 80, ECinv- is [0..79] and ECinv+ is [81..65535]. If there is to be a specific value or range to be given, it will be written as ECv(i), ECinv-(j) or ECinv+(k). Following the above example: ECv(0) is 80, ECinv-(0) may be 79, ECinv-(1) may be [0..79], ECinv+(0) may be [81..1024] and ECinv+(1) may be 65535.

The Equivalence Class Partitioning can be supplemented by another method called Boundary Value Analysis. Values close to the edges of the input are selected, so that the test case covers both upper and lower edges of an equivalence class [5]. Following the above example, boundary values for ECinv- are 0 and 79, and for ECinv+ 81 and 65535.

Equivalence classes and boundary test values are generated for each field of each rule predicate. Test cases for each rule predicate are composed of the combination of test values of each field. The algorithm for the construction of equivalence classes and choice of boundary values is shown in the next section. Examples are presented in the following section.

4 Test Case Generation Algorithm

The algorithm is built on the model stated by *Pressman* and the approaches of Equivalence Class Partitioning and Boundary Value Analysis. It assumes that a policy tree can be generated from firewall policy rules so that the model stated by *Pressman* is applicable. The algorithm is given in Figure 2.

```
for <protocol>
            ECv is protocol_value
            ECinv- is (protocol_value – 1)
            ECinv+ is (protocol_value + 1)
for <ip>
            if ip is a host
                        ECv is ip
                        ECinv- is [0.0.0.0 ... ip-1]
                        ECinv+ is [ip+1 ... 255.255.255.255]
            if ip is a network_range
                        ECv is network_range
                        ECinv- is [0.0.0.0 ... lower bound of network_range-1]
                        ECinv+ is [upper bound of network_range+1 ... 255.255.255.255]
            if ip is " *.*.*.* "
                        ECv(0) is 10.0.0.1, ECv(1) is 172.16.0.1, ECv(2) is 192.168.0.1,
                        ECv(3) is 127.0.0.1, ECv(4) is 88.241.34.41, ECv(5) is 215.15.568.23
            for <network_range>
                        EC_(0) is center value of network_range
                        EC_(1) is bottom value of network_range
                        EC_(2) is top value of network_range
                        EC_(3) is range that has the center value within network_range
                        EC_(4) is range that has the bottom value within network_range
                        EC_(5) is range that has the top value within network_range
for <port>
            if port is an integer
                        ECv is port
                        ECinv- is (port – 1)
                        ECinv+ is (port + 1)
            if port is " any "
                        ECv(0) is 0, ECv(1) is 65535, ECv(2) is 23, ECv(3) is 80
```

Figure 2: Test Case Generation Algorithm for Firewall Policies

As mentioned above, the <protocol> field may have one of these values: IP, ICMP, TCP or UDP. These values are represented as positive integers. For instance, TCP is 6 and UDP is 17. For <protocol> field ECv will always be a unique integer n and there will be ECinv- and ECinv+. By taking BVA into consideration, ECinv- is decided to be (n-1) and ECinv+ to be (n+1).

The <src_ip> and <dst_ip> fields can be a host (e.g., 193.140.248.11), or a network address range (e.g., 193.140.248.*). If the field is a host then ECv is that IP address, ECinv- is the range of lower addresses and ECinv+ the range of higher addresses. If the IP address field is a network address range then ECv is that address range, ECinv- is the range of lower addresses and ECinv+ the range of higher addresses. For an address range in any EC, the following six test values will be used: center value, bottom value, top value, range that has the center value, range that has the bottom value, and range that has the top value. The ECv of 193.140.248.* address range may be constructed as ECv(0) = 193.140.248.127, ECv(1) = 193.140.248.1, ECv(2) = 193.140.248.254, ECv(3) = 193.140.248.32: 193.140.248.192, ECv(4) = 193.140.248.1: 193.140.248.63, and ECv(5) = 193.140.248.160: 193.140.248.254. Notice that for bottom and top values last octet is 1 and 254 respectively. Since 0 and 255 are special addresses, they are not included into the ECv.

If the IP address field is *.*.*.* then ECv is all possible IP addresses and there are no invalid equivalence classes for <src_ip> and <dst_ip> fields. For this ECv, the following six test values will be used: 10.0.0.1, 172.16.0.1, 192.168.0.1, 127.0.0.1, 88.241.34.41, and 215.15.568.23. The first four test values are selected from four different address ranges which are local use only addresses. They have higher probability to uncover an error than any address. The last two addresses are believed to be from two different ISPs.

The <src_port> and <dst_port> fields are either a unique integer (port) or "any". If the field value is a unique integer then ECv is that unique integer (port), ECinv- is the range of lower ports and ECinv+ the range of higher ports. In this work, only the top value of ECinv- and the bottom value of ECinv+ are taken into consideration. If ECv is 80 then ECinv-(0) is 79 and ECinv+(0) is 81.

If the field value is "any" then ECv is all possible ports and there are no invalid equivalence classes for <src_port> and <dst_port> fields. For this ECv, the following four test values will be used: 0, 65535, 23, and 80. The number of values can be easily incremented as foreseen with respect to future work described in the Conclusion section.

5 Test Case Generation Example

The application of the test case generation algorithm given above is presented with one comprehensive example. The rule predicate to be covered is <tcp, 193.140.248.*, any, *.*.*.*, 25> of rule <1: tcp, 193.140.248.*, any, *.*.*.*, 25 accept>. Auto-generated test cases are shown in Table 1.

Table 1: Auto-Generated Test Cases for Rule Predicate <tcp, 193.140.248.*, any, *.*.*.*, 25>

Test Case No	protocol	src_ip	src_port	dest_ip	dest_port
1	6 (tcp)	193.140.248.*	any	*.*.*.*	25
2	5	193.140.248.*	any	*.*.*.*	25
3	7	193.140.248.*	any	*.*.*.*	25
4	6 (tcp)	193.140.248.127	any	*.*.*.*	25
5	6 (tcp)	193.140.248.1	any	*.*.*.*	25

6	6 (tcp)	193.140.248.254	any	*.*.*.*	25
7	6 (tcp)	193.140.248.64:193.140.248.191	any	*.*.*.*	25
8	6 (tcp)	193.140.248.1:193.140.248.63	any	*.*.*.*	25
9	6 (tcp)	193.140.248.160:193.140.248.254	any	*.*.*.*	25
10	6 (tcp)	193.140.247.127	any	*.*.*.*	25
11	6 (tcp)	193.140.247.1	any	*.*.*.*	25
12	6 (tcp)	193.140.247.254	any	*.*.*.*	25
13	6 (tcp)	193.140.247.64:193.140.248.191	any	*.*.*.*	25
14	6 (tcp)	193.140.247.1:193.140.248.63	any	*.*.*.*	25
15	6 (tcp)	193.140.247.160:193.140.248.254	any	*.*.*.*	25
16	6 (tcp)	193.140.249.127	any	*.*.*.*	25
17	6 (tcp)	193.140.249.1	any	*.*.*.*	25
18	6 (tcp)	193.140.249.254	any	*.*.*.*	25
19	6 (tcp)	193.140.249.64:193.140.248.191	any	*.*.*.*	25
20	6 (tcp)	193.140.249.1:193.140.248.63	any	*.*.*.*	25
21	6 (tcp)	193.140.249.160:193.140.248.254	any	*.*.*.*	25
22	6 (tcp)	193.140.248.*	23	*.*.*.*	25
23	6 (tcp)	193.140.248.*	80	*.*.*.*	25
24	6 (tcp)	193.140.248.*	0	*.*.*.*	25
25	6 (tcp)	193.140.248.*	65535	*.*.*.*	25
26	6 (tcp)	193.140.248.*	any	10.0.0.1	25
27	6 (tcp)	193.140.248.*	any	172.16.0.1	25
28	6 (tcp)	193.140.248.*	any	192.168.0.1	25
29	6 (tcp)	193.140.248.*	any	127.0.0.1	25
30	6 (tcp)	193.140.248.*	any	88.241.34.41	25
31	6 (tcp)	193.140.248.*	any	215.15.568.23	25

| 32 | 6 (tcp) | 193.140.248.* | any | *.*.*.* | 24 |
| 33 | 6 (tcp) | 193.140.248.* | any | *.*.*.* | 26 |

Since a test case contains values for all fields in the predicate, the algorithm aims to generate test values for a field while keeping the values of the rest of the fields unchanged. The very first test case is exactly what the predicate is. Then 2nd and 3rd test cases are generated for the <protocol> field while keeping the rest unchanged.

Test cases through 4th to 21st are for <src_ip> field. Test cases from 4 to 9 are generated from ECv. The next 6 test cases belong to ECinv- and then the following 6 test cases belong to ECinv+. <src_port> will be tested through test cases 22 to 25. *.*.*.* <dest_ip> field will be tested through 26th and 31st test cases. The last two test cases are for <dest_port>.

6 Conclusion

A test case generation approach which is critical to firewall implementation testing is presented in this paper. First, the algorithm is described and then its operation is explained with an example. Depending on the type of the rule predicate, the number of test cases for each predicate is in multiple of tens. Assuming that a typical firewall rule sequence does not exceed 40 rules, some number of test cases around 1000 will be generated. When those test cases are converted into test packets and injected into the firewall, the extra traffic generated by the test packets will be insignificant.

The next step in the future plan is to work on different firewall rule sequences and find out the number of test cases necessary to test those firewalls. Moreover in the future, test case generation approach will be extended by including knowledge of the protected network, by generating test cases using recorded real network traffic, by integration of penetration testing goals (e.g. checking for vulnerabilities, spoofing, etc.), and by considering stateful connections.

References

1. Heidi, H. (2004). Specification Based Firewall Testing, Master of Arts Thesis, Texas State University-San Marcos, May 2004.
2. Zaugg, G. (2005). Firewall Testing, Diploma Thesis, ETH Zürich, 2005.
3. Pressman, R.S. (1994). Software Engineering: A Practitioner's Approach, European 3Rev.ed., 1994, UK.
4. Al-Shaer, E. & Hamed, H. (2004). Discovery of policy anomalies in distributed firewalls. In IEEE INFOCOM'04, March 2004.
5. Burnstein, I. (2003). Practical Software Testing: a Process Oriented Approach, Springer-Verlag, New York, 2003.
6. Myers, G. J. (2004). The Art of Software Testing, John Wiley & Sons Inc., 2004.

Performance Study of Secure IEEE 802.11g Networks

Zeynep Gurkas Aydin, A.Halim Zaim, M.Ali Aydin

Department of Computer Engineering, Faculty of Engineering
Istanbul University, Avcilar, Istanbul, Turkey
{zeynepg, ahzaim, aydinali}@istanbul.edu.tr

Abstract. Wireless Local Area Networks are rapidly growing. They also continue to gain popularity in business and computer industry. This increased productivity and growing popularity caused many improvements on both physical configuration and security mechanisms. In this study we investigated the experimental performance of different security mechanisms on IEEE 802.11g networks with multiple clients.

1 Introduction

Wireless local area networks (WLANs) are of great importance in network technologies. WLAN systems like IEEE 802.11 networks became common access networks in private and public environments. Unlike a traditional wired LAN, users have much more freedom for accessing the network. Such benefits also come with several security considerations. Security risks in wireless environments include risks of wired networks plus the new risks as a result of mobility. To reduce these risks and protect the users from eavesdropping, organizations have been adopted several security mechanisms [1-4].

The traditional WLAN security mechanism is Wireless Equivalent Privacy (WEP). WEP is an encryption scheme designed in 1999 along with 802.11b standard to provide wireless security. It employs RC4 (Rivest Cipher 4) algorithm from RSA Data Security [4]. However, several serious weaknesses were identified by cryptanalysts and WEP was superseded by Wi-Fi Protected Access (WPA) in 2003, and then by the full IEEE 802.11i standard (also known as WPA2) in 2004 [5]. Despite the serious security flaws WEP still provides a minimal level of security [2].

In this paper we investigated the performance of the implemented IEEE 802.11g networks with multiple clients for different security mechanisms available for WLANs. We focused on different types of traffic and packet lengths and different bandwidth issues such as the ones mentioned in [7] and extended the results to 802.11g networks.

2   Wireless LAN Security Issues

Number of committees have made an effort including the IEEE's 802.11 Task Group i (TGi), the Internet Engineering Task Force (IETF) and the National Institute of Standards (NIST) about the security of wireless networks. However, the group that plays the most powerful role in the development of WLAN security standards is the TGi [6]. The security mechanisms for 802.11 wireless networks have been developed in the following chronological order: Wired Equivalent Privacy (WEP), Wi-Fi Protected Access (WPA), IEEE 802.11i (WPA2). WEP employs RC4 algorithm for encryption and uses 64 bit secret key. WEP uses two key sizes: 40 bit and 104 bit; to each is added a 24-bit initialization vector (IV) which is transmitted directly.

2.1   802.1x Authentication

802.1x, port-based network access control makes use of the physical access characteristics of IEEE 802 LAN infrastructures in order to provide a means of authenticating and authorizing devices attached to a LAN. EAP (Extensible Authentication Protocol) is a transport layer protocol designed especially for authentication. EAP-TLS authentication method is widely used method and requires RADIUS (Remote Authentication Dial-In User Service) server to provide necessary options of 802.1x. [8].

3   Experimental Studies

In this study we implemented our 802.11g network in laboratory environment. Our network consists of two clients and an access point (AP) and we investigated the performance of this 802.11g network for different security mechanisms and for TCP and UDP protocols. We generated the network traffic by special tools to provide congested and uncongested networks and collected the desired data with network analyzer programs.

The network which we implemented our tests consist of a server, two clients and an access point. There is 54 Mbps connection between clients and AP and 100 Mbps local area connection between AP and server.

We have chosen the following six different security configurations for our experiments including the basic encryption and authentication mechanisms of wireless LANs and IEEE 802.1x.

1) No Security: The default security setting of wireless networks
2) 64 bit WEP Authentication and WEP Authentication
3) 128 bit WEP Authentication and WEP Authentication
4) EAP-TLS Authentication (802.1x)
5) 64 bit WEP Authentication and EAP-TLS Authentication
6) 128 bit WEP Authentication and EAP-TLS Authentication

3.1 Traffic Generation, Data Collection and Implementation of 802.1x

IP Traffic software was used to generate traffic in our experiments [9]. This tool is able to generate suitable traffic for 802.11g networks using either TCP or UDP protocol, replay the traffic, manage the size of packets and choose the bandwidth of traffic. IP Traffic also supports several connections at the same time. Ethereal network analyzer software was used to collect data from the traffic flow in our experiments [10]. Ethereal works on the server computer and is able to collect required data. To implement the security mechanisms including 802.1x authentication, we used LucidLink RADIUS Server on the server computer to provide EAP-TLS authentication mechanisms [11].

3.2 Performance Measurements and Parameters

Response time and Throughput were measured in our experiments to examine network performance. Response Time is the required total time for the completion of traffic flow between end users. Throughput is the total number of bytes (or kilobytes) transmitted over the traffic during response time.

a.Total Packet Number : We generated 43000 packets (a random value) with IP Traffic software in session .

b.Bandwidth : The maximum bandwidth supported by our 802.11g AP is 54 Mbps. We selected three different bandwidth values for our tests to consider the congestion status of the network. These values are 2000 Kbps and 25000 Kbps for uncongested networks and 55000 Kbps for congested networks.

c.Protocols : Traffic have been generated using both TCP and UDP protocols.

d.Packet Length : Random length packets have been generated to support the network's characteristic like a real network traffic.

4   Results

The performance criteria that we have chosen have been obtained separately for TCP and UDP protocols. All tests have been implemented with one client and two clients respectively and have been repeated five times, the results and statistics have been determined according to the average values. Fig.1 and Fig.2 express the results for the tests on uncongested networks. Total throughput and total response time values of TCP and UDP traffic types against different security configurations are shown in Fig.1. The performance of UDP protocol is greater than TCP protocol because of the connection mechanism TCP uses. On the other hand, as the level of security configurations increase, throughput decreases for both TCP and UDP. The increase of the complexity on each level caused this situation. TCP is significantly slower than UDP for all security configurations. Connection-oriented structure of TCP causes this great difference between response times. Fig.3 and Fig.4 express the results for the tests on congested networks. Total throughput values of TCP and UDP traffic types against different security configurations are shown in Fig.3. The difference between throughput values of TCP and UDP is significant especially for the security configurations that include WEP encryption. In congested networks, unlike TCP, UDP's lack of congestion control cause greater throughput performance than TCP. The performance of fourth security configuration that contains no WEP encryption is better than other security configurations. WEP encryption affect the network performance in a negative way in congested networks for both TCP and UDP traffic types. In Fig.4, for security configurations 2, 3 and 5, 6, total response time of the TCP network is highly different from UDP. As we mentioned above, WEP encryption takes a bit long time in existence of congestion. The security configuration 4 differs from the others in its affects on response time. In the lack of WEP, total time is not so much different from UDP.

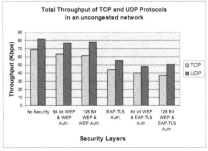

Fig. 1. Total Throughput in uncongested networks

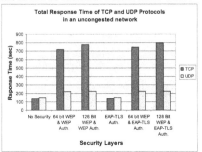

Fig. 2. Total Response Time in uncongested networks

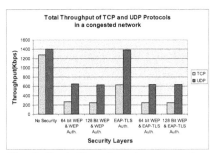

Fig. 3. Total Throughput in congested networks

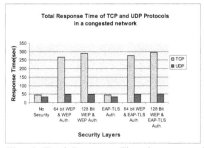

Fig. 4. Total Response Time in congested networks

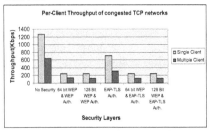

Fig. 5(a). Average per–client throughput of congested TCP networks

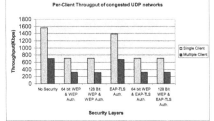

Fig. 5(b). Average per-client throughput of congested UDP networks

All of the tests illustrated in Fig. 5(a) & 5(b) have been implemented for both single and multiple client networks, unlike the previous tests. Fig.5(a) and Fig.5(b) show the difference between TCP and UDP per-client throughput values of single client and multiple client congested networks. As we can see from these graphics, as the number of clients increased, per-client throughput has been decreased due to the congestions. Per-client throughput decreased by 50.55 % for TCP and decreased by 45.6% for UDP in congested networks.

5 Conclusion

In this study, we investigated the experimental performance results of IEEE 802.11g networks for different security configurations and obtained the expected results. [7]

Our results showed that the complexity in security configurations affect the network throughput and response time differently according to the network situation. Security configurations' effects on congested and uncongested networks are different. In congested networks, WEP has decreased the network performance more than 802.1x authentication. TCP is significantly slower than UDP especially in congested networks. UDP has no congestion control mechanisms, so UDP's throughput performance is greater than TCP. Connection-oriented structure of TCP also increases the total response time. The results also showed that the number of clients in a network affect the network performance especially for congested networks. The increase in the number of clients causes an increase in total response time and a decrease in per-client throughput.

Finally, the existence of security absolutely decreases the performance in all types of networks, but this decrease in performance must be ventured to provide the integrity and security of communication through wireless medium.

References

1. GURKAS, G.Z., 2003, A Comparative Analysis of Wireless Security Protocols, Thesis (MS), Computer Engineering Program, Institute of Science, Istanbul University
2. WLANA Organization and Education, What is a Wireless LAN? (online), http://www.wlana.org/learn/educate1.htm
3. POLLIN, D., MAXIM, M., 2002, Wireless Security, RSA Press, McGraw Hill. ISBN 0-07-222286-7
4. WONG, J., 2003, Performance Investigation of Secure 802.11 Wireless LANs: Raising the Security Bar to Which Level? , Thesis (MS), Master of Commerce in Accountancy, Finance, and Information Systems, University of Canterbury
5. RSA Security, http://www.rsasecurity.com/.
6. IEEE 802.11 Task Groups, http://grouper.ieee.org/groups/802/11/
7. IEEE 802.11 wireless LAN security performance using multiple clients, Baghaei, N.; Hunt, R. Networks, 2004 (ICON 2004) Proceedings. 12th IEEE International Conference on Volume 1, 16-19 Nov. 2004 Page(s):299 - 303 vol.1
8. 802.1x Port Based Authentication, http://www.tldp.org/HOWTO/html_single/8021X-HOWTO/
9. IP Traffic, http://www.zti-telecom.com/pages/iptraffic-test-measure.htm.
10. Etheral, http://www.ethereal.com/
11. LucidLink RADIUS Server, http://www.lucidlink.com

Configuration of Microsoft ISA 2004 Server and Linux Squid Server and Evaluation of Some Security Tools

Dr. Hedaya Alasooly[1]

[1] Palestine Ministry of Telecommunication and Information Technology, Government Computer Center, Omar Al-Mokhtar Street, Gaza, Palestine
hasooly@gov.ps

Abstract. The paper concerns about basic Microsoft ISA server and Linux Squid Server configuration. The paper covers also briefly some of the security tools.

Keywords: Microsoft ISA server, Linux Squid Serve, Security Tools.

1 Introduction

The paper concerns about basic Microsoft ISA server and Linux Squid Server configuration. It covers also in brief manner some of the security tools. The paper is composed from three parts

2 Microsoft ISA Server 2004 Configuration

Because this paper is supposed to be short paper, I suggest for complete understanding of Microsoft ISA Server 2004 configuration to refer to my full published paper at ISOneWorld 2007 conference proceeding, which can be found at http://www.isy.vcu.edu/~gdhillon/secconf/secconf07/papers.htm.

All of the network rules and access rules make up the firewall policy. The firewall policy is applied in the following way:
1. A user using a client computer sends a request for a resource located on the Internet.
2. If the request comes from a Firewall Client computer, the user is transparently authenticated using Kerberos or NTLM if domain authentication is configured. If the user cannot be transparently authenticated, ISA Server requests the user credentials. If the user request comes from a Web proxy client, and the access rule requires authentication, ISA Server requests the user credentials. If the user request comes

from a Secure NAT client, the user is not authenticated, but all other network and access rules are still applied.

3. ISA Server checks the network rules to verify that the two networks are connected. If no network relationship is defined between the two networks, the request is refused.

4. If the network rules define a connection between the source and destination networks, ISA Server processes the access rules. The rules are applied in order of priority as listed in the ISA Server Management interface. If an allow rule allows the request, then the request is forwarded without checking any additional access rules. If no access rule allows the request, the final default access rule is applied, which denies all access.

5. If the request is allowed by an access rule, ISA Server checks the network rules again to determine how the networks are connected. ISA Server checks the Web chaining rules (if a Web proxy client requested the object) or the firewall chaining configuration (if a Secure NAT or Firewall Client requested the object) to determine how the request will be serviced.

6. The request is forwarded to the Internet Web server.

There are two types of network relationships:
  a) Route: ISA server routes traffic between network sources and destinations (no network translations are used). Routed relationships are directional.
  b) NAT: ISA server hides the source computers by replacing their network IP address of its outgoing traffic by its external IP address.

All the HTTP requests pass through the web proxy component on ISA server, regardless of the client type. To configure the ISA server as proxy, expand the Configuration, and then Networks, and click the network whose web access properties you would like to configure, this is usually the internal network. Click Edit the selected Network, and in the Web Proxy Tab, ensure that Enable Web Proxy Clients is selected, and you can configure the ISA Server to listen on the HTTP connections on specified port, i.e. 80 or 8080. In Firewall Client Tab, ensure that Enable Firewall client support for this network is selected, and use a Web proxy server is selected, and configure the ISA server name or IP address.

2.1 ISA Server Templates:

For best understanding of the firewall configuration, take a look at the rule base constructed automatically for the ISA 2004 server templates. There are five templates for the ISA Server.
  a) ISA server with single adapter: Specify that the ISA server will be used in a single network adapter configuration inside your internal or perimeter network In this case the ISA server will be used for web proxying, caching, Web publishing and OWA server publishing.
  b) Edge firewall template: ISA server with two adapters, one is connected to internal network and the other to the internet. ISA Server will connect your internal network to internet and protect it from intruders.

c) Three leg perimeter template: Connect your internal network to internet and protect it from intruders, and publishes services in the perimeter network to internet.

d) Front firewall template: Use the ISA server as front line of defense in a back to back perimeter network configuration. Use this configuration when you have two firewalls between the protected internal network and internet.

e) Back firewall template: Use the ISA server as the back line of defense in a back to back perimeter network configuration. Use this option when you have two firewalls between the protected internal network and internet.

You can refer to my published paper at http://www.isy.vcu.edu/~gdhillon/secconf/secconf07/papers.htm for the rule base configuration of each template. I will just show here the configuration of the three leg perimeter template in order to allow limited web access and allow access to network services on perimeter network

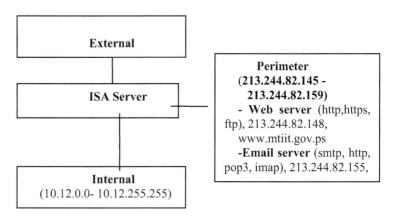

Fig. 1. Three leg perimeter network

1. Allow HTTP, HTTPS, FTP from Internal Network and VPN Clients Network to Perimeter Network and External Network (Internet).
2. Allow DNS traffic from Internal Network and VPN Clients Network to Perimeter Network.
3. Allow all protocols from VPN Clients Network to Internal Network.

Table 1. The rule base to allow limited web access and allow access to network services on perimeter network

Rule Name	From	To	Protocol	Network Relation	Users	Action

Local Host Access	Local host	All networks		Route		
VPN Clients to Internal Network	Quarantined VPN Clients, VPN Clients	Internal		Route		
Internet Access	Quarantined VPN Clients, VPN Clients, Internal	External		NAT		
Perimeter Configuration	Quarantined VPN Clients, VPN Clients, Internal	Perimeter		NAT		
Perimeter Access	Perimeter	External		Route		
Web Access Only	Internal, VPN Clients	External, Perimeter	Ftp, Http, Https		All	Allow
VPN Clients to Internal Network	VPN Clients	Internal	All Outbound Traffic		All	Allow
Allow DNS to perimeter	Internal, VPN Clients	Perimeter	DNS		All	Allow
Default Rule	All Networks	All Networks	All Traffic		All	Reject

To publish the web site www.mtiit.gov.ps in the web server 213.244.82.148, a web server publishing rule must be created which will cause the following in the rule base:

Table 2. The rule base caused by the web server publish rule

Rule Name	Public Name	From	To	Traffic	Listener	Action
Web server publish rule	Request to www.mtit.gov .ps	Anywhere	Net Address of the server to publish 213.244.8 2.148	http	External interface: 80	Allow

To publish the mail server 213.244.82.155, a mail server publishing rule must be created which will cause the following in the rule base:

Table 3. The rule base caused by the mail server publish rule

Rule Name	From	To	Traffic	Network	Action
SMTP Publish rule	Anywhere	Net address of the server to publish 213.244.82.155	SMTP Server	External interface:	Allow
POP3 publish rule	Any where	213.244.82.155	Pop3 Server	External interface	Allow
IMAP4 publish rule	Any where	213.244.82.155	IMAP4 Server	External interface	Allow
Mail Exchange RPC Server	Any where	213.244.82.155	Exchange RPC Server	External interface	Allow
SMTP publish rule	Any where	213.244.82.155	SMTP Server	External interface	Allow
NNTP publish rule	Any where	213.244.82.155	NNTP Server	External interface	Allow
IMAP4 publish rule	Any where	213.244.82.155	IMAP4 Server	External interface	Allow
For exchange, HTTP publish rule, path: - /exchange/* - /exchweb/*/OMA/* -/public/* -/Microsoft-Server-ActiveSync/*	Any where	213.244.82.155 Public name: webmail.gov.ps	HTTP Server	External interface: 80	

If the ISA server shall forward the requests to upstream server, then a web chaining rule will be required to be created.

2.2 Intrusion Detection

By default, the ISA Server is configured with most of the intrusion detection options already enabled. In ISA Server Management, click General, then click Enable Intrusion Detection and DNS Attack detection, then select the attack options that you wish to enable. The attack options: Windows out-of-band attack, Land attack, Ping of death attack, Port scan, IP half scan, UDP bomb, DNS attacks (DNS hostname overflow, DNS length overflow, DNS zone transfer).

2.3 Application and Web Filtering

HTTP filters are applied on a per-rule basis. To configure the HTTP filters on particular access rule or Web publishing rule, modify the HTTP policy for that rule. Right click the rule and choose configure http. You can configure ISA to allow or deny requests based on General Properties, HTTP Method, Extensions, HTTP Headers, and HTTP Signatures.

There are also filters for other types of applications, i.e.

- DNS filter: Allows organization to screen incoming DNS communications for malicious commands and data before they reach to DNS server.
- FTP filter: Enables FTP protocols (client and server)
- MMS filter: Enables Microsoft Media Streaming protocol
- PNMP filter: Enables Real Networks Streaming Media protocol
- POP intrusion detection filter: Protects POP email servers from buffer overflow attacks.
- PPTP filter: Enables PPTP tunneling through ISA Server
- PRC filter: Enables publishing of RPC servers
- RTSP filter: Enables Real Time Streaming Protocol
- H.323 filter: Enables H.323 protocol
- Socks version 4 filter: Enables SOCKS 4 communication
- SMTP filter: Uses content inspection to examine SMTP commands and ensure that they are not harmful to organization email. Filters SMTP traffic based on keywords, users/domains, attachments, SMTP commands.

3 Linux Squid Configurations

Because this paper is supposed to be short paper, I suggest for complete understanding of Squid Server Configuration is to refer to my full published paper at ISOneWorld 2007 conference proceeding, which can be found at http://www.isy.vcu.edu/~gdhillon/secconf/secconf07/papers.htm. I used webmin for Squid server configuration to make things easy. Configuring Squid as a proxy server lets you access HTTP, FTP, Gopher, and Wide Area Information Server (WAIS) sites. You can also set up Squid to block certain types of traffic. After you configure your Squid cache in Squid Webmin module, you need to set the ports and IP addresses on which you want Squid to listen for proxy requests.

To specify the ports that Squid uses, click *Ports and Networking* on the Squid module's main page. In the *Proxy Addresses and Ports* table, enter a port number such as 8080 in the first empty field in the Port column. In the Hostname/IP address column, either select All to accept connections on any of your system's interfaces or select the second option to enter an IP address in the adjacent text box. Squid uses also the Inter-Cache Communication Protocol (ICP) to communicate with other proxies in a cluster. The default is 3130 for ICP.

In my test environment, I set the proxy IP address to 10.12.1.149 and the proxy port to 8080 in Webmin Squid module. Configuring these settings made the following changes in the Squid configuration file (i.e., /etc/squid/squid.conf):

http_port 10.12.1.149:8080
icp_port 3130

Next, you must create ACLs that you'll use to create proxy restrictions. An ACL basically serves as a test that you apply to a client request to see whether it matches. Then, based on the ACLs that each request matches, you can choose to block the request, prevent caching, force the request into a delay pool, or hand the request off to another proxy server. Many types of ACLs exist—for example, one type checks a client's IP address; another matches the URL being requested; and others check the destination port, web server hostname, authenticated user, and so on. Squid has many ACL types, although not all of them are available in all the versions of Squid. To create an ACL, select Access Control List from the Webmin Squid module. The table in my paper in http://www.isy.vcu.edu/~gdhillon/secconf/secconf07/papers.htm   lists the ACL types you can create for Squid. For testing, I created the ACLs that Table 4 lists. There are also some predefined ACLS

Table 4. The added ACLS

ACL Name	ACL Type	Included Elements	
Denyed-sites	Web Server Hostname	.msn.com, .download.com	.webmin.com,
FTP	URL Protocol	ftp	
Internal	Client IP Address	10.12.0.0 through 10.12.255.254	

After you generate some ACLs, you can use them to create, edit, and move proxy restrictions. Squid includes several predefined proxy restrictions by default. Squid compares each request to all the defined proxy restrictions in order and stops when it finds a restriction that matches the request. The action you previously set for the restriction determines whether Squid allows or denies the request. To create a proxy restriction, click the Access Control icon on the Webmin Squid module's main page. Next, click *Add proxy restriction* below the list of existing restrictions to go to the creation form. In the Action field, select either Allow or Deny depending on whether you want Squid to process matching requests. You can use the *Match ACLs* list to trigger an action if a request matches the ACLs that you select in the list. The *Don't match ACLs* list lets you select ACLs that must *not* match for an action to trigger. In addition, you can make selections from both lists to indicate that an action should occur only if all the ACLs on the *Match ACLs* list match and none of the ACLs on the *Don't match ACLs* list match. By default, Squid has an ACL named All that matches all requests. You can use this ACL to create restrictions that allow or deny everyone. Use the arrows in the proxy restrictions table to move a restriction to the correct location. Ensure that the *Deny all* entry is on the bottom so that the restriction you just set will have an effect. For testing purposes, I created a proxy restriction that gives access to internal network clients, prevents access to some Web sites (denyed-sites: .webmin.com, .msn.com, .download.com), and prevents access to FTP sites. I set the action to Allow. From the *Match ACLs* list, I selected internal ACL; from the *Don't match ACLs* list, I selected ftp and denyed-sites ACLs. Then, I moved my ACL above the last default *Deny all* rule. There are also some predefined restrictions. These proxy restrictions caused several changes to the Squid configuration file (i.e., /etc/squid/squid.conf).

```
acl all src 0.0.0.0/0.0.0.0
acl manager proto cache_object
acl localhost src 127.0.0.1/255.255.255.255
acl to_localhost dst 127.0.0.0/8
acl SSL_ports port 443 563
acl Safe_ports port 80 # http
acl Safe_ports port 21 # ftp
acl Safe_ports port 443 563 # https, snews
acl Safe_ports port 70 # gopher
acl Safe_ports port 210 # wais
acl Safe_ports port 1025-65535 # unregistered ports
acl Safe_ports port 280 # http-mgmt
acl Safe_ports port 488 # gss-http
acl Safe_ports port 591 # filemaker
acl Safe_ports port 777 # multiling http
acl CONNECT method CONNECT
acl internal src 10.12.0.0-10.12.255.254
acl denyed-sites dstdomain .download.com .msn.com .webmin.com
acl ftp proto ftp
http_access allow manager localhost
http_access deny manager
http_access deny !Safe_ports
http_access deny CONNECT !SSL_ports
http_access allow localhost
http_access allow internal !denyed-sites !ftp
http_access deny all
```

To configure your server to use another proxy for requests other than those to a certain network or domain, go to the Squid module's main page in Webmin and click the Access Control icon. Create a Web Server Hostname or Web Server Address ACL that matches the Web servers you want your proxy to fetch directly. Give the ACL a descriptive name (e.g., Direct). Go back to the main page and click Other Caches. Click *Add another cache* to go to the cache host creation form. In the Hostname field, enter the master cache server's fully qualified host name (e.g., bigproxy.example.com). Select *parent* from the Type menu to indicate to Squid that the other proxy is at a higher level (and thus has more cached pages) than the original proxy. In the *Proxy port* field, enter a port number that the other proxy is listening on, such as 8080. In the *ICP port* field, enter the port that the proxy uses for ICP requests, which is typically 3130. Click Save at the bottom of the page to return to the list of other caches. At the bottom of the page that opens is a form that contains a section called *ACLs to fetch directly*. This section is actually an ACL table similar to the proxy restrictions table that I mentioned previously. Use this section's *Add ACLs to fetch directly* link to first add an entry to allow your *Direct* ACL. Then, add an entry to deny the All ACL. These settings tell Squid to directly fetch pages from local Web servers but to pass all other requests on to the parent proxy. You can configure Squid to contact other caches in a cluster for each request and check if a page is already cached. If a page is already cached, Squid retrieves the page from the other proxy instead of from the originating Web server. To configure multiple proxies to use ICP to communicate, click the Other Caches icon on the module's main page. Next, select *Add another cache* to open the cache host creation form. In the Hostname field, enter the full host name of one of the other caches. From the Type menu, select *sibling* to indicate that the other cache is at the same level as the original cache. In the *Proxy port* field, enter the HTTP port on which the other proxy listens. In the *ICP port* field, enter the port number that the other proxy uses for ICP (usually 3130). Click Save to add the other proxy and to return to the Other Caches list. Repeat these steps for each host in the cluster. Finally, click Apply Changes at the top of the page.

I set up my Squid server to consider the computer center proxy server 10.12.0.32:80 as its parent proxy server. My configuration caused the following change in the Squid configuration file (i.e., /etc/squid/squid.conf):
cache_peer 10.12.0.32 parent 80 3130

After you configure Squid, you can start (or restart) Squid from the command line.
service squid restart

4   Evaluation of Some Security Tools

Top security tools can be found in http://sectools.org/. Most important vulnerabilities in Windows and Linux can be found in www.sans.org/top20/. There is a good course that covers most of the hacking and security issues, which is the "Certified Ethical Hacking" course. I published one book named " Quick Evaluation of Some Security Tools: GFI LANguard, Nessus, Snort, Base, Rman, OSSEC, Sguil, SnortCenter " at http://www.lulu.com/content/821501.

Generally the security tools are divided to vulnerability scanners and intrusion detection tools.

4.1   Vulnerability Scanners

Nessus is one of the best free network vulnerability scanner available, and can run on UNIX and Windows. It is constantly updated, with more than 11,000 plugins for the free. You can also create users with Nessus and use NessusWX Client as its output is more orginized.

GFI LANguard scans IP networks to detect what machines are running. Then it tries to discern the host OS and what applications are running. It also tries to collect

Windows machine's service pack level, missing security patches, wireless access points, USB devices, open shares, open ports, services/applications active on the computer, key registry entries, weak passwords, users and groups, and more. Scan results are saved to an HTML report, which can be customized/queried. It also includes a patch manager which detects and installs missing patches.

I tried to test both of the tools when scanning a Windows and Linux machines. It seems that both tools provides some special type of information, so I don't want to tell which tool is better. I just advice to try both tools when checking the vulnerabilities. You cant scan a machines behind firewall when using GFI LANguard, but Nessus can do. GFI LANguard requires administrative privileges to give the best results. Please refer to my book at Lulu.com http://www.lulu.com/content/821501 for installation guide and to have a look at the scan results for both security scanners. Other security scanners that can be tried also: Tenable NeWT , Shadow Security Scanner, and Microsoft Baseline Security Analyzer.

4.2 Intrusion Detection Tools

Under windows, I tested the BlackICE for intrusion detection and prventation. You   can   download   the   evaluation   version   from www.iss.net/issEn/DLC/blackiceevaluation.jhtml.

Snort is lightweight network intrusion detection and prevention system excels at traffic analysis and packet logging on IP networks. Through protocol analysis, content searching, and various pre-processors, Snort detects thousands of worms, vulnerability exploit attempts, port scans, and other suspicious behavior. Snort uses a flexible rule-based language to describe traffic that it should collect or pass, and a modular detection engine.

You can use BASE or ACID as web based front end for snort alert analysis. They are a PHP-based analysis engine to search and process a database of security events generated by Snort. Their features include a query-builder and search interface for finding alerts matching different patterns, a packet viewer/decoder, and charts and statistics based on time, sensor, signature, protocol, IP address, etc.

You can use SnortCenter for "Snort IDS Rule & Sensor Management". It is a web based front end multi-sensor management and analysis console. Rman is another tool that can be used for the same purpose. I just felt that Rman is easier to be used than SnortCenter.

OSSEC HIDS is an Open Source Host-based Intrusion Detection System. It performs log analysis, integrity checking, rootkit detection, time-based alerting and active response. If you have one system to monitor, you can install the OSSEC HIDS locally on that box and do everything from there. However, if you are administering a few systems, you can select one to be your OSSEC server and the others to be OSSEC agents, forwarding events to the server for analysis. You can use /ossec-wui as a web interface for OSSEC.

Sguil (pronounced sgweel) is built by network security analysts for network security analysts. Sguil's main component is an intuitive GUI that provides real-time events from Snort/barnyard. It also includes other components which facilitate the practice of Network Security Monitoring and event driven analysis of IDS alerts. The Sguil client is written in tcl/tk and can be run on any operating system that supports tcl/tk (including Linux, *BSD, Solaris, MacOS, and Win32).   Sguil should be installed in the following order:

Step 1: Install mysql and create the sguil database.
Step 2: Install the GUI server (sguild).
Step 3: Install the GUI client (sguil.tk).
Step 4: Install the sensor.

My special note is that installing Sguil sensor is not easy, as you should patch and install both Snort and Barnyard packages, and then start them, but I was getting errors through my work.

Please refer to my book at Lulu.com http://www.lulu.com/content/821501 for complete installation guide of Snort, BASE, ACID, SnortCenter, Rman, OSSEC, Sguil.

5   Conclusion

The paper talked about basic Microsoft ISA server and Linux Squid Server configuration My conclusion is that, the ISA server can be implemented as a firewall and proxy with wide range of topologies, application filters and intrusion detections. Also the ISA server supports many server publishing scenarios. Squid server can be implemented as proxy server to access http, ftp, gopher and wais sites, although it can block some type of traffics according to the given restriction rules. The paper covered also in brief manner some of the security tools.

References

1. http://www.webmin.com web site.
2. http://www.microsoft.com web site.
3. http:// www.squid-cache.org website.
4. http://sectools.org/ web site.

Personal Data Protection in Turkey: Technical and Managerial Controls

Yalcin Cebi[1], Osman Okyar Tahaoglu[2]

[1]Dokuz Eylul University, Computer Engineering Dept. Kaynaklar Campus Buca, Izmir, Turkey
[2]Global Bilgi Inc., Fatih Avenue. Dereboyu Str. No:8 Halkali, Istanbul, Turkey
[1] yalcin@cs.deu.edu.tr, [2] okyar.tahaoglu@global-bilgi.com.tr

Abstract. In the European Union and some other countries legal directives are in progress which consist of regulations about personal data protection, processing such personal data resulting with any expression of opinion about the individual. Also in Turkey a draft Personal Data Protection Act which was prepared in 2003 and waiting to be approved includes new practices for individuals, public bodies and corporates. During the identification of these practices and compliance gaps, how much the bodies are ready for this act, how will the requirements of this act be supplied by the bodies, and the differences between sectoral practices must be investigated for Turkey. This paper consists of the new applications which individuals, public bodies and corporates will face as a result of the draft act, the compliance with the current international standards from a general perspective, ways in which the system security countermeasures may ensure the act's requirements.

Keywords: Personal data protection, data security legislation, information security.

1 Introduction

A survey done by the Council of Europe in 2003 shows that 70% of the European population has no information about that is being done in their own countries to protect personal data [1]. In consequence of this, the Council of Europe has decided to celebrate the 28th of January as *Data Protection Day* in order to raise the awareness of individuals [2].

The need for data protection has risen to protect the individual's rights and human rights against governments, and data transfers between countries and territories. Data sharing and transmission is now fast and easy because of new developing technologies however this also causes easy sharing of private data for improper purposes and by inadequate methods. As a result of increasing computer crimes, the Turkish National Police have constructed a special unit for Internet and Cyber Crimes in 2002.

Privacy is a fundamental human right that must be protected as a primary defense. Personal data can be defined as all of the information that can express any opinion about an individual or corporate. Privacy in the context of personal data can be divided into the following separate but related concepts: [3]

Information privacy, which involves the establishment of rules governing the collection and handling of personal data such as credit information, and medical and government records. It is also known as "data protection";

Bodily privacy, which concerns the protection of people's physical selves against invasive procedures such as genetic tests, drug testing and cavity searches;

Privacy of communications, which covers the security and privacy of mail, telephones, e-mail and other forms of communication; and

Territorial privacy, which concerns the setting of limits on intrusion into domestic and other environments such as the workplace or public space.

It is obvious that corporations and several parties process data for several operational purposes and with request of the data owner. At the end of such process, data can be indexed, classified, stored, transferred or made anonymous. Making data anonymous is formatting the data so that the output information cannot pin-point the individual (data subject), cannot be associated with the data subject directly or indirectly and the source of raw data cannot be identified. Permission for processing data can only be given to the authorized Database System Controller (DBSC) who collects the data from the data subject. The DBSC is the individual or corporate party which has taken permission from the data subject to process the data in a relevant filling system for pre-defined purposes and by pre-defined methods. DBSC is the competent authority to specify the processing methods and can outsource the processing to a Database System Controller Representative (DBSR) agent.

The Personal Data Protection Act in draft is still being prepared in Turkey [4]. The draft includes a series of rules about data privacy, data security and protection. In this Act sensitive personal data means personal data consisting of information as to name, surname, the racial or ethnic origin of the data subject, his political opinions, his physical or mental health or condition, his sexual life and financial profile.

If this information can be used to identify the individual owner or the corporate owner then it can be defined, as the data, and the data subject is associated. If the data is associated with a corporate it may be possible to identify the managers or that corporate and therefore it will also be possible to associate an individual. Therefore the act covers the persons and corporations.

2 Data Protection Legislation

2.1 Data Protection in Europe

The genesis of modern legislation in this area can be traced to the first data protection law in the world enacted in the Land of Hesse in Germany in 1970. The Council of Europe's 1981 *"Convention for the Protection of Individuals with regard to the Automatic Processing of Personal Data"* is a reference for today's data protection legislation. The national laws of the member countries have been developed following this directive [5].

Other current directives force the member states of the European Union (EU) to prepare and deploy their own data protection laws. These legislations aim to keep the security level of data used and shared among the states for commercial, legislative and social objects. The baseline of the security level is set by the Directive 108. Each country must look for a data protection act from the other member while sharing personal data. If the DBSR is located in a member country but the DBSC is out of the borders of EU, this condition is also required [6].

EU and Council of Europe have supported the Directive 108, by enacting several regulations for telecommunication, private and public sectors.

Especially the data that falls in the definition "sensitive personal data" is identified to give a direction for the members. The United Kingdom(UK) Data Protection Act defines the sensitive data as; (a) the racial or ethnic origin of the data subject, (b) his political opinions, (c) his religious beliefs or other beliefs of a similar nature, (d) whether he is a member of a trade union, (e) his physical or mental health or condition, (f) his sexual life, (g) the commission or alleged commission by him of any offence, or (h) any proceedings for any offence committed or alleged to have been committed by him, the disposal of such proceedings or the sentence of any court in such proceedings [7].

2.2 Data Protection Authorities

Each data protection act includes a statement for establishing a regulator authority. This authority is responsible for building an infrastructure to make this act possible by preparing the supporting regulations, registry system and the audit mechanism. Each EU member gives different names for this authority like, regulator, commissioner, supervisor or commissioner. For example EU, UK and Greece call their central authorities European Data Protection Supervisor, The Information Commissioner, and Data Protection Authority respectively.

The United States of America (USA) does not have an act directly for data protection and does not have an authority at present. Privacy Rights Clearinghouse (PRC), a non profit organization located in California, San Diego, reports the chronology of data losses and identity thefts according in USA annually. According to the 2006 reports, 327 events took place where a hundred records have been affected[8]. There is no such organization in EU that takes records of incidents. The percentage of data loss and vandalism events reported from the private sector is 40% for notebook theft, 20% personnel errors and software malfunctions. Personnel errors and software malfunctions take the first order in public sector with a 44% ratio and computer theft follows by 21%.

The absence of a data protection act in the USA should be taken into account and the relation of existence of such an act and the incidents occurred should be investigated. One of the duties of the regulator bodies should be to investigate this correlation.

3 Data Protection in Turkish Laws and Regulations

3.1 General Situation

Turkey has signed the Directive 108 in the same year it has been approved by the EU. Every Turkish citizen has rights protected by the organic law about protection of private and family life [9]. Article 20 of the organic law identifies the immunity of personal life with its exceptions for national security while the articles 21 and 22 describe the immunity of houses and freedom of communication. These articles cover the main idea of Directive 108. Some regulations and rules currently in progress in Turkey and Europe are listed in Table 1. Details and history of regulations of each EU member country can be reached from the official web page of the Council of Europe [5].

Table 1. Data Protection Laws and Regulations.

Regulation	Type/Article	Definition
Organic Law [9]	Article 20-22	Privacy of private life and freedom of communication.
Turkish Penal Law [10]	Article 132-140	Privacy of private life, recording, sharing and deleting personal information.
Convention for the Protection of Human Rights and Fundamental Freedoms [11]	Article 8	Protection of family right.
Personal Information Processing and Protection of Privacy in The Telecommunications Sector [12]	Regulation	
Council of Europe Directive No: 108 [2]	Directive	Protection of Individuals with regard to the Automatic Processing of Personal Data
UK Data Protection Act [7]	Act, 1998 Chapter 29	Data Privacy Act 1998.

Turkey while walking through the information age continues to develop new regulating infrastructures to protect fundamental rights, to penalize the nonconformities for every special sector. Especially the cases and events that took place in financial sector, private sector and in courts have taken on a role for the revisions and updates in the penal laws to prevent such event reoccurring. These different updates cover the privacy of communication, recording private data and publishing it. These developments concern not only individuals but associations, institutes and corporations also. Corporations and the legal sector will be required to review their organizations and applications and deploy additional countermeasures to comply with the new coming rules.

3.2 Draft Law on the Protection of Personal Data

The draft "Personal Data Protection Law" has not yet become law. The draft law is a complement of the legislations indicated above. The draft is a regulation that draws the boundaries of usage and processing practices of data.

The draft law consists of five parts and 14 chapters. The first part describes the boundaries of objective, scope, definitions and processing of data with adequate and acceptable purposes for data processing. Second part includes the article about the rights of the data subject, necessary controls for processing the data and data transit to the third parties in or out of the borders. The third part includes the registry of DBSC to a system managed by the Personal Data Protection Authority (PDPA) and audit methods.

The organization and responsibilities of the PDPA and the relationship with the data controller are defined in the fourth and the fifth parts.

3.2.1 Purpose and Convenience for Processing of Data

The article no. 4 of the act determines that the DBSC is required to inform the data subject during the collection of the information about his purposes. DBSC must also have an authorization from the data subject. The statements used in such contract must be definitive and clear. The context of the gathered data must be sufficient and well proportioned with the service taken by the data subject from the DBSC. In accordance with this definition, a merchant requesting the home phone number or e-mail from his customer during a purchase may be discussed as an "insufficient data" for the service. Similarly, a bank's credit customer may be asked to give extra information other than financial and guarantees that may not be necessary also.

Article no. 5 describes the suitable and appropriate legal circumstances of data processing. An articulate allowance and agreement is required at this point. The way of declaring and approval of such an agreement includes detailed legal definitions and is out of the scope of this paper. However it is clear that there must be a statement describing "the purpose of data processing, and the identities of the responsible data controllers" [13] in the agreement. In addition to this, an opportunity must be given to the data subject to choose to agree or disagree.

3.2.2 Scope for Individuals and Corporations

According to the act personal data is the all data that can be associated with an individual or corporate. Therefore each case of nonconformity will have different results and effects. A disclosure of an individual's data will be a privacy problem where a disclosure of a company data may result in financial loss and image loss.

3.2.3 Duties and Responsibilities of the Both Parties

The data subject has the right to know whether the DBSC has a record about him or not. In practice a customer of a merchant will be able to reach the details of the database of the store or the chain store and will be able to request to update or delete it. And the merchant will have to process this request except for the record that has to keep according to the labor laws. Merchant will establish a channel to accept the requests and will announce and operate the channel free of charge or for an acceptable fee.

3.2.4 Data Controller's Duty and Information Security

DBSC is responsible for deploying managerial and technical controls to protect personal data, process it with privacy and integrity, protect it away from misuse, unauthorized alteration, deletion and modification. The controls will be relevant technical and managerial countermeasures.

The relevant controls indicate computing a cost-benefit analysis, handling the data as an asset and analyzing the risk associated with this asset and ensuring the baseline of security level. The actions and countermeasures described in the act include parallel descriptions with the international standards *"ISO17799:2005 Information Technology-Code of Practice for Information Security Management"* and *"ISO27001:2005 Information Security Management System"*.

The definition of information security in the introduction of ISO27001 is; information security is achieved by implementing a suitable set of controls, including policies, processes, procedures, organizational structures and software and hardware functions. These controls need to be established, implemented, monitored, reviewed and improved, where necessary, to ensure that the specific security and business objectives of the organization are met. This should be done in conjunction with other business management processes [14].

The essence of information security management system is based on risk management. The assessment of data, its context and importance for the corporation, the vulnerability analysis, identifying the effect and probability of a compromise on the data, measuring and controlling the risk, all include managerial and technical countermeasures. As a result of the assessment, the controls will be chosen from a set of practices, technologies and procedures. Some can be classified as;

- Protection of data from unauthorized access,
- Training personnel responsible for the processing of private data,
- Guaranteeing the secrecy of data and the responsibilities by signed contracts in the case of an outsource of operations or sharing of data with third parties,
- Other technical and managerial controls.

Assuming that ISO27001 standard will be enough to comply with the act may be wrong. But this internationally accepted standard should be analyzed. Since risk management forms the framework of the standard, corporations must prepare a scope within their main business activity and aim to protect the data used in this scope. The applied countermeasures will increase the security level up to a certain point. The act also estimates similar solutions as the international standard and foresees the application of countermeasures by taking into account current technology and costs. Therefore if the corporations establish a security system and develop their systems by referencing the standard, they will make a progress to comply. On the other side, each corporation will decide on a different acceptable level of risk and this situation may bring new concurrency problems.

3.2.5 The Authority in Turkey
According to the act an authority called Personal Data Protection Authority will be established. PDPA will be responsible for operating registry system and auditing the DBSC. PDPA will force the DBSC to register to the inventory of data controllers. PDPA will prepare the regulations to guide the rules of the registry system, the attributes of the inventory. According to the EU directives; the registry inventory includes the name and surname of the DBSC and DBSR, purpose for the data processing, the authorized parties, plans and the measures taken for the transmission of the data to the third countries [15].

3.2.6 Complaints and Public Bodies
According to the act the PDPA will start an investigation in the case of a complaint of a data subject. PDPA requests the data controller to take action to change the applications and comply with the act immediately but if the data controller is a public body a 30 day period is given to improve the practices. This is a reactive approach but information security needs more proactive approaches.

4 Conclusion and Future Work

Studies on the Turkish Personal Data Protection Draft Act that includes rules and boundaries of private data and data security still continue. The scope of the act includes the protection of data owned by individuals and corporations which aims to audit the attempts to reach the origin of data subject from an anonymous data. The draft act in progress determines new regulations to the public and private sector to secure their systems with new practices.

The act will aim to protect the subject data in a boundary but it does not give any suggestion or direction for the open databases which can be used together and to reach the subject data. Open databases can be associated and used to find the source of information. Several vulnerabilities may arise and these must be investigated. In order to solve these possible problems our recommendations are;

- Corporations must cooperate,
- An authority must be responsible for monitoring the open databases and services and computing the risk level of the databases,
- Data sharing and transit between corporations must be monitored.

A code of ethics for every professional body must be prepared to take into account the possibility of misuse of information of information technology staff, experts, consultant and auditor. The existence of data protection legislation and the cases of data theft must be explored. A risk management framework can be modeled in a national wide scope. Different approaches are necessary for every sector like insurance, finance, banking, telecommunication, health, etc. We will work on a benchmark model for the current compliance status and the practices of these sectors with the draft act and will endeavor to suggest a model.

References

1. European Opinion Research Group, "Data Protection", *Eurobarometer Survey on the Protection of Privacy,* (2003).
2. Council of Europe, *Data Protection Web Page,*
   <http://www.coe.int/T/E/Legal_affairs/Legal_co-operation/Data_protection> (Feb,17 2007).
3. Privacy International, "Overview of Privacy", *Privacy and Human Rights 2005 Report, 29 Şubat 2006,* <http://www.privacyinternational.org> (Jan 26, 2007).
4. Ministry of Justice, "Current Draft Acts to be approed", Official Web Page, Nov. 9 2005, <http://www.kgm.adalet.gov.tr/basbakanliktabulunanlar.htm> (Feb 22, 2007).
5. Council of Europe Data Protection Web Page, "National Laws", <http://www.coe.int/t/e/legal_affairs/legal_co-operation/data_protection/documents/national_laws/nationallaws_en.asp> (Jan 10, 2007).
6. HUET, J., Study contracts involving the transfer of personal data between Parties to Convention Ets 108 and third countries not providing an adequate level of protection, (2001).
7. UK Data Protection Act 1998, *Sensitive Personal Data,* Chapter 29, Part I, Section 2, (1998).
8. Privacy Rights Clearing House, *Chronology of Data Breaches 2006:Analysis,* 1 Şubat 2007, <http://www.privacyrights.org/ar/DataBreaches2006-Analysis.htm> (26 Şubat 2007).
9. Grand National Assembly of Turkey, Organic Law of Turkish Republic, Nov 7, 1982, <http://www.tbmm.gov.tr/Anayasa.htm> (Feb 10, 2007).
10. Grand National Assembly of Turkey, *Turkish Penal Law*, Sep 26, 2004, Approved: Apr 1, 2005, <http://www.tbmm.gov.tr/kanunlar/k5237.html> (Feb 14, 2007).
11. Council of Europe, *Convention for the Protection of Human Rights and Fundamental Freedoms,* Nov 4, 1950, <http://www.avrupakonseyi.org.tr> (Feb 28, 2007)
12. Turkish Telecommunication Authority, "Ordinance on Personal Information Processing and Protection of Privacy in The Telecommunications Sector", Laws and Regulations, Feb 6, 2004, <http://www.tk.gov.tr/Duzenlemeler/Hukuki/yonetmelikler/Yonetmelikler.htm> (Feb 28, 2007)
13. Başalp, N., Kişisel Verilerin Korunması ve Saklanması, Yetkin Yayınevi, 2004, ISBN: 975-464-301-6 *(9754643016).*
14. International Standard ISO17799:2005 Code of Practice for Information Security Management, (2005) viii.
15. Bilişim Şurası Hukuk Çalışma Grubu, Türkiye 2. Bilişim Şurası (2004) 129.

Integrating Security Assurance within the (Rational) Unified Process

Mohammad Reza Razzazi[1], Yashar Heydari[1]

[1]Computer Engineering Department, Amirkabir University of Technology,
242 Hafez Ave., Tehran, Iran.
{Razzazi, Heydari}@ce.aut.ac.ir

Abstract. Security is turning into one of the most important issues in software and information systems. Developing security-critical solutions and assuring and certifying them are gaining a vast concern and importance in the software industry. In this paper we investigate the integration of software Security Assurance activities (as defined by the Common Criteria for Information Technology Security Evaluation, 1999 ed.) within the Unified Process. We compare and contrast these two frameworks, and identify the aspects of security assurance activities and disciplines that are directly supported, partially supported or not supported by the Unified Process.

Keywords: Unified Process, Security, Assurance, Software Development Process, Security Assurance, Common Criteria.

1 Introduction

Over the past few years, agile methods for software development have captured much attention. Agile methods require that software development be iterative and incremental, which is not a new idea [7], in order to achieve effective communication and feedback. An increasingly popular framework for iterative and incremental development is known as the Unified Process (UP) [2].

Agile methods strive to achieve a quicker development of software at the expense of increased, yet acceptable security and general risks. For safety and security critical software, however, risk mitigation is a very serious concern. Many security assurance methods have evolved over the years to ascertain software projects being flawless and secure. Security assurance methods are a set of guidelines and activities that lead the developer throughout the development lifecycle to achieve a high-quality, low risk software with certain security aspects being developed.

Common Criteria is a recognized and utilized standard for Security Evaluation and Assurance [1]. Almost all of U.S. government projects require to be certified by CC organization, in which, an organization independent from the one developing the software, performs the evaluation and certification activities during and after the software development lifecycle. If the developing organization were using the UP, how would the CC assurance activities integrate into the overall process? This is what we are trying to address throughout this article.

In this article, we explore the use of the UP in the context of a project whose software lifecycle specifically includes CC assurance activities according to ISO/IEC Std. 15408. First we discuss the background of the ISO/IEC Std. 15408, the Unified Process and its variants, as well as some related works. Then, we explain our approach to integrate CC assurance activities in the context of the UP. Finally, we present the partial results of our analysis, followed by a discussion of some conclusions and future works.

2   Background

In this section we present the ISO/IEC Std. 15408 for software security assurance as well as the framework of the UP. We also discuss related works that examine a phase-oriented iterative process in the context of CC framework, and an analysis of agile processes in conjunction with security assurance methods.

2.1   Security Evaluation and Assurance Philosophy (ISO/IEC 15408)

Software security assurance is a set of activities whose purpose is to raise software security during the development lifecycle. Although there are several approaches (or models) for planning the lifecycle of software development, but there are clearly certain disciplines that exist in the phases within any lifecycle. These disciplines include requirements engineering, analysis, design, implementation and testing. At each step along the development lifecycle, mistakes can be made, which in turn can affect the quality and security of the final software product as well as the schedule and cost of the project. Security evaluation and assurance activities strive to lead the developers and project managers to follow some pre-matured disciplines and guidelines in order to prevent these mistakes and establish a way to let a third party organization to evaluate the product. This Evaluation, in turn, will result to a determination and assurance of the correctness and completeness of the realization of security requirements throughout the lifecycle, and will lead to detect any mistakes or insecure activities, which have the potential to lead into an insecure acquisition. These activities are not sufficient, however, to assure that the final software product is a totally flawless system which can be called a secure system.

The CC philosophy is that the threats to security and organizational security policy commitments should be clearly articulated and the proposed security measures be demonstrably sufficient for their intended purpose [1]. The CC does not exclude, nor does it comment upon, the relative merits of other means of gaining assurance. Research continues with respect to alternative ways of gaining assurance. As mature alternative approaches emerge from these research activities, they will be considered for inclusion in the CC, which is so structured as to allow their future introduction.

ISO/IEC Std. 15408 has been used by several private and government organizations to structure the security assurance activities performed on various projects. The assurance activities result in an Evaluation Assurance Level (EAL), which is assigned to the product as a certification of a level of assurance. Microsoft Corporation, for example, was certified in October of 2002 at EAL-4+ by CC security evaluation and assurance organizations [4].

2.2 Common Criteria Overview

In December 1999, ISO/IEC 15408, Parts 1-3 (Criteria for IT Security Evaluation), was approved as an international standard. According to this standard the part of an IT system or product which should be evaluated based on the CC is called Target of Evaluation (TOE) and has to fulfill different security requirements which are verified by an evaluation authority.

The security requirements of the CC are divided into security functional requirements (requirements on the product) and security assurance requirements (requirements on the process) and are structured into classes. The functional requirements are realized in the functions of the system in order to achieve the security objectives of the TOE. The assurance requirements contain measures to be undertaken during development in order to keep the risk for weak points low. They are necessary for guaranteeing the confidence that the TOE meets its security objectives. The number and strictness of the assurance requirements to be fulfilled depends on the Evaluation Assurance Level (EAL) one has chosen for the TOE.

The following section contains all measures and assurance classes independent of a chosen EAL which are important for a development based on the CC:

Security Target (class ASE): The Security Target (ST) is the core document of system development based on the CC. It contains the security analysis, a description of the TOE and its boundaries, a definition of the assets of the TOE and the threats against the assets, and the security objectives corresponding to the threats. The ST is concluded with a TOE summary specification, describing the security functions of the TOE. For further information, see [1].

Configuration management plan (class ACM): All assurance requirements concerning the configuration management should ensure that the integrity of the TOE is preserved. Developers have to write a configuration management plan which contains a description of the configuration management system used in the project.

Design and representation (class ADV): The assurance requirements about design and representation require correct and consistent specifications and designs on different levels of abstraction and formality (e.g., a semi-formally specified high-level design).

Lifecycle documentations (class ALC): The assurance requirements for the lifecycle support are important for controlling the development and maintenance of the TOE. For example, in this class there are requirements on the documentation of development tool usage.

Test documentation (class ATE): In this class all assurance requirements belong to test activities like test documentation, test depth and extent of testing. Test activities are used to validate that the TOE satisfies all security functional requirements defined in the ST.

Vulnerability assessment (class AVA): Activities concerning this class correspond to an analysis of the vulnerabilities (design-specific weaknesses) of a system.

Guidance documents (class AGD): The CC contains assurance requirements which refer to the content of the user and administration guidance. Both documentations have to be understandable, consistent and complete. The aim is to show users and administrators how to operate with the system and its security functions in a secure manner.

Delivery and operation documentations (class ADO): The CC contains assurance requirements to ensure the security during delivery, installation, start-up and operational use of the TOE.

Fig. 1 shows the CC Evaluation Levels and needed assurance classes and components for each level captured by [5].

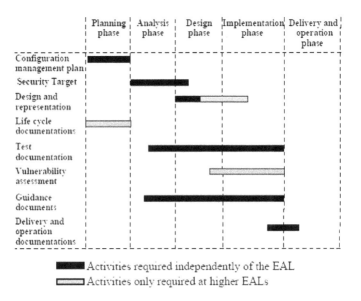

Fig. 1. Integration of CC Requirements into process

2.2 Unified Process

The Unified Process (UP), also known as the Unified Software Development Process, is a process framework for software development, based on the principles of iterative and incremental development. The UP supports projects that are large or small scale, that are technically and managerially complex or simple. Concrete, product variants of the Unified Process exist, namely the Rational Unified Process (RUP) [9, 12] and the Unified Process for Education (UPEDU) [8].

One of the strengths of the UP is its extensive use of visualization to describe the software development process. An overview of the Unified Process is available at

UPEDU web site. This overview shows how a software engineering process is broken down into phases, each of which consists of one or more iterations. Within the iterations, various disciplines are practiced. These include the traditional disciplines such as Requirements, Analysis and Design, Testing, etc. There is a relative level of activity within the lifecycle of the process. For example, it is reasonable to assume that, early in the project, work on requirements is done at a higher level than testing. It is also reasonable that the testing activities tend to be synchronized with the incremental releases of software, at the end of each cycle or iteration.

The UP, as defined in UPEDU and RUP, is comprised of the following key elements – roles, activities, workflows and artifacts, which describe respectively who, how, when, and what of a software development process. For example, in the Requirements discipline an analyst (role) participates in the Understand Problem Workflow, by eliciting stakeholder requests (activity) using the information from the project's vision document (artifact) to produce a use-case model (artifact). In another example, the Test discipline, workflows include Plan and Design Test, Implement Test, and Execute Test. A tester (role) plans tests (activity) to produce a test plan (artifact) based on the use-case model and the supplementary specifications (artifacts). Another important aspect of the UP is the existence of templates and case-study examples of the various UP artifacts.

The UP is a flexible framework, and as such can (and usually must) be tailored to fit a particular project's needs. The latest version of the RUP software provides software tools and plug-ins that allow the tailoring and customization of RUP.

2.3 Related Works

There are some studies in security assurance literature in which one examines RUP-like, iterative processes in the context of other security assurance models and frameworks. Vetterling et al. have employed an iterative process in acquisition of mobile money exchange software in accordance with EAL-2 of CC [5]. The authors show that their process should be tailored to support the appliance of CC assurance activities into it and present their experiments. They also discuss the artifacts that should be provided extra to their process context and consequent effort needed.

Beznosov and Kruchten have studied the agile processes and their pain points in context of various security assurance methods [3]. The authors categorize assurance methods in four categories according to their conformity with agile development processes. The pain points mentioned by the authors affecting this categorization are 1) Tacit Knowledge/Documentation, 2) Lifecycle, 3) Re-factoring and Iteration, and 4) Test Philosophy. These issues are the cornerstones of the agile methods and mostly have some basis in RUP disciplines and best practices. The authors believe that CC, is a mismatch method being used in agile methods context.

Mellado et al. have proposed a Common Criteria based security requirements engineering [13], which uses a security resources repository and integrates the Common Criteria into the software development lifecycle to handle the software security requirements.

3   Analysis Approach

Our goal in this article is to determine the level of support that the UP offers with respect to security assurance defined by ISO/IEC 15408. Firstly, we consider a comparison of the key concepts of each framework. Secondly, we attempt to identify which of the specific assurance tasks are supported by the UP.

3.1   Key Concepts of UP and CC

The UP defines roles, workflows, activities and artifacts. CC has concepts that are analogous to these. Activities in the UP are very similar to activities in the CC. Workflows in the UP are close to the Components in CC. However, the UP is more explicit about the order and concept of activities within workflows than the CC is about the order of activities within components. The various required inputs and outputs for CC activities defined in the standard, mostly have equivalent artifacts in the UP. In fact, many of them, e.g., Software Requirements Specification, Test Plan Documentation and etc. map directly to UP artifacts.

Both CC and UP, define the notion of role. However, CC mentions three specific roles and it does not require (nor provide guidance about) the specifics of different roles related to evaluation and assurance activities. The UP tends to do this for all of its activities and workflows. This can be seen as a potential benefit to considering CC in the context of the UP.

3.2   Identification of CC Assurance Activities in the UP

In the second part our analysis, we studied each CC assurance class and its components proposed in CC documentation part three and identified in which UP phase(s) it appears (if at all). Then, we determined whether or not the UP explicitly supports the component. In this analysis, a CC component is considered as supported by the UP, if there is a similar activity in terms of produced artifacts in it. Any component from CC that has not been clearly specified or described in the UP is considered as a missing activity and some tailoring in the UP would be required in order to support that CC activity.

4   Results

In this section, we summarize the results and discuss some of details of the UP support (or lack) for CC. Almost most of the activities defined by CC are supported by the UP. Some are partially supported and others are not supported at all.

We ignore the ASE and APE classes of the CC assurance class series, these classes try to evaluate and assure the two primary documents of CC, namely Protection Profile (PP) and Security Target (ST), which are security requirements of the TOE and their main solutions accordingly. These two documents deal with security requirements specification and designing solutions for those requirements, which are traditional ways to handle the system requirements. In the UP requirements emerge along the lifecycle first iterations, and the solution (system architecture and design) emerge and evolve through the next iterations in the Elaboration and Construction phases. This brings out a different idea of treatment while facing the PP and ST requirements in context of RUP methodology. While iterative lifecycle is based on the successive enlargement and refinement of a system through multiple iterations, with cyclic feedback and adaptation as core drivers to converge upon a suitable system [11], requirements specification and design emergence should be considered as a successive and incremental progresses too, which postpones the PP and ST creation from the very fist steps of software acquisition lifecycle to very late steps of it. We believe that CC assurance strategy and concept is mostly based on traditional methodological approaches in software development (i.e. linear approaches), where all requirements are analyzed and frozen in the first phases of development lifecycle. This is an evidence of lack of methodology independence paradigm that CC claims, which is under further investigation by the authors.

4.1 Direct support of CC Assurance Classes in the UP

We consider the following CC Assurance Classes and it's components to be mostly supported by the UP: ACM Class (for Configuration and change management), ADO Class (for the Delivery & Operation Activities), AGD Class (for Guidance Documentation Generation) and ATE Class (for Test Activities).

ACM class is supported specifically with automated tools such as Clear Case and Clear Quest, part of the RUP suite of tools, or other CM tools available. This class can be directly supported by Configuration and Change Management (CCM) workflow in the RUP, although some of ACM components need some artifacts which are not directly supported by it, but most of these artifacts are part of CM tools documentation or can be produced by roles defined in the CCM Workflow.

ADO class as defined in the CC deals with software secure deployment, which is supported by the Deployment workflow in the RUP. The purpose of this class is to assure secure and correct delivery of the product and secure installation, generation and start-up of the TOE, which are done through documented procedures that guide users and deployment personnel for secure delivery. The generation of these guidelines and artifacts is supported by the Deployment workflow in the RUP, although, it is possible to employ the deployment tools available (e.g. Install-Shield) to enforce these procedures automatically for secure deployment.

AGD class requirements, as specified in the CC, are to provide the required guidance documents for administrators and end-users. However, these requirements can be addressed by the Deployment workflow in the UP, as there is a specific role called "Course Developer" [9], who is responsible for providing guidance documents. In order to fulfill the content requirements of the AGD class documents, these documents can be prepared under the control of "Technical Writer" (another UP role) in order to be consistent with the security functions and their functionality descriptions.

ATE class is another assurance class defined by the CC, which is fully supported by the UP Test workflow. This assurance class deals with different aspects of testing the TOE, such as Test Coverage (covering all security related parts in test), Test Depth (for testing the system in different abstract levels and layers), Functionality Testing, and Independent Testing (which concerns the third party tests). All of test aspects mentioned above can be provided through the test plan and design activities in the UP. A complete suite of automatic test tools are available in the RUP to help the developers and independent testers to address different test scenarios and test-cases.

4.2   Partial Support of CC Assurance Classes in the UP

We use the term "partial support" for those assurance classes which are supported by the UP workflows partially, or in other words, they match from general point of view, but some role definitions or activity tailoring is required in order to fulfill all assurance requirements and artifacts need. According to this, the following assurance classes of CC, and their components are partially supported by the UP: ALC Class (for Life-Cycle support) and ADV Class (for Development guidelines).

ALC class, as defined by CC, is an assurance class which deals with software development lifecycle. Some parts (families) of this class deal with lifecycle definition, its requirements and development tools and techniques applied during lifecycle. These requirements are addressed directly by the RUP itself and its documents and tools. Some other parts (families) are dedicated to development security (i.e. development environment security) and flaw remediation techniques and procedures applied during lifecycle to make the lifecycle and product secure and flawless. Although these latter parts are not directly supported by the UP, but they can be tailored to its activities very easily. For example, a flaw remediation scenario or activity can be defined in beginning of each iteration and some systematic/automatic approaches can be placed in too. It would be worthy to mention that iterative programming, which is one of the most valuable and best practices in the UP, can be very helpful in flaw remediation practices because of several tests and feedbacks to developers in each one of iterations.

ADV class is about development guidelines and requires developers to implement the system in several abstract levels and as simple as possible from the code complexity and coupling point of view to achieve a simpler code for evaluation. The purpose is to make a structured implementation from the most abstract level (functional spec.) down to least abstract level (source code), and correspondence evidence between these levels to assure that all security requirements have been met and implemented correctly and completely. The overall UP process deals with several abstraction levels from Use-Cases to source code and according to object-oriented approach the design and implementation is divided into classes and components to make the system simpler and much more understandable, which is very important for evaluation of codes. According to [11], requirements management tools, provide support in capturing requirements and organizing them in documents and in a requirements repository along with important attributes that are used for managing requirement scope and change, which makes the correspondence activities simpler and much more systematic.

4.3 Missing Aspects of CC Assurance Activities in the UP

As a missing aspect of CC assurance activities, we could mention the AVA class, which concentrates on Vulnerability Assessment and Analysis. According to its security analyzing nature, this class can not be mapped into UP activities or workflows, but generally it is possible to define some activities to measure the vulnerabilities during development lifecycle.

Some parts of this assurance class (AVA), which are about the reduction of misuse of TOE, require that the administrator and user guidance documentation to be much more descriptive in the sense of possible vulnerabilities and insecure states of system, their consequences and implications about maintaining secure operation. These aspects could be addressed in Deployment workflow in the UP, where the guidance documents are provided. Other parts (families) of CC assurance class AVA, like covert channel analysis, or strength evaluation of security functions, etc. are not supported or mentioned in the UP workflows and they should be added, defined and tailored in various workflows as necessary.

Generally, CC assurance mechanism is to evaluate and assure the single-iteration, phase-base, documentation-driven development methodologies like the traditional waterfall. Modern methodologies like UP and agile methods, which are mostly iterative and individual-centric, would experience some difficulties or even unexpected blockades while being applied in the context of conventional security assurance methods.

5 Conclusions and Future Work

Results presented in the last sections show that, although the UP with its vast workflows address almost every aspect of software engineering requirements and can fulfill any development requirement, but it needs to significantly be tailored to support other standards like ISO/IEC 15408.

It is important to notice that CC assurance classes and their approach to evaluate and assure a system, is based on product documentation evaluation, and these classes do not use the practices and strengths of development methodologies for assessment, which leads to a series of pain points and problems in this conjunction.

Since there is not a specific role definition in CC, tailoring of the UP to support the CC could be a beneficial step in an organization that wishes to integrate the CC activities within the UP methodology.

In security-critical software systems, applying security evaluation and assurance activities in a most explicit way in the UP is an important issue in order to achieve a secure system. While in these systems security is the most critical aspect and is of the highest concern, time and cost factors are also crucial. As in most of evaluation cases, the evaluated software is almost out of date at the time of release. "Good security is not some an add-on or after-the-fact concern."[4], so it should be very integrated into the software acquisition process.

The UP's highly iterative approach may seem too radical for traditional, safety critical development environments. We feel that this is not necessarily the case, as long as the assumptions about what is expected to change during the lifecycle, is well understood. The relatively short iterations in the UP are designed to minimize risks due to uncertainties related to technology, user needs, human resources, etc. Many of these risks exist as well in projects that are safety critical. An agile approach based on the UP that has been tailored for support of security assurance methods can be beneficial where used appropriately. The authors, at the moment, are working on the methodological aspects of integration of security assurance methods, especially CC, into various software development processes such as agile methods, specifically XP method, which seems has the most contrast with conventional security assurance processes and frameworks.

Acknowledgments. This paper is supported by Iran Telecommunication Research Center (ITRC).

6 References

1. CC, "Common Criteria for Information Technology Security Evaluation, 2.1 ed.", available at: http://www.commoncriteriaportal.org/ (1999) (Accessed: 2006).
2. I. Jacobson, G. Booch, and J. Rumbaugh, The unified software development process. Addison Wesley (1999).
3. K. Beznosov, P. Kruchten, Towards Agile Security Assurance, workshop on new Security Paradigms, Software Development Practices (2004) 47-54.
4. E. H. Spafford, Exploring Common Criteria: Can it Ensure that the Federal Government Gets Needed Security in Software?, USA House Government Reform Committee Sub-committee on Technology, Information Policy, Intergovernmental Relations and the Census, Testimony, available at: http://homes.cerias.purdue.edu/~spaf/usgov/ (2003) (Accessed:2005).
5. M.Vetterling, G. Wimmel, Secure Systems Development Based on Common Criteria: The PalME Project, in Proceedings of ACM Workshop SIGSOFT (2002).
6. Information Technology – Security Techniques, A Framework for IT security Assurance, ISO/IEC JTC 1/SC 27, DIN Deutches Institut fur normung e.V., Burggranfenstr. 6, 10772 Berlin, Germany, available at: http://www.ni.din.de/sc27 (2003) (Accessed:2004).
7. C. Larman and V.R. Basili, Iterative and incremental developments - a brief history, Computer Journal, vol. 36, (2003) 47-56.
8. école Polytechnique de Montréal, "Unified Process for EDUcation (UPEDU)," http://www.upedu.org/upedu/, (2002) (Accessed: 2007).
9. P. Kruchten, The Rational Unified Process An Introduction, Second Edition, Addison Wesley, (2000).
10. J. Wayrynen, M. Boden, and G. Bostrom, Security Engineering and eXtreme Programming: an Imposibble Marriage? , Extreme Programming and agile Methods XP/Agile Universe 2004, C. Zannier , H. Erdogmus, and L. Lindstorm, Eds. LNSC3134, Berlin: Springer-Verlag, (2004) 117-128.
11. C. Larman, Applying UML and Patterns, An Introduction to Object-Oriented Analysis and Design and the Unified Process, Second Edition, Prentice Hall, (2001).
12. P. Kroll and P. Kruchten, The Rational Unified Process made easy: a practitioner's guide to the RUP, Addison Wesley, (2003).
13. D. Mellado, E. Fernández-Medina, M. Piattini, A Common Criteria based security requirements engineering process for the development of secure information systems, Computer Standards & Interfaces, vol. 29, (2007) 244-253.

Secure Communication and Access Control for Mobile Web Service Provisioning

Satish Narayana Srirama[1], Anton Naumenko[2]

[1]RWTH Aachen University, Informatik V (Information Systems)
Ahornstr 55, 52056 Aachen, Germany
[2]Industrial Ontologies Group, Department of Mathematical Information Technology,
P.O. Box 35, 40014 University of Jyväskylä, Finland
srirama@cs.rwth-aachen.de, annaumen@cc.jyu.fi

Abstract. It is now feasible to host basic web services on a smart phone due to the advances in wireless devices and mobile communication technologies. While the applications are quite welcoming, the ability to provide secure and reliable communication in the vulnerable and volatile mobile ad-hoc topologies is vastly becoming necessary. The paper mainly addresses the details and issues in providing secured communication and access control for the mobile web service provisioning domain. While the basic message-level security can be provided, providing proper access control mechanisms for the Mobile Host still poses a great challenge. This paper discusses details of secure communication and proposes the distributed semantics-based authorization mechanism.

Keywords: Access Control, Communication system security, Mobile Communication, Mobile web services.

1 Introduction

The high-end mobile phones and PDAs are becoming pervasive and are being used in variety of applications like location based services, banking services, ubiquitous computing etc. The higher data transmission rates achieved with 3G and 4G technologies also boosted this growth in the wireless market. The situation brings out a large scope and demand for software applications for such high-end mobile devices. To meet this demand and to reap the benefits of the fast growing web services domain and standards, the scope of the mobile terminals as both web services clients and providers is being observed. While mobile web service clients are common these days, we have studied the scope of mobile web service provisioning, in one of our previous projects. [5]

Mobile web service provisioning offers many of its applications in domains like collaborative learning, social systems, mobile community support etc. While the applications are quite welcoming, the ability to provide secure and reliable communication in the vulnerable and volatile mobile ad-hoc topologies is vastly becoming necessary. Moreover with the easily readable mobile web services, the complexity to realize security increases further. Secure provisioning of mobile web services needs proper identification mechanism, access control, data integrity and confidentiality.

In our current research, we are trying to provide proper security for the mobile web service provider ("Mobile Host") realized by us. The security analysis suggests that proper message-level security can be provided in mobile web service provisioning with reasonable performance penalties on the Mobile Host. While the basic message-level security can be provided, the end-point security comprising proper identity and access control mechanisms, still poses a great challenge for the Mobile Host. Here we propose to utilize distributed architectures of semantics-based authorization mechanism to ensure pro-active context-aware access control to mobile web services.

The rest of the paper is organized as follows: Section 2 discusses the concept and analysis of mobile web service provisioning domain. Section 3 addresses the issues of securing the communication for mobile web services. Section 4 presents our research ideas towards implementation of semantics-based access control for mobile web services and section 5 concludes the paper with future research directions.

2   Pervasive Mobile Web Service Provisioning

Traditionally, the hand-held cellular devices have many resource limitations like limited storage capacities, low computational capacities, and small display screens with poor rendering potential. Most recently, the capabilities of these wireless devices like smart phones, PDAs are expanding quite fast. This is resulting in quick adoption of these devices in domains like mobile banking, location based services, social networks, e-learning etc. The situation also brings out a large scope and demand for software applications for such high-end wireless devices.

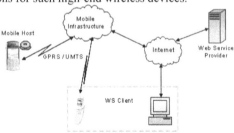

Fig. 1. Mobile terminals as both web service providers and clients

Moreover, with the achieved high data transmission rates in cellular domain, with interim and third generation mobile communication technologies like GPRS, EDGE and UMTS [2], mobile phones are also being used as Web Service clients and providers, bridging the gap between the wireless networks and the stationery IP networks. Combining these two domains brings us a new trend and lead to manifold opportunities to mobile operators, wireless equipment vendors, third-party application developers, and end users [4, 5]. While mobile web service clients are quite common these days, and many development tools are available from major vendors, the research with mobile web service provisioning is still sparse [3, 4]. During one of our previous projects, a small mobile web service provider ("Mobile Host") has been developed for resource constrained smart phones. Figure 1 shows the scenario with mobile terminal as both web service provider and client.

Mobile Host is a light weight web service provider built for resource constrained devices like cellular phones. It has been developed as a web service handler built on top of a normal web server. The web service requests sent by HTTP tunneling are diverted and handled by the web service handler. The Mobile Host was developed in PersonalJava [7] on a SonyEricsson P800 smart phone. The footprint of our fully functional prototype is only 130 KB. Open source kSOAP2 [8] was used for creating and handling the SOAP messages.

The detailed evaluation of the Mobile Host clearly showed that service delivery as well as service administration can be performed with reasonable ergonomic quality by normal mobile phone users. As the most important result, it turns out that the total WS processing time at the Mobile Host is only a small fraction of the total request-response invocation cycle time (<10%) and rest all being transmission delay [5]. This makes the performance of the Mobile Host directly proportional to achievable higher data transmission rates. Thus the high data transmission rates achievable, in the order of few Mbps, with advanced mobile communication technologies in 2.5G, 3G and 4G, help in realizing these Mobile Hosts also in commercial applications [9].

Mobile Host opens up a new set of applications and it finds its use in many domains like mobile community support, collaborative learning, social systems and etc. Many applications were developed and demonstrated, for example in a distress call; the mobile terminal could provide a geographical description of its location (as pictures) along with location details. Another interesting application scenario involves the smooth co-ordination between journalists and their respective organizations. From a commercial viewpoint, Mobile Host also renders possibility for small mobile operators to set up their own mobile web service businesses without resorting to stationary office structures. [5]

The Mobile Hosts in an operator proprietary network can also form a P2P network with other mobile phones and can share their individual resources and services. P2P offers a large scope for many applications with Mobile Host. Not just the enhanced application scope, the P2P network also offers better identification and discovery mechanisms of huge number of web services possible with Mobile Hosts. [10]

While the applications possible with Mobile Host are quite welcoming in different domains, the ability to provide the secured communication and access control in the vulnerable and volatile mobile ad-hoc topologies is quite challenging.

3 Securing Message Communication of Mobile Web Services

Once a mobile web service is developed and deployed with the Mobile Host, the service and the provider are prone to different types of security breaches like denial-of-service attacks, man-in-the-middle attacks, intrusion, spoofing, tampering etc. As web services use message-based technologies for complex transactions across multiple domains, traditional point-to-point security paradigms fall short. Potentially, a web-service message traverses through several legitimate intermediaries before it reaches its final destination. The intermediaries can read, alter or process the message. Therefore, the need for sophisticated end-to-end message-level security becomes a high priority and is not addressed by existing security technologies and standards in the wireless domain.

At the minimum, the mobile web service communication should possess the basic security requirements like proper authentication/authorization, and confidentiality/Integrity. Secure message transmission is achieved by ensuring message confidentiality and data integrity, while authentication and authorization will ensure that the service is accessed only by the legitimate service requestors. Even though a lot of security specifications, protocols like WS-Security [11], SAML [12] etc., exist for web services in traditional wired networks, not much has been explored and standardized in wireless environments. Our study contributes to this work and tries to bridge this gap, with main focus at realizing some of the existing security standards in the mobile web services domain.

The WS-Security specification from OASIS is the core element in web service security realm in wired networks. It provides ways to add security headers and tokens, insert timestamps, and to sign and encrypt the SOAP messages. To adapt the WS-Security in the mobile web service communication, the web service messages were processed with different encryption algorithms, signer algorithms and authentication principles, and were exchanged according to the standard. The performance of the Mobile Host was observed during this analysis, for reasonable quality of service. The main parameters of interest were the extra delay and variation in stability of the Mobile Host with the launched security overhead.

For the analysis, a SonyEricsson P910i smart phone was used. The device supports J2ME MIDP2.0 [13] with CLDC1.0 [14] configuration. For cryptographic algorithms and digital signers, java based light weight cryptographic API from Bouncy Castle crypto package [15] was used. kSOAP2 was modified and adapted according to WS-Security standard and utilized to create the request/response web service messages. The Mobile Host was redesigned with J2ME and the adapted kSOAP2.

To achieve confidentiality, the web service messages were ciphered with symmetric encryption algorithms and the generated symmetric keys were exchanged by means of asymmetric encryption methods. The messages were tested against various symmetric encryption algorithms, along with the WS-Security mandatory algorithms, TRIPLEDES, AES-128, AES-192 and AES-256 [16]. The PKI algorithm used for key exchange was RSA-V1.5 with 1024 and 2048 bit keys. Upon successful deployment of confidentiality, we considered data integrity on top of confidentiality. The messages were digitally signed and were evaluated against two signature algorithms, DSAwithSHA1 (DSS) and RSAwithSHA1. [6]

Figure 2 shows the total times taken at the Mobile Host for processing the web service requests. The timestamps consider the added security overload, for different symmetric algorithms on Mobile Host, but exclude the transmission delays. The test configuration considered here was RSA-1024 key exchange and RSA signature. This test is conducted against varied soap message sizes ranging from 1 to 10 KB. The extra delay for highly secured communication, AES-256 bit ciphered was approximately ~3 sec with RSAwithSHA1 signature for reasonable message sizes of 5KB. The delay was reasonable with respect to the Mobile Host's processing capability. [5]

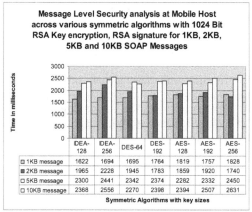

Fig. 2. Mobile Host's processing times for various message sizes and symmetric key algorithms using RSA signature

The detailed performance analysis suggested that not all of the WS-Security specification can be adapted to the mobile web service communication with today's smart phones. But with latest developments in speed and performances of processor chips for mobile devices, the scenario is going to change soon. With our security study for today's smart phones, we are recommending that the best way of securing SOAP messages in mobile web service provisioning is to use AES symmetric encryption with 256 bit key for encrypting the message and RSAwithSHA1 to sign the message. The symmetric keys are to be exchanged using RSA 1024 bit asymmetric key exchange mechanism. The cipher data and the keys are to be incorporated into the SOAP message according to WS-Security specification.

4 Semantics-based Access Control Mechanisms

For the trusted and distributed management of access control to protect mobile web services, we propose to use Semantics-Based Access Control (SBAC). SBAC is the result of adoption of the Semantic Web vision and standards [22] to the access control research and development field. Administration and enforcement of access control policies based on semantics of web services, clients, mediating actors and domain concepts comprise the most suitable approach to handle openness, dynamics, mobility, heterogeneity, distributed nature of environments that are involved in the mobile web service provisioning. We have previously defined the SBAC research framework [17], the SBAC model [18] and ontologies [19], the SBAC abstract architecture [17-21], conducted quantitative evaluation of the prototype of the SBAC policy enforcement function [19], and described adoption of SBAC for Semantic Web Services [20] and Multi-Agent Systems [21]. This paper covers the adoption of SBAC for the mobile SOA providing the analytical feasibility study of SBAC deployment options. This analytical evaluation takes into account results of experiments with the prototype of the SBAC enforcement function [19]. Before the analysis, we give a short description of the SBAC research framework, model, ontologies, and abstract architecture.

The SBAC research framework defines the SBAC research and development outcomes and their interrelations [17]. The whole framework reuses achievements in the Semantic Web research and development area like standards, ontologies, methodologies, frameworks, tools, platforms, etc. For example, the model-theoretic semantics of SBAC [18] is an extension of the direct model-theoretic semantics defined in the Web Ontology Language (OWL) standard [23] and Semantic Web Rule Language (SWRL) [24]. The SBAC model has been expressed in the form of ontologies [19]. Thus ontologies are the key part of the SBAC framework [17]. The ontology engineering constitutes the traditional domain modeling.

The abstract SBAC architecture is the upper view on components of SBAC and interactions between them to provide authorized access to protected resources. Basically, the abstract architecture is a bridge between theoretical findings and adoption of the SBAC into practice. Figure 3 shows elements of the SBAC abstract architecture encompassing the SBAC enforcement function only.

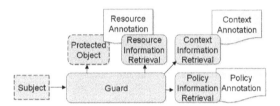

Fig. 3. The SBAC abstract architecture for the enforcement function.

The elements of the SBAC enforcement mechanism are the pro-active guard, the unified information retrieval components for policies (PIR), contextual data (CIR), and resource annotations (RIR). The guard is a proxy for both protected resources and information retrieval components. The guard pro-actively collects all relevant data and enforces an access decision based on the iterative reasoning over the semantically encoded access control policies and the semantic annotations of a subject, an operation, an object and a context of access [18, 20].

For the authorized mobile web service provisioning with the help of Mobile Host, the meanings of some crucial generic elements of the SBAC architecture are:

-- A subject of access is a human user who invokes the protected web services on the Mobile Host using a mobile phone or regular computer through Internet and mobile networks.

-- An operation of access is a web service's operation itself.

-- An object of access represents a protected web service deployed on the Mobile Host together with the input values as they determine the information outputs and/or physical effects on the state of the world as a result of service enactment.

-- A guard mediates access to the protected web service and enforces rules of corresponding access control policies. The guard must evaluate all requests, correctly evaluate policies, be incorruptible, and nonbypassable. Guard might be a standalone proxy or embedded as a wrapper of web service. Peculiarities of the mobile web service provisioning are in favor of the intermediate proxy implementation. This is a more general case and the functionality of guards can be integrated with resources.

-- A policy has rules that define which users may access the mobile web service. The goal is to isolate policy decision logic from resource and enforcement code. In the SBAC the policy is always an ontology or a set of ontologies that define semantic profiles of users, mobile web services, context, and policy rules of access.

-- A subject and object descriptors are well known patterns that provide access to the relevant attributes of subject and objects of access. A representation of descriptors for users, mobile web services, policies and context in the form of semantic annotations for mobile environments is reasonable because checking of attributes is independent from establishing them; there are different sources of attributes; different attributes are needed in different contexts; etc.

-- Context is a container for data that are relevant for access control decisions and enforcement. Different temporal and special characteristics like the time and location are traditionally considered as contextual information in mobile environments.

The components of the SBAC abstract architecture have to be deployed to the Mobile Host and middleware nodes depending on characteristics and requirements of use cases. There are several reasonable options of deployment of the SBAC components for protected mobile web service provisioning with unique characteristics and implications on the level of security and quality of mobile web service.

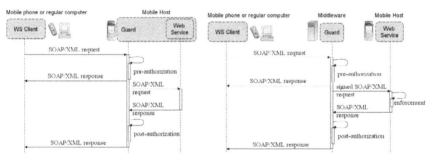

Fig. 4. Deployment options with 1. Embedded guard to the Mobile Host and 2. Middleware guard that is a proxy web service for the mobile web service

Figure 4 represents the first option with the embedded guard to the Mobile Host as a wrapper of the mobile web service. This is the most natural option for the pervasive mobile web service provisioning with the P2P communication between mobile clients and Mobile Hosts [10]. The clients access the service in the regular way. SOAP/XML messaging between the guard and service can be done using native RPC calls without delays of wireless or wired communication as other options have. One crucial advantage of this option is the opportunity to perform post-authorizations i.e. procedures of access control that must be performed after service enactment e.g. filtering of the content of response. This option supports the principle of end-to-end security in contrast to other options. However computational limitations of mobile phones demand light-weight functionality of the guard that prohibits use of complex semantics-based algorithms of reasoning for authorization decision making process.

Figure 4, with option 2, illustrates the deployment option where the guard is a middleware component and intermediate web service proxy that provides the same interface as the original mobile web service, decorates web service invocation with the SBAC policy enforcement mechanism, and delegates authorized requests to the mobile web service. The middleware guard is deployed in the Internet or mobile infrastructure. When the guard is in the Internet, clients are able to access it in a traditional way. Moreover Mobile Hosts receive a less number of requests or in other words only authorized requests, thus improving the scalability of the Mobile Host. The post-authorization is still possible. The middleware guard can aggregate and serve several Mobile Hosts and web services. Mobile-to-mobile requests experience delays of wireless communication twice when the guard is not an embedded but middleware component. There is a need to implement and deploy an enforcement component on the Mobile Host to validate signatures of the guard.

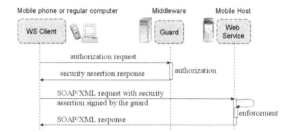

Fig. 5. Deployment option where guard is a third-party authorization authority.

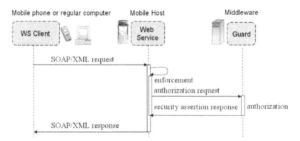

Fig. 6. Deployment option with the delegation of authorization to the middleware guard.

The deployment option, shown in figure 5, where the guard is a third-party authorization authority, creates additional inconveniences for clients to get authorization assertions prior to access protected mobile web services. Then enforcement components deployed on Mobile Hosts validate security assertions provided with requests in addition to verification of digital signatures in the previous option. The validation of security assertions is necessary to check that a security assertion corresponds to an operation for which a request is obtained when mobile web services provide more than one operation. Although this case might look too complex, however this is probably the most suitable option for the industrial, commercial or professional use of mobile web services when clients can get security tokens with long period of validity on the basis of their memberships in or subscriptions to different organizations, social networks, commercial services, etc. This option allows direct multiple requests to mobile web services using the same security token over time without overheads of the authorization decision making process for each request.

Delegation of authorization, when mobile web services initially receive all requests directly from clients and then generate requests for the access control decision to the middleware guard, is the last option we consider in this paper. Figure 6 depicts sequence diagram of communication between actors for this option. While such kind of deployment is possible, it has several significant shortcomings without clear advantages compared to above mentioned options. There are following needs: to embed the enforcement component for authorization messaging with all possible time overheads; to verify signatures of the guard; to process all requests from clients; etc. The only advantage is the shift of demanding functionality to the middleware guard.

5   Conclusion and Future Research Directions

This paper mainly discussed issues in providing secured communication and access control for the mobile web service provisioning. The paper first introduced the concept of mobile web service provisioning and then discussed the security breaches for the developed Mobile Host. It later presented the analysis of message-level security for the Mobile Host. The detailed performance analysis suggested that basic message-level security can be provided for the Mobile Host, even though not all the standards can be adapted to the mobile web service communication.

The paper later discussed the SBAC mechanism and adapting this mechanism in the provisioning of mobile web services, to achieve trusted and distributed management of access control for protecting mobile web services. Conducted analysis of deployment options reveals that they all are reasonable for realization and have different implications to security and QoS for the Mobile Host. However, future research in this domain, mainly addresses realizing the integrated security infrastructure. Further research on SBAC for MWS demands to assess all possible threats and attacks especially for the middleware guard. There is a need also to adopt SAML for the exchange of signed security assertions between WS clients, SBAC guards and Mobile Hosts. The adoption of SBAC for real-world application of MWS would also exemplify and align SBAC with practical concerns. Detailed performance analysis of the Mobile Host is again important, so that the extra load caused by the security mechanisms will not have serious impedances on the battery life of the devices and smart phone's basic purposes like making normal telephone calls.

Acknowledgments. This work is supported by the Research Cluster Ultra High-Speed Mobile Information and Communication (UMIC) at RWTH Aachen University (http://www.umic.rwth-aachen.de/) and partly funded by the grant from the Rector of the University of Jyväskylä, Finland.

References

[1]   Web Services Activity, http://www.w3.org/2002/ws/ (2006)
[2]   GSM   World: GSM   -   The   Wireless   Evolution,
      http://www.gsmworld.com/technology/index.shtml (2006)
[3]   JCP: J2ME Web Services Specification, JSR 172, http://jcp.org/en/jsr/detail?id=172
      (2006)
[4]   Balani, N.: Deliver Web Services to mobile apps, IBM developerWorks (2003)
[5]   Srirama, S., Jarke, M., Prinz, W: Mobile Web Service Provisioning, Int. Conf. on Internet
      and Web Applications and Services, ICIW06, IEEE Computer Society, pp. 120-125 (2006)
[6]   Srirama, S., Jarke, M., Prinz, W., Pendyala, K.: Security Aware Mobile Web Service
      Provisioning, In Proceedings of the Int. Conf. for Internet Technology and Secured
      Transactions, ICITST'06, e-Centre for Infonomics, pp. 48-56 (2006)
[7]   Java support in SonyEricsson mobile phones P800 and P802, Developer guidelines from
      SonyEricsson Mobile CommunicationsAB, www.SonyEricssonMobile.com (2003)
[8]   KSOAP2, A open source SOAP implementation for kVM, http://kobjects.org/ (2006)
[9]   4G Press: World's First 2.5Gbps Packet Transmission in 4G Field Experiment,
      http://www.4g.co.uk/PR2006/2056.htm (2005)
[10] Srirama, S.: Publishing and Discovery of Mobile Web Services in P2P Networks, Int.
      Workshop on Mobile Services and Personalized Environments (MSPE '06) (2006)
[11] OASIS: WS-Security version 1.0, http://www.oasis-open.org/specs/#wssv1.0 (2006)
[12] OASIS: Security   assertion   markup   language   SAML   V2.0,   http://docs.oasis-
      open.org/security/saml/v2.0/saml-2.0-os.zip (2006)
[13] JCP: Mobile Information Device Profile 2.0, JSR 118, http://jcp.org/en/jsr/detail?id=118
      (2006)
[14] JCP: Connected   Limited   Device   Configuration   Version   1.0,   JSR   30,
      http://jcp.org/en/jsr/detail?id=30 (2006)
[15] Bouncy   Castle: Bouncy   Castle   lightweight   cryptography   API,
      http://www.bouncycastle.org/documentation.html (2006)
[16] Security algorithms, http://www.rsasecurity.com/ (2006)
[17] Naumenko, A.: A Research Framework towards Semantics-Based Access Control,
      International Journal of Network Security, (Submitted for review January 2007).
[18] Naumenko, A.: Contextual rules-based access control model with trust, In Proceedings of
      Int. Conf. for Internet Technology and Secured Transactions, ICITST 2006, 11-13
      September, e-Centre for Infonomics, pages 68-75 (2006)
[19] Naumenko A.: Semantics-Based Access Control - Ontologies and Feasibility Study of
      Policy Enforcement Function, In Proceedings of the 3rd Int. Conf. on Web Information
      Systems and Technologies (WEBIST), INSTICC Press, pp. 150-155. (2007)
[20] Naumenko, A. and Luostarinen, K.: Access control policies in (semantic) Service-Oriented
      Architecture, In Proceedings of the Semantics 2006 conf., Austrian Computing Society
      (OCG), 28-30 November, Vienna, Austria, pp 49-62 (2006)
[21] Naumenko A., Katasonov A., Terziyan V.: A Security Framework for Smart Ubiquitous
      Industrial Resources, In: Enterprise Interoperability II: New challenges and Approaches,
      Proceedings of the 3rd Int. Conf. on Interoperability for Enterprise Software and
      Applications (IESA-07), March 28-30, 2007, Portugal, Springer, 183-194 pp. (2007)
[22] Berners-Lee T., Handler J., Lassila, O.: The Semantic Web, Scientific American. (2001)
[23] Patel-Schneider P., Hayes P. and Horrocks I. (eds.): OWL Web Ontology Language
      semantics and abstract syntax, W3C Recommendation, www.w3.org/TR/owl-absyn/
      (2004)
[24] Horrocks I., Patel-Schneider P., Boley H., Tabet S., Grosof B. and Dean M.: SWRL: A
      Semantic Web Rule Language combining OWL and RuleML, W3C Member Submission,
      W3C, www.w3.org/Submission/SWRL/ (2004)

Creating Application Security Layer Based on Resource Access Decision Service

Mehmet Özer Metin[1], Cevat Şener[1], Yenal Göğebakan[2]

[1] Department of Computer Engineering, Middle East Technical University, Ankara, Turkey
[2] Cybersoft, Ankara, Turkey
(ozermetin@gmail.com, sener@ceng.metu.edu.tr, yenal.gogebakan@cs.com.tr)

Different solutions have been implemented for different security aspects (access control, application security) of enterprise web applications. However combining "enterprise-level" and "application-level" security aspects in one layer could give great benefits such as reusability, manageability, and scalability. In this paper, we propose adding a new layer to n-tier web application architectures, which use RAD service implementations to execute enterprise and application security policies. Proposed architecture enables applications not only benefit from "enterprise-level" security policies provided by RAD, but also implements "application-level" security based on RAD services to eliminate web application attacks including but not limited to those based on cross-site scripting, SQL injection, forceful browsing, cookie poisoning, invalid input and most importantly session stealing.

1. Introduction

As more business requirements are embedded in web applications, the requirement of complex enterprise-level security policies for authorization, and handling these policies in application logic of web application reduce reusability and manageability of whole system [9]. Enterprise-level security policies have dynamic manner and can not be easily handled by existing general purpose security mechanisms. Implementations of Resource Access Decision (RAD) specification can also be used for authorization problem of complex web applications.

On the other hand, web applications suffer from broad kind of security attacks and the most dangerous of these attacks are the ones that target application-level vulnerabilities. Application-level web security refers to vulnerabilities inherent in the code of a web-application itself. Traditional firewall based security techniques fail to prevent these kinds of attacks and it is almost impossible to investigate whole source code of web application to find the vulnerable part. So it is much harder to detect the source of vulnerability and defend the system.

Executing enterprise-level security policies for requests that suffer from application vulnerability can probably result in error-prone access decisions. In order to decide on enterprise-level security policies, web requests must be free of application-level security vulnerabilities. A correct access decision can be granted only if a request satisfies both "enterprise-level" and "application-level" security policies. Besides, combining "enterprise-level" and "application-level" security aspects in one layer in a transparent manner can result in great benefits such as reusability, manageability, scalability.

2. Access Control Problems in Enterprise Web Applications

The Internet is forcing enterprises to implement collaborative business and governmental solutions that integrate internal systems. To ensure security, these enterprise applications must implement complex access control rules that originate from both business logic and integration of business transactions. At this point access control rules become so called "enterprise-level security policies". However as access control logic becomes closer to enterprise level, policy rules become more dynamic, more domain-specific, and more context dependent. These rules are enforced organization widely, and are usually embedded in application systems. As a result, the traditional approaches in execution of these rules are costly and error-prone.

Beznosov [2] defines domain (application)-specific factors in security decisions as follows; "An application-specific factor is a certain characteristic or property of an application's resource, produced, modified and processed in the course of normal application execution and not for the sole purpose of a security policy decision." According to this point of view, all business objects in enterprise applications can be a source of access policies with their underlying business rules. And collection of these domain-specific access policies defines the "enterprise-level security" policies. According to the separation of concerns principle [3] "enterprise-level security" policies should be handled by a uniform, fine-grained, and transparent way.

The employment of domain-specific factors in security decisions is not new; one of the earliest examples can be found in OSI access control framework [1]. From there on, various distributed application systems try to encapsulate domain-specific factors. These affords can be classified into three categories:

- **Middleware infrastructures**: Most common distributed application technologies, such as J2EE, .NET, COM and CORBA [2] integrated access control engines that manage object interactions. However, they all suffer from low expressiveness for controlling enterprise applications that execute business transactions and business services that require much more abstraction to be controlled by an object interaction access control.
- **Access control frameworks:** The major aim of access control frameworks [4-6] is to supply a centralized authorization engine, which is a uniform access control interface that asks for access permissions. Authorization engines are able to interpret and execute enterprise-policy rules that are defined by policy specification languages such as Ponder [8] and eXtensible Access Control Markup Language (XACML) [7]. These frameworks are powerful choices for expressing enterprise-

level security policies. However they are not transparent solutions. So they are error-prone and are hard to be organization-wide.

- **Commercial Access Managers:** Most of the commercial application server vendors [12-14] have access manager products and also other vendors [15][16] have products that can integrate into variety of application servers. The common strategy of these products is managing user identities and roles assigned to appropriate privileges based on RBAC [11]. However, RBAC fails to separate enforcement function and decision function that is needed to evaluate domain-specific access policies [2].

On the other hand, as organizations have been increasing their reliance on web applications and are confronted with steadily maturing network-layer defenses, attackers are presumably turning their attention to the application layer and the corresponding business applications that are being served. According to SANS Institute [17] statistics; from 1Q05 to 1Q06 there has been a 20% rise in the number of application-specific vulnerabilities identified, and over 50% of these are based on web applications. The statistics also show that over 80% of all malfunctions that emerged in the past year were focused on exploiting application-layer vulnerabilities.

The Open Web Application Security Project (OWASP) [18] is one of the foundations that is dedicated to find and classify possible web application attacks, and offers countermeasures for them. OWASP publishes "Top Ten Most Critical Web Application Security Vulnerabilities" list to inform the public about the most dangerous vulnerabilities. According the latest list [19], the most critical one is considered as invalidated input. Since request parameters is the only input source for web application, sniffing HTTP request and validating each parameter is the most critical step toward securing web applications.

The most common solution is web application firewalls. According to web application security consortium (WASC), a web application firewall is "An intermediary device, sitting between a web-client and a web server, analyzing OSI Layer-7 messages for violations in the programmed security policy. A web application firewall is used as a security device protecting the web server from attack." [20]. Nowadays there are both academic proposals for web application firewalls [21], as well as open-source [22] and commercial ones [23][24].

Access control decisions can easily be manipulated or even bypassed, if the demander application is vulnerable. In other words, to decide on enterprise-level security policies, web requests must be free of application-level security vulnerabilities. A correct access decision can only be granted if a request satisfies both "enterprise-level" and "application-level" security policies. Most web applications use different solutions to provide "enterprise-level" and "application-level" security. However, this solution reduces manageability, reusability and scalability of the whole system. RAD implementations can be extended to apply both enterprise-level and application-level security policies in a single solution so that complex access control rules that originate from enterprise-level policies, and security rules that originate from application-level policies can be evaluated at the same place.

3. Access Control and Security Solution Based On RAD

The Resource Access Decision (RAD) specification released by the Object Management Group (OMG) is a mechanism for obtaining authorization decisions and administrating access decision policies [10]. In Beznosov's work this facility is rated as one of the best solutions that can be used by security-aware applications [2].

Our solution EYEKS ("Erişim, YEtkilendirme ve Kişiselleştirme Sistemi" in Turkish, meaning "Access, Authorization and Personalization System") uses the authorization engine, CSAAS, which is in fact a RAD implementation with RBAC [11] capabilities (presented in Akademik Bilişim Conference 2005 [9]). As mentioned before, RAD implementation does not force applications for authorization. It is the responsibility of the application to invoke authorization function and to take appropriate actions. However, EYEKS forces each request to be authorized and takes responsive actions to satisfy whole security policies in a transparent way.

Mapping Policies to RAD

RAD specification requires resources and their valid operations to be well defined. Resource can be any entity in the computer system, and an operation defines a valid procedure performed on any resource of the system. Every resource-operation pair can be associated with a number of "policies" that define access policies to do the requested operation on that resource. Access is granted only when that operation satisfies the associated policy rules on the specified resource. Policies are evaluated using attributes of an operation. These attributes can be dynamic (attribute value is evaluated at the time of the request) or static (parameters that are passed directly with the operation.) A policy grants or denies an access, based on the values of these attributes. A conceptual model of these relations is given in figure 1.

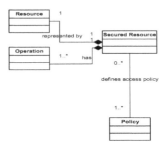

Fig. 1.

3.1.1 Enterprise Policy Mapping

RAD specification suits well for security requirements of web applications. Resources and operations can be defined to fulfill the application requirements, and a central access control layer can govern access rights of the whole web application. Request parameters can be mapped to attributes of security policies, and policy rules can be chained together to control access decisions of the web application. Determination of resources and operations is a critical step before designing an access control system that depends on RAD. They can be chosen according to the behavior of the web application. Two different approaches can be taken:

- Valid URL's can be chosen as resources and HTTP methods (PUT, GET, POST, DELETE) can be chosen as valid operations on that resource.
- Web Forms can be chosen as resources and two conceptual operations can be defined as operation on that resource; VIEW and SUBMIT.

It is possible to go a step further in abstraction on this subject. If the web application is served on more than one web context, or the whole system consists of more than one web application, web applications or web contexts can be handled as resources and valid URLs can be chosen as operations. This approach is more suitable for web farms. Either way, all HTTP request parameters must be passed to policy evaluator to be evaluated at each policy chain. After successfully naming the resources and possible operations, the web application becomes directly mapped to RAD domain. At this step, business access rules, so called enterprise-level security policies, can be added to the system. An example is given in Table-1.

Table 1. An example of enterprise policy mapping.

Resource	Operation	Parameters	Policy
docft.jsp	VIEW	USERID ACCOUNT_INFO	EFTTimeCheckPolicy VIEWAccountPolicy
	SUBMIT	USERID ACCOUNT_INFO TRANS_ACC_INFO TRANS_AMOUNT	EFTTimeCheckPolicy VIEWAccountPolicy TransAmountPolicy

According to Table 1, *doeft.jsp* is responsible for an EFT operation and the possible operations on the page are VIEW and SUBMIT. <doeft.jsp,VIEW> resource-operation pair can be linked with an enterprise security policy (EFTTimeCheckPolicy), which defines when a view operation is allowed (for example between working hours eg. 9 am-5 pm), and also with another enterprise security policy VIEWAccountPolicy, which checks whether the account really belongs to the specified user. <doeft.jsp,SUBMIT> pair can also be controlled by the same policies as <doeft.jsp,VIEW>, and additionally, can be linked with a security policy that checks whether the transfer could be allowed (TransAmountPolicy) (the policy that checks whether the transfer amount is less than the upper limit of user defined). As seen in the example, all enterprise-level security policies can be associated with any related resource-operation and they are reusable. These policies can be implemented by using policy evaluators defined in RAD specification [10] and can be managed by RAD implementation. Policies can be added, removed or changed dynamically without altering the enterprise web application code.

3.1.2 Application Security Policy Mapping:

Nearly 80% of web application attacks are because of parameter manipulation, which are generally caused by data validation vulnerabilities. A careful centric design of data validation would free web application from these vulnerabilities. However checking against possible vulnerability exploits and validating input at every point of entry to web application is costly, error-prone, and unmanageable. These "application-level" security policies that eliminate web attack risks must be taken into consideration. An access control system, that depends on RAD specifications could possibly work for evaluating "application-level" security policies. Since it is guaranteed that all HTTP request parameters are passed to the RAD facility, the security officer can define DoEftViewSecurityPolicy on <doeft.jsp,VIEW> pair, and DoEftSubmitSecurityPolicy on <doeft.jsp,SUBMIT>, which define possible parameters and their expected values for each pair.

For general use, some policies that check for known security exploits has been already implemented and built into the system. Security officers can link these policies to any <resource, operation> pair in the application. These pre-implemented policies start with SECURITY tag, and target injection and input validation attacks. These policies can be extended or altered according to application security needs.

Carefully designed security policies, which define safe values for each parameter to be free from web vulnerabilities, can be written and forced to be used for every possible resource-operation pair. If any parameter for an operation on any web page is found to be malicious, that access would automatically be denied by the RAD system before actual execution.

System Architecture

The architecture of the proposed solution, which incorporates RAD facility to be used for access control, is shown in figure 2. A specific layer, so called "application security layer", is created and placed in frontend. Backend layers consist of real web applications and databases and have no direct access to the outside world. All communications from outside world to backend web application must be intercepted and authorized from application security layer. RAD service has placed in this layer and can only communicate with EYEKS. This layer is also responsible for handling session security. It eliminates the risk of information disclosure and implements safe authentication methods.

EYEKS is designed as a logically layered structure. All user requests are captured at the uppermost layer and are processed through the inner layers, and are then dispatched to the backend web applications. The logical layers are (from uppermost to innermost layer):

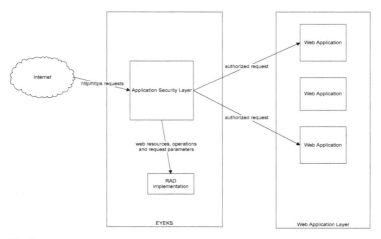

Fig. 2.

- **Request Listener Layer** is the entrance point of all client requests. This layer consists of two Java Servlets and can be configured to listen to any port by editing server configurations of the application server. The first servlet, i.e. the request listener servlet, is configured to hold root address (/), so that all requests to application security layer can be intercepted. All page and login requests are captured and handled by this servlet. The second servlet is a file upload servlet. All file upload operations are handled by this servlet.
- **Request Parser Layer** is where all requests are parsed to find out to which backend web application it refers, and also to find out to which scenario it belongs. Parameter-value pairs and backend context information (which web application does the request belongs to) are passed to the next layer.
- **Security Layer** is responsible for authentication and authorization. Authentication and authorization module checks if the user has enough credentials to make the coming request, by using a RAD implementation. Access to any page can be controlled by more complex policies like policies about allowed parameter-value pairs. Any request that does not satisfy these policies are denied.
- **Session Management Layer** is responsible for creating and handling user sessions. User sessions are created by the login scenario and are handled by the page request scenario. User sessions are stored in database or LDAP to be retrieved on every request. This layer also provides a unique encrypted token to be sent to client. The system provides these tokens that will be sent back by overwriting the HTML file generated by backend web applications. So the token mechanism for handling session security is fully transparent.
- **Request Dispatcher Layer** is the last layer of EYEKS structure. This layer consists of two modules and handles HTTP tunneling, requesting the original page from backend web application, modifying and returning it back to client. Log Manager is the other module which logs every access to the database or file. So detailed access logs of the whole system are handled by a single source.

System Details

EYEKS executes 4 main scenarios; *Login, Page Request, File Upload* and *Logout*. Login can be configured to work with hardware-token authentication, LDAP authentication, or password based authentication. *Page Request* and *File Upload* scenarios are fully controlled by RAD implementation and distributed session management semantics. *Logout* basically deletes user session.

Authorization strategy of EYEKS is based on combination of RBAC (Role Based Access Decision) and RAD (Resource Access Decision) standards, which are implemented by CSAAS [10]. Access policies written on resource and operation pairs give full force to the system on closing security breaches of the backend web applications. In these policies, all parameters can be checked for their allowed values. Also time based access policies can be implemented to control the permissions. For example, the web application can be closed to users having a specific role for some time period. IP based policies are other common types of policies that can be frequently used. Authentication module of EYEKS combines RAD with RBAC standard. So users can be assigned to system defined *roles*. And the roles a user assigned to define which *permissions* of RAD standard the user have. These roles can be hierarchically arranged. The authorization system of EYEKS can be fully configured by the system administrator according to the security needs of the system.

HTTP tunneling is an important strategy for securing a backend application. Client requests come only to application security layer, and then, the application security layer maps this request to the appropriate web application, and creates a new connection to the backend server. Firewall between application security layer and backend servers must be configured only to allow requests from application security layer and to deny everything else. Request dispatcher module, described in System Architecture section, is responsible for HTTP tunneling. The mapping between coming requests to backend web applications is described in a property file.

EYEKS offers a new more secure method for handling user sessions. The method not only considers security but also considers distribution execution so that web applications can be installed in clustered servers. Session management is handling by using encrypted token that hold user credentials such as user id and request sequence number. Using sequence information avoids session hijacking. Even if this encrypted token is hijacked by a malicious user, sending it back to the application security layer will not work. This token is inserted in every response of user request and it is granted that it will send back with the next request of user.

4. Test Results

EYEKS has been used in one of the biggest e-government projects of Turkey. The system became online on October 2004, and has been used for nearly 2.5 years. Table 2 gives statistics about how many business transactions, login and page request has been done per month in last year respectively. The last column stands for the total request numbers on a peek day.

Table 2. EYEKS Usage Statistics

	Start	Average	Peek	Peek Day
Transactions	34,534	1,856,324	3,124,236	649,024
Login	131,448	3,559,883	6,910,198	833,670
Page Request	975,481	39,046,279	63,808,771	4,128,295

Project is started with 13,466 registered users and by January 2007, 181,747 users have been registered. 4 resources (project consists of four different web applications), 67 operations, and 8 different roles are defined in RAD implementation. Enterprise security rules of the project were implemented by 11 user defined policies and by 210 permission mappings to policies.

EYEKS is installed on 3 servers. Two machines have four Solaris Ultra SPARC CPUs with 8GB RAM and one machine has two Solaris Ultra SPARC CPU with 2GB RAM. EYEKS scales and responses well though the heavy load as stated in table 2. Average peek CPU usage on peek days is 27%, where backend servers (8 servers) usage is 92%. On the other hand, artificial load test showed that under the load of 300 concurrent users, EYEKS payload is only 8 %. EYEKS logs also showed that in last 3 months, total number of 865,327 requests, and 938,787 incorrect password tries per month were classified as malicious and were denied.

5. Discussion & Conclusion

In this paper, a fully implemented solution, EYEKS, to secure a web application is presented. Adding a new layer to web application, which deals with all security aspects, frees application developers from thinking security aspects of their applications. This approach also results in a more functional and structured system.

EYEKS is designed to eliminate the threats stated in problem definition section. Creating a separate layer, using HTTP tunneling and making service proxy for backend applications, eliminates information disclosure of information type of attacks such as common file query, link traversal, directory enumeration, and path truncation. On the other hand, parameter filtering and input validation mechanisms, implemented by application security policies that executes on RAD solution, eliminate invalidated input type of attacks such as cross site scripting, SQL injection, parameter passing, and forcing a parameter. User credentials are protected from session stealing attacks such as session hijacking and cookie tampering by a secure implementation of distributed session handling mechanisms.

A centralized view of security aspects enables the web application to be more manageable. Business-depended enterprise-level security policies and protection mechanisms (application-level security policies) can be added together to form a full security policy chain that can be managed by RAD specification. RAD implementations offer high available, fine-grained, extensible, and dynamic access control mechanism which suits well for web application authorization needs.

As a result, adding an application security layer that controls organization-wide security policies, can give great benefits such as reusability, manageability, scalability to all kinds of web applications.

6. References

[1] OSI, "Information Technology -- Open Systems Interconnection -- Security frameworks in open systems -- Part 3: Access control," ISO/IEC JTC1 10181-3, 1994.

[2] K. Beznosov. Object Security Attributes: Enabling Application-Specific Access Control in Middleware. In DOA'02, pages 693-710, London, UK, October 2002.

[3] H. Ossher and P. Tarr. Using multidimensional separation of concerns to (re)shape evolving software. Commun. ACM, 44(10):43-50, 2001.

[4] S. Jajodia, P. Samarati, M. L. Sapino, and V. S. Subrahmanian. Flexible support for multiple access control policies. ACM Trans. Database Syst., 26(2):214–260, 2001.

[5] T. Ryutov and C. Neuman, "Access Control Framework for Distributed Applications (Work in Progress)," IETF, Internet Draft draft-ietf-cat-acc-cntrl-frmw-03, March 9 2000.

[6] T. Ryutov and C. Neuman, "Generic Authorization and Access control Application Program Interface: C-bindings," IETF, draft-ietf-cat-gaa-bind-03, March 9 2000.

[7] OASIS. Core Specification: eXtensible Access Control Markup Language (XACML) Version 2.0.

[8] N. Damianou, N. Dulay, E. Lupu, and M. Sloman. The Ponder Policy Specification Language. LNCS, 1995:18-28,2001.

[9] Y. Göğebakan. Cok Katmanlı Internet Uygulamalarında Yetkilendirme Problemi, Akademik Bilişim 2005

[10] OMG Security Specifications, Resource Access Decision (RAD) V1.0

[11] NIST Standards, Role Based Access Control (RBAC)

[12] BEAWebLogic Enterprise Security V4.2, (last accessed April 28 2007) http://edocs.bea.com/wles/docs42/index.html.

[13] Oracle Access Manager.

[14] IBM WebSphere Application Server.

[15] Securant, (last accessed April 28 2007) "Unified Access Management: A Model for Integrated Web Security," http://www.cleartrust.com.

[16] WebDeamon, Integrative Security Management for Web-Based Enterprise Applications

[17] SANS Institute, (last accessed April 28 2007) http://www.sans.org/top20/.

[18] The Open Web Application Security Project, http://www.owasp.org/index.php/Main_Page

[19] OWASP Top Ten Project, (last accessed April 28 2007) http://www.owasp.org/index.php/Category:OWASP_Top_Ten_Project.

[20] Web Application Security Consortium Glossary.

[21] D. Scott and R. Sharp, Abstracting Application-Level Web Security

[22] ModSecurity, (last accessed April 28 2007) Open Source Web Application Firewall, http://www.modsecurity.org.

[23] SecureSphere, (last accessed April 28 2007) Web Application Firewall http://www.imperva.com/products/securesphere/web_application_firewall.html.

[24] Citrix Application Firewall, (last accessed April 28 2007) http://www.citrix.com/English/ps2/products/product.asp?contentID=25636

More Secure Authentication using Multiple Servers (MSAMS)

Belgin Bilgin, İbrahim Sogukpinar

Gebze Institute of Technology, Kocaeli, TURKEY
{belgin, ispinar}@bilmuh.gyte.edu.tr

Abstract. Using remote user authentication is an important necessity in many applications. There are many methods to provide security in these applications. Some of them use smart cards and one secret key stored server. In this work we have proposed an authentication scheme (MSAMS) using multiple servers. MSAMS is more secure than other solutions because the main goal of MSAMS is to provide its security by distributing parts of secret. MSAMS use (k,n)-threshold scheme to generate parts of the key and to recompose the secret. The system consists of secure authentication, mutual confidence and security of all transaction.

Keywords: Security, Remote Authentication, Smart Card, Sharing Secret.

1 Introduction

Nowadays, we use remote user authentication in many applications such as software of mobile phones, e-commerce, e-banking, e-government, remote login and database management systems. These operations are very important and should be secure. Therefore, we have to provide the security of them during the transmission over insecure networks. Many schemes based on hash functions [9, 15, 20, 26 and 32] and based on public key [2, 5, 11, 13, 18 and 25] have been suggested for user authentication. Some schemes [2, 3, 4, 5, 6, 7, 8, 10, 13, 18 and 25] use smart cards. All of them consist of the same weakness: the security of the secret key. Because, the scheme has only one secret and this secret key is stored in one server. If someone corrupts the server, he learns the secret and could generate new illegal users' smart cards. In this paper, we have proposed a new solution for secure authentication system using smart cards and multiple servers.

In a banking application, transaction is done by using POS machine or kiosk. There is the Internet connection between the client machine and authentication server. However, user and remote system need secure communication on the Internet. For this purpose, we must supply mutual authentication and generate a unique session key for using in the following operations. Some schemes support mutual authentication and session key agreement. All of them have same risks about the server's secret key. If the secret key is stolen, the system is not secure anymore.

In this work, we have proposed a solution for the security of similar systems. The proposed system has three parts: Initialization Process, Registration and

Authentication & Session Key Agreement Phases, and Password Changing Process. The security of the proposed system depends on the Shamir's secret sharing algorithm [24] and security of one-way hash function and it is nonce-based. The nonce is a generated random number that its value has not been used before. This is used to avoid replay attack and the time synchronization problem.

The remainder of this paper is organized as follows. In section 2, we review some fundamental information and related works. In section 3, we introduce the proposed solution. In section 4, we analyze the security and the performance of the proposed scheme. Finally, we conclude this paper in section 5.

2   Background Information and Related Works

There are many solutions for remote authentication. Some schemes [17, 21, 33, 31] use password or verification tables stored remote system for checking the users. These schemes will suffer from the many security risks because the tables are maintained in the remote server. For instance, if an attacker modifies the verification table, the system will be broken. To keep this table in the remote server is also costly.

Later, schemes using smart cards has proposed. In these schemes, servers have own secret key and do not need to keep any password or verification tables. Some of them are related to smart cards. We can classify them as follows [28]. The first class is RSA-based authentication schemes, such as Yang-Shieh Scheme [29], Fan-Li-Zhu Scheme [10] and Yang-Wang-Chang Scheme [30]. Another class is ElGamal-based authentication schemes. For example: Hwang-Li Scheme [13], Awasthi-Lal Scheme [2] and Kumar's Scheme [16]. The last class is Hash-based authentication schemes. There are some example for this class: Sun's Scheme [27], Hwang-Lee-Tang Scheme [12], Chien-Jan-Tseng Scheme [8], Chen-Lee-Horng Scheme [6], Liao-Lee-Hwang Scheme [22],[28], Yoon-Ryu-Yoo Scheme [34], Lee-Kim-Yoo Scheme [19].

An ideal user authentication scheme using passwords should have following properties defined by [28].

P1. The scheme does not need password or verification tables.
P2. The users can chose and change freely their passwords.
P3. The password cannot be revealed by the administrator of the system.
P4. The password is not transmitted in plain text over the insecure network.
P5. The length of the passwords must be as short as possible for memorizing.
P6. The scheme must be efficient and practical.
P7. Any unauthorized login attempts can be quickly detected.
P8. At the end of the authentication, a unique session key should be generated.
P9. The ID should be dynamically changed for every session.
P10. The scheme is still secure even if the server's secret key is stolen.

Also, it should have capability of these security properties in [28].

S1. Denial of Service Attacks
S2. Forgery Attacks (Impersonation Attacks)
S3. Forward Secrecy
S4. Mutual Authentication
S5. Parallel Session Attacks
S6. Password Guessing Attacks

S7. Replay Attacks
S8. Smart Card Loss Attacks
S9. Stolen-verifier Attacks

Some banking applications work on the Internet. The secret information of a client (user) is transferred over the insecure network. When a user wants to pay the cost of his shopping, he could use his credit card with chip, which is a smart card. He enters your password. After the user authentication, the main job,payment, is realized. There are three important matter: secure authentication, mutual confidence and security of all transaction. Firstly, we have to provide a secure authentication. Two sides must be able to trust each other. Also, we need security during transaction. MSAMS covers them.

3   More Secure Authentication using Multiple Servers (MSAMS)

MSAMS includes five servers and two firewalls. One of the servers is the registration server (RegSer), another is the authentication server (AuthSer) and the others are storage servers storing the parts of the secret key. The servers settled behind a firewall are secure network except AuthSer. AuthSer is in the demilitarized zone (DMZ).

RegSer generates the secret key and the special function for sharing secret. Computing the parts of the key, storing the first part and distributing the other parts of the key are the jobs of RegSer. (k,n)-threshold scheme [24] is used for sharing the key. We suggest a new method, which has some modifications on Juang's scheme [14]. All system is represented in Figure 1.

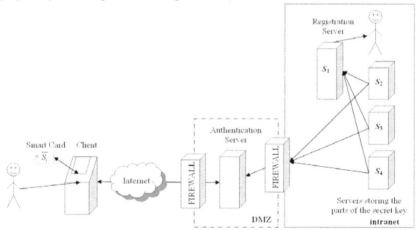

Figure. 1. System Block Diagram

MSAMS has all of the aforementioned properties except P6 and P9.

3.1 Initialization Process

Firstly, we need to generate a (s) secret key. This job is processed by RegSer. The secret key is used in registration and authentication phases. We use (k,n)-threshold scheme for sharing secret [24] to provide the key security. RegSer generates a special function $f(.)$mod p. Where the degree of the function is $k-1$, $d(f)=k-1$, p is a big prime number and the constant value of the function is s. In our solution, the value of k is selected as four, so $d(f)$ is 3. It means that four parts, called *share*, is used to recompose secret key. This value also defines the number of storage servers. So, the system uses three storage servers. Three shares are kept in the storage servers. Other two shares are stored in the user's smart card. When the user wants to login the system, AuthSer will recompose the secret key by taking two shares from the user's smart card and other two shares from storage servers. The remaining share in the storage server is auxiliary for the failing possibility of any server. If one of the storage servers fails then the auxiliary comes to action. Thus, computing the secret key is successful even if one of the servers is down or inaccessible. If we use less storage servers, such as two, RegSer can not compute secret key by using Lagrange Interpolation Formula [1]. Also, using more storage servers will cause maintenance problem. It also gives opportunity to an attacker to compose secret key without user's smart card.

Each share includes three values: $s_i=(x_i,y_i,l_i)$, where $y_i=f(x_i)$ and l_i is priority level of share $(l_i =\{0,1\},l_i=1$ if share's owner is user or RegSer, $l_i=0$ otherwise). RegSer generates x_1, x_2, x_3, x_4 and computes y_1, y_2, y_3, y_4 for these values. RegSer stores $s_1=(x_1,y_1,1)$ and function $f(.)$ mod p. Then it distributes $s_i=(x_i,y_i,0)$, for $i=2,3,4$, storage servers. The storage servers store these values to use in registration and authentication phase. Then, the values of s, s_2, s_3 and s_4 are removed.

3.2 Registration Phase

When a user wants to register the system, RegSer calls $\{s_2, s_3, s_4\}$ from storage servers. It recomposes the secret key s by using s_1, which is kept on RegSer, and $\{s_2, s_3, s_4\}$. After that, RegSer deletes the values of $\{s_2, s_3, s_4\}$ quickly. Afterwards, U_i : $v_i=h(ID_i,s)$, which is the user's secret, is computed by RegSer. ID_i identifies user U_i, s is a secret number computed in RegSer and $h(.)$ is a secure one-way function. RegSer generates x_{i1} and x_{i2}, and calculates y_{i1} and y_{i2}. s_{i1} consists of the values of x_{i1}, y_{i1} and $l=1$. Also, s_{i2} consists of the values of x_{i2}, y_{i2} and $l=1$. After the computation, the value of s is removed. The value of $\overline{s_i}$, which represented in (1), is computed by encrypting the values of s_{i1} and s_{i2}.

$$\overline{s_i} = E'_{SPB}\left(s_{i1},s_{i2}\right) \tag{1}$$

Later, the user enters his password and $w_i = v_i \oplus PW_i$ is computed, where the symbol "$\oplus$" denotes an exclusive operation. After these computations, ID_i, $\overline{s_i}$ and w_i are written on the smartcard of user U_i. Finally, the value of s_i is deleted.

3.3 Authentication and Session Key Agreement Phase

This phase consists of three steps described as follows:

Step 1: ru_j, which is the jth random number to use in computing session key, and N_1, which is a nonce, are generated in smartcard. Smartcard computes $v_i = w_i \oplus PW_i$ after the user enters his password. Later, it computes $EncMsg$ in (2).

$$EncMsg = E_{v_i}\left(ru_j, h(ID_i \parallel N_1)\right) \tag{2}$$

Where; E denotes symmetric encryption and the symbol "$\parallel$" presents string concatenation operation.

After these computations, N_1, ID_i, $EncMsg$ and $\overline{s_i}$ are sent AuthSer. Receiving the message, AuthSer decrypts $\overline{s_i}$ by using own secret key (SPR: Server's PRivate key) (3). Thus, it gets the first part of the secret key. Then, it requests other shares from storage servers. It checks value of all l. AuthSer recomposes secret key by using Lagrange Interpolation Formula [1] if there are shares such that two shares, which have $l=1$ from user, and two shares, which have $l=0$ from storage servers.

$$\left(s_{i1}, s_{i2}\right) = \left((x_{i1}, y_{i1}, 1), (x_{i2}, y_{i2}, 1)\right) = D'_{SPR}\left(E'_{SPB}\left(s_{i1}, s_{i2}\right)\right) \tag{3}$$

$$v_i = h(ID_i, s) \tag{4}$$

AuthSer generates v_i by using s (4). Later, it destroys s, s_{i1}, s_{i2}, s_2, s_3, s_4 and it decrypts $EncMsg$ and it gets ru_j. It checks whether the message contains $h(ID_i \parallel N_1)$ and N_1 is fresh. If both of these controls are yes, AuthSer computes and sends $E_{v_i}(rs_j, N_1+1, N_2)$ to user, where; rs_j, which will be used to create a session key k_j with ru_j, is a random number generated in AuthSer, like ru_j.

Step 2: After receiving the message, smartcard decrypts this message by using $D_{v_i}(E_{v_i}(rs_j, N_1+1, N_2))$. Then, it checks whether N_1+1 is fresh or not. If it is yes, it computes j^{th} session key k_j in (5).

$$k_j = h\left(rs_j, ru_j, v_i\right) \tag{5}$$

Then, smartcard computes and sends the new message in (6) to server.

$$E_{k_j}\left(N_2 + 1\right) \tag{6}$$

Step 3: AuthSer generates k_j and decrypts the message received by using (7). Then, it checks whether N_2+1 is fresh or not. If this value is fresh, secure communication can be started between user U_i and server by using k_j known by user and server.

$$D_{k_j}\left(E_{k_j}\left(N_2 + 1\right)\right) \tag{7}$$

3.4 Password Changing Process

MSAMS solves also the changing password problem. RegSer or AuthSer is not used in this phase. User can change his password himself on client systems. After user types his old password, the client system computes v_i (8). Later, the user chooses a new password and enters this password. The client system computes the new w_i (9) and updates the value of w_i on user's smartcard.

$$v_i = w_i \oplus PW_i \tag{8}$$

$$w_i' = v_i \oplus PW_i' \tag{9}$$

4 Security and Performance Analysis and Comparison

The properties of MSAMS and the security of the system are explained in this part of the article.

1. MSAMS contains no verification or password table.
2. Users can choose their passwords freely and also can change them.
3. *Password Changing Process:* In the password changing process, there is no check whether the old password is valid or not. It means that the smart card may be unusable if the user enters the wrong password.
4. *Time Synchronization Problem:* There is no time synchronization problem because the MSAMS system does not use timestamps. The system is the nonce-based scheme.
5. *Collisions in Hashes:* Hashing is used in this scheme. The output bit size is longer than the original messages in MSAMS. So, hash collision does not occur. This property is related to the hashing algorithm. Chosen strong one-way hash function, $h{:}x{\rightarrow}y$, must provide following properties [23]:
    1. x can be any size
    2. The output must be fixed size
    3. When f and x is known, it is easy to compute $f(x)$
    4. It must be infeasible to find any pair of x and x', where $x{\neq}x'$ and $f(x){=}f(x')$. Also, this function has two more properties:
    5. The output size of this one-way has function must be 128 bits.
    In additionally, the sizes of *ID* and *N* must be fixed. This property prevents hash collision like $h(ID_i' \parallel N_1'){=}h(ID_i \parallel N_1)$.
6. *Denial of Service Attacks:* MSAMS is secure against denial of server attacks.
7. *Forgery Attacks (Impersonation Attacks):* Random numbers cannot derive from *EncMsg*. The attacker cannot create another valid message.
8. *Forward Secrecy (Stolen Secret Key):* MSAMS protects secret key by using (k, n)-threshold scheme. The secret key is not stored in only one computer. Three shares are in storage servers. Users have their share. Another share is in RegSer, which is in the secure zone and is closed to remote access. Nobody can recompose the secret key without user side even other parts are known. Two users would have

four shares, but they could not recompose secret because each user has two shares encrypted in his smart card. MSAMS is important because of providing security of the secret key.

9. *Mutual Authentication:* The AuthSer and the users can authenticate each other.
10. *Parallel Session Attacks:* This attack does not work on our scheme because the attacker need secret key to compute *EncMsg* including N_1.
11. *Password Guessing Attacks:* MSAMS is secure against password guessing attack. The system uses very simple method for this attack. The user can try to login to system maximum three times. After the third wrong password, the system does not accept new connection request of the user.

The size is 32 bits for passwords, which are length 4 characters. The number of various passwords is 2^{32}. An attacker could find the password after 2^{31} attempts. The probability of this is $3/2^{31}$. When we choose the length of the passwords is 8 characters, 64 bits, the number of attempts is 2^{63}. The probability in here is $3/2^{63}$. We prefer 8 characters passwords because its probability is less, $(3/2^{63}) << (3/2^{31})$.

12. *Replay Attacks:* An attacker cannot create different message by using new nonce value. The attacker must know user's secret v_i. Using nonce prevents replay attacks.
13. *Smart Card Loss Attacks:* The attacker need right password to change password even if the user loss smart card. Because, the system does not check entered password. If attacker enters wrong password as old one, it will make the smart card unusable.
14. *Stolen-verifier Attacks:* Password-verifiers are not stored in servers. So, this attack is not successful on MSAMS.

The following tables (Table 1 and Table 2) compare MSAMS to some other schemes.

Table 1. Security Comparisons [28] (Si: Security requirements, Y: Supported, N: Not Supported)

Scheme	S1	S2	S3	S4	S5	S6	S7	S8	S9
MSAMS	Y	Y	Y	Y	Y	Y	Y	Y	Y
Yoon et al. [29]	Y	Y	Y	Y	Y	Y	Y	N	Y
Juang [32]	Y	Y	Y	Y	Y	Y	Y	N	Y
Chien et al. [14]	Y	Y	Y	Y	N	Y	Y	N	Y
Sun [26]	Y	Y	N	N	Y	Y	Y	Y	Y

Table 2. Properties Comparisons [28] (Y: Achieved N: Not Achieved)

Scheme	P1	P2	P3	P4	P5	P6	P7	P8	P9	P10
MSAMS	Y	Y	Y	Y	Y	N	Y	Y	N	Y
Yoon et al.[29]	Y	Y	N	Y	Y	Y	Y	N	N	Y
Juang [32]	Y	N	Y	Y	Y	Y	N	Y	N	N
Chien et al. [14]	Y	N	N	Y	Y	Y	N	N	N	N
Sun [26]	Y	N	N	Y	Y	Y	N	N	N	N

4.1 Performance Analysis

MSAMS consists of four part: initialization process, registration phase, authentication phase and password changing process.

In the following table, each part of the computational costs are demonstrated. MSAMS is preferable although the computational cost is much. Because, it provides secure system which is aforementioned in this section.

Table 3. Performance Analysis

Stage	Time Complexity
Initialization Process	$4T_f+T_{Gf}$
Registration Phase	$T_{LIF}+T_h+T_f+T_\oplus+T_S$
Authentication & Session Key Ag. Phase	$T_\oplus+4Th+3T_S+4T_S+T_{LIF}$
Password Changing Process	$2T_\oplus$

T_f: computational cost of f (.)mod p, T_{Gf}: cost of generating f (.) mod p function, T_{LIF}: computational cost of Lagrange Interpolation Formula, T_h: cost of hashing function, $T_\oplus$: cost of exclusive operation, T_S: cost of symmetric encryption/ decryption

5 Conclusion

There are many applications which need remote authentication. Providing the security of them is an important necessity. MSAMS proposed in this work includes secure authentication, mutual confidence and security of all transaction. It provides secrecy of remote authentication via secret sharing. Also secrecy of the following operations is increased by using session key. Although the computational cost of this solution is high, the system is preferred for more security. However, it needs to improve the performance of proposed solution.

References

1. Abramowitz, M. and Stegun, I.A. (Eds.):   Handbook of Mathematical Functions with Formulas, Graphs, and Mathematical Tables", 9th printing. New York: Dover (1972) 878-879 and 883
2. Awasthi, A.K. and Lal, S.: A remote user authentication scheme using smart cards with forward secrecy, IEEE Transactions on Consumer Electronics, Vol. 49, no. 4 (2003) 1246–1248
3. Chan, C.K. and Cheng, L.M.: Cryptanalysis of a remote user authentication scheme using smart cards, IEEE Transaction on Consumer Electronics, Vol. 46 (2000) 992–993
4. Chang, C.C. and Wu, T.C.: Remote password authentication with smart cards, IEE Proceedings-E, Vol. 138 (1991) 165–168
5. Chang, C.C. and Liao, W.Y.: A Remote Password Authentication Scheme Based upon ElGamal's Signature Scheme, Computers & Security, Vol. 13, no. 2 (1994) 137–144
6. Chen, T.H., Lee, W.B., and Horng, G.: Secure SAS-like password authentication schemes, Computer Standards & Interfaces, vol. 27 (2004) 25–31

7. Chien, H.Y., Jan, J.K., and Tseng, Y.M.: A modified remote login authentication scheme based on geometric approach, Journal of Systems and Software, Vol. 55 (2001) 287–290

8. Chien, H., Jan, J., Tseng, Y.: An efficient and practical solution to remote authentication: smart card, Computer&Security, 21(4) (2002) 372-375

9. Das, M.L., Saxena, A. and Gulati, V. P.: A Dynamic ID-based Remote User Authentication Scheme, IEEE Trans. On Consumer Electron., Vol. 50, no.2 (2004) 629-631

10. Fan, L., Li, J.H., and Zhu, H.W.: An enhancement of timestamp-based password auhentication scheme, Computers & Security, Vol. 21 (2002) 665–667

11. Jablon, D. P.: Strong Password-only Authenticated Key Exchange, ACM Computer Communications Review, Vol. 26, no. 5 (1996) 5-20

12. Hwang, M.S., Lee, C.C. and Tang, Y.L.: A simple remote user authentication scheme, Mathematical and Computer Modelling, Vol. 36 (2002) 103–107

13. Hwang, M.S. and Li, L.H., A New Remote User Authentication Scheme Using Smart Cards, IEEE Trans. On Consumer Electron., Vol. 46, no. 1 (2000) 28-30

14. Juang, W.S., Efficient password authenticated key agreement using smart cards, Computer&Security, Vol. 23, no. 4 (2004) 167-173

15. Ku, W.C.: A Hash-Based Strong-Password Authentication Scheme without Using Smart Cards, ACM Operating Systems Review, Vol. 38, no. 1 (2004) 29-34

16. Kumar, M.: New remote user authentication scheme using smart cards, IEEE Transactions on Consumer Electronics, Vol. 50, no. 2 (2004) 597–600

17. Lamport, L., Password authentication with insecure communication, Comm. ACM 24 (1981) 770–772

18. Lee, C.C., Hwang, M.S., and Yang, W.P.: A flexible remote user authentication scheme using smart cards, ACM Operating Systems Review, Vol. 36, no. 3 (2002) 46-52

19. Lee, S.W., Kim, H.S. and Yoo, K.Y.: Improved efficient remote user authentication scheme using smart cards, IEEE Transactions on Communications, Vol. 50 (2004) 565–567

20. Lee, C.C., Li, L.H., Hwang, M.S.: A Remote User Authentication Scheme Using Hash Functions, ACM Operating Systems Review, Vol. 36, no. 4 (2002) 23-29

21. Lennon, R.E., Matyas, S.M., Mayer C.H.: Cryptographic authentication of time-invariant quantities, IEEE Trans.Commun. COM-29 (6) (1981) 733–777

22. Liao, I-En, Lee, C.C., and Hwang, M.S.: Security enhancement for a dynamic id-based remote user authentication scheme, in IEEE CS Press, International Conference on Next Generation Web Services Practices (NWeSP'05), Seoul, Korea (2005) 437–440

23. Merkle, R. C.: One-way hash functions and DES, in Advances in Cryptology, CRYPT0'89, Lecture Notes in Computer Science, Vol. 435 (1989) 428-446

24. Shamir, A.: How to share a secret, Communications of the ACM 22, (1979) 612-613

25. Shen, J.J., Lin, C.W., and Hwang, M.S.: A Modified Remote User Authentication Scheme Using Smart Cards, IEEE Trans. On Consumer Electron., Vol. 49, no. 2 (2003) 414-416

26. Shimizu, A., Horioka, T., and Inagaki, H.: A Password Authentication Method for Contents Communications on the Internet, IEICE Trans. On Commun., Vol. E81-B, no. 8 (1998) 1666-1673

27. Sun, H.: An efficient remote user authentication scheme using smart cards, IEEE Trans Consumer Electron, Vol. 46, no. 4 (2000) 958-961

28. Tsai, C.S., Lee, C.C., and Hwang M.S.: Password Authentication Schemes: Current Status and Key Issues, International Journal of Network Security, Vol.3, No.2 (2006) 101–115

29. Yang, W.H. and Shieh, S.P.: Password authentication schemes with smart cards, Computers & Security, Vol. 18, no. 8 (1999) 727–733

30. Yang, C.C., Wang, R.C., and Chang, T.Y.: An improvement of the Yang-Shieh password authentication schemes, Applied Mathematics and Computation, Vol. 162 (2005) 1391–1396

31. Wang, S.J., Remote table-based log-in authentication upon geometric triangle, Comp. Stand. Inter. 26 (2004) 85–92

32. Yeh, T.C., Shen, H. Y., and Hwang, J. J.: A Secure One-Time Password Authentication Scheme Using Smart Cards, IEICE Trans. on Commun., Vol. E85-B, no. 11 (2002) 2515-2518
33. Yen S.M., Liao, K.H.: Shared authentication token secure against replay and weak key attack, Inform. Process. Lett.(1997) 78–80
34. Yoon, E.J., Ryu, E.K., and Yoo, K.Y.: An improvement of Hwang-Lee-Tang's simple remote authentication scheme, Computers & Security, Vol. 24, no.1 (2005) 50-56

Research Challenges in Cryptology
(Extended Abstract)

Bart Preneel

Katholieke Universiteit Leuven, Dept. Electrical Engineering-ESAT/COSIC,
Kasteelpark Arenberg 10, B-3001 Leuven, Belgium
bart.preneel@esat.kuleuven.be

Abstract. During the last decade, cryptology has become a commodity. Many present day on-line applications are unthinkable without cryptographic technology. For example, cryptographic algorithms and protocols have been integrated into our bank cards, mobile phones and software systems to protect information and transactions and to offer privacy. These success of cryptology may lead us to believe that the cryptography problem is "solved." However, this extended abstract demonstrates that there are still many challenges ahead, both in the area of foundations and applications of cryptology.

Keywords: cryptology, information security

1 Introduction

The science of cryptology studies mathematical techniques in order to provide secrecy, authenticity and related properties for digital information [12]. It also allows to establish trust relationships over open networks and enables the collaboration of mutually distrusting parties towards achieving a common goal. Cryptology is a fundamental enabler for security, privacy and dependability in an on-line society. Cryptographic techniques can be found at the core of computer and network security, of digital identification and digital signatures, digital rights management systems, etc. Their applications vary from e-business, m-business, e-voting and on-line payment systems to wireless protocols and ambient intelligence. Today we find cryptology in our GSM or 3G mobile phones, on our bank or credit cards, on our desktop (browser), on WLAN connections, and for a growing number of European citizens even in our electronic identity cards.

From an implementation point of view, cryptography has evolved from manual methods to mechanical devices in the first half of the twentieth century. From the 1960s to the 1980s cryptography was mainly implemented in electronic *hardware* devices. Around 1990 general purpose hardware became fast enough to allow for *software* implementations of cryptology. We are now witnessing a trend in which cryptographic hardware will be integrated into every general purpose processor, while an ever increasing number of small processors (hundreds or thousands of devices per user) that bring cryptology *everywhere*.

For outsiders, who have a limited understanding of the complexity of the field, the widespread deployment of cryptology may give the impression that the cryptography problem has been "solved." We have cryptographic algorithms and protocols available that can be called as a "black box" by security engineers; consequently, one may believe that research efforts in security should be focused exclusively on building trust infrastructures, addressing application security issues and integrating security into applications. This (incorrect) impression is strengthened by the (correct) observation that security systems fail usually due to other reasons than cryptographic flaws (such as incorrect specifications or implementations, malicious software, incorrect configurations, social engineering attacks, ... [1]).

While cryptology is a scientific discipline that is becoming more mature, there is a strong need for research, both in the area of foundations and in the area of applied cryptology. On the one hand, the world in which cryptosystems are deployed is changing and the threats to their security increase. This calls for continuous monitoring of state-of-the art cryptanalytic methods in order to assess the security of deployed systems. Maintenance of their security is crucial for making our information infrastructure secure. On the other hand, future developments (e.g., ambient intelligence) present new challenging applications, which need to be addressed by different or better cryptologic methods than the ones we know today.

2 Research Challenges

Three main research challenges in cryptology have been identified as a priority by the ECRYPT Network of Excellence [6].

2.1 Improved trade-offs

Cryptographic algorithms and protocols offer a trade-off between cost (footprint in hardware or software, power and/or energy consumption), security and performance. Achieving any two of these three is straightforward. For example, if cost is not an issue, it is rather easy to make a very fast and very secure solution. Similarly, if a system can be very slow, it can offer an acceptable security at a low cost. The real challenge is to improve these three parameters at the same time, in particular for specific extremes beyond the current state of the art, namely:

Extremely low cost solutions are essential to get cryptography everywhere (ambient intelligence, sensor networks and RFIDs). A specific target is encryption with less than 1500 gates or an entity authentication protocol that consumes less than 50 mJoules.

Extremely fast solutions for applications such as bus encryption, and authenticated encryption for Terabyte storage devices and Terabit networks.

Need for high security solutions: for applications such as e-voting, e-health and national security we need cryptographic algorithms that provide guaranteed protection for 50 years or more. While this is conceivable for symmetric cryptography, this goal is currently a major challenge for public key cryptography, in view of progress in research to attack hard mathematical problems and the anticipated development of quantum computers.

An example of the areas in symmetric cryptography in which substantial progress is needed are hash functions and stream ciphers. For hash functions, more work is needed to develop a basic understanding of the required security properties and how these properties can be achieved by generic and specific constructions. This has become apparent after the breakthroughs of Wang et al. [18, 19] on MD5 [16] and SHA-1 [8]. This has prompted NIST to start an open competition for an advanced hash function standard. For general purpose applications, block ciphers are probably more suited than stream ciphers and we seem to have a better understanding of how to design a good block cipher. However, if we want to encrypt 3 to 5 times faster than AES [4, 9] or to develop very compact hardware, stream ciphers may offer a specific advantage. The goal is to push the limits without compromising long term security. The eSTREAM initiative [7] organized by ECRYPT tries to make progress towards achieving this goal. The widely used public key algorithms depend on a small set of problems in algebraic number theory (factoring [15], discrete logarithm in $\mathbb{Z}_p$ [5] and discrete logarithm on an elliptic curve over a finite field [2]). If a breakthrough would be made in solving any of these problems, or if quantum computers could be built [17], we would have to abandon all these schemes. There exists a small number of alternatives based on coding theory and lattices. These alternative public-key schemes typically require more computation and/or memory than the schemes used in practice; in addition, more security analysis is needed before they can be widely adopted. For a more detailed overview of recent progress in cryptanalysis, see [14].

2.2   Advanced protocols

While substantial progress has been made in proving the security of simple building blocks such as authenticated encryption or two-party authenticated key agreement, the development and analysis of more complex cryptographic protocols for distributing trust is a major challenge. These protocols allow to reduce the requirement of trusting a centralized system and/or specific machines or hardware components by distributing this trust over a larger number of entities. This is very important for privacy sensitive applications such as voting, auctions and mining of medical data. The creation of complex cryptographic protocols has been based on an approach for provable security based on assumptions for cryptographic algorithms; due to the complexity of the proofs and methods, there is a need for advanced tools to develop and verify such proofs. A second goal is to take into account more realistic deployment models such as concurrent composition, asynchronous interactions, complex privacy models and protocols in which the players are not "malicious" or "honest" but rather "rational"; the latter approach results in a novel game theoretic approach.

2.3   Secure implementations

The need for the development of secure hardware and software implementations of cryptographic components has been understood only in the past decade. By now it is clear that the impact of physical attacks (such as timing attacks [10, 13], power analysis [11] and fault attacks [3]) on implementations is much larger than anticipated; every year several new side channels are being discovered, and there appears to be no uniform way to counter these. A substantial research effort is required to develop a solid theory and countermeasures against physical attacks at all layers (circuit, logic gate, algorithm protocols); this problem appears to be very difficult.

3   Conclusion

The basic research questions in cryptology deal with questions such as: which problems are hard, is it hard to factor the product of two large primes, or how difficult is it to solve large non-linear systems of equations, even if these a generated from a cipher with a regular structure. All of these problems are known to be very difficult, and researchers understand that so far we have only scratched the surface in this area.

The goal of cryptography is to build on these foundations by constructing algorithms, primitives, protocols and applications. The long term target is to develop clear security definitions and to create a deep understanding of the relation between the security properties of the applications and those of all the building blocks. This may make it easier to develop stable and long term solutions, that may stay secure for 10 years or more. In addition, we should develop systems in which building blocks can be replaced in an efficient and transparent way.

Acknowledgements. This author's work was supported in part by the Concerted Research Action (GOA) Ambiorics 2005/11 of the Flemish Government, by the IAP Programme P6/26 BCRYPT of the Belgian State (Belgian Science Policy), and by the European Commission through the IST Programme under Contract IST-2002-507932 ECRYPT.

References

1. R.J. Anderson, "Why cryptosystems fail," *Communications ACM*, Vol. 37, No. 11, November 1994, pp. 32–40.
2. R.M. Avanzi, H. Cohen, C. Doche, G. Frey, T. Lange, K. Nguyen, F. Vercauteren, *Handbook of Elliptic and Hyperelliptic Curve Cryptography*, H. Cohen, G. Frey, Eds., Chapman & Hall/CRC, 2005.
3. D. Boneh, R. DeMillo, R. Lipton, "On the importance of checking cryptographic protocols for faults," *Advances in Cryptology, Proceedings Eurocrypt'97, Lecture Notes in Computer Science 1233*, W. Fumy, Ed., Springer-Verlag, 1997, pp. 37–51.

4. J. Daemen, V. Rijmen, *The Design of Rijndael. AES – The Advanced Encryption Standard,* Springer-Verlag, 2001.
5. W. Diffie, M.E. Hellman, "New directions in cryptography," *IEEE Transactions on Information Theory,* Vol. IT–22, No. 6, 1976, pp. 644–654.
6. ECRYPT Deliverable D.SPA.22, *Challenges for Cryptology Research in Europe for 2007-2013,* revision 1.0, 26 May 2006, http://www.ecrypt.eu.org.
7. ECRYPT eSTREAM initiative, http://www.ecrypt.eu.org/estream
8. FIPS 180-2, *Secure Hash Standard,* Federal Information Processing Standard (FIPS), Publication 180-2, NIST U.S. Dept. of Commerce, August 26, 2002 (Change notice 1 published on December 1, 2003).
9. FIPS 197, *Advanced Encryption Standard,* Federal Information Processing Standard, NIST, U.S. Dept. of Commerce, November 26, 2001.
10. P. Kocher, "Timing attacks on implementations of Diffie-Hellman, RSA, DSS, and other systems," *Advances in Cryptology, Proceedings Crypto'96, Lecture Notes in Computer Science 1109,* N. Koblitz, Ed., Springer-Verlag, 1996, pp. 104–113.
11. P. Kocher, J. Jaffe, B. Jun, "Differential power analysis," *Advances in Cryptology, Proceedings Crypto'99, Lecture Notes in Computer Science 1666,* M. Wiener, Ed., Springer-Verlag, 1999, pp. 388–397.
12. A.J. Menezes, P.C. van Oorschot, S.A. Vanstone, *Handbook of Applied Cryptography,* CRC Press, 1997.
13. D. Osvik, A. Shamir, E. Tromer, "Cache attacks and countermeasures: The case of AES," *Topics in Cryptology – The Cryptographers' Track at the RSA Conference 2006, Lecture Notes in Computer Science 3860,* D. Pointcheval, Ed., Springer-Verlag, 2006, pp. 1–20. Extended version at www.wisdom.weizmann.ac.il/ tromer/papers/cache.pdf
14. B. Preneel, "A survey of recent developments in cryptographic algorithms for smart cards," *Computer Networks,* Vol. 51, No. 9, 2007, pp. 2223-2233.
15. R.L. Rivest, A. Shamir, L. Adleman, "A method for obtaining digital signatures and public-key cryptosystems," *Communications ACM,* Vol. 21, No. 2, 1978, pp. 120–126.
16. R.L. Rivest, "The MD5 message-digest algorithm," *Request for Comments (RFC) 1321,* Internet Activities Board, Internet Privacy Task Force, April 1992.
17. P.W. Shor, "Algorithms for quantum computation: discrete logarithms and factoring," *Proceedings 35nd Annual Symposium on Foundations of Computer Science,* S. Goldwasser, Ed., IEEE Computer Society Press, 1994, pp. 124–134.
18. X. Wang, H. Yu, "How to break MD5 and other hash functions," *Advances in Cryptology, Proceedings Eurocrypt'05, Lecture Notes in Computer Science 3494,* R. Cramer, Ed., Springer-Verlag, 2005, pp. 19–35.
19. X. Wang, Y.L. Lin, H. Yu, "Finding collisions in the ful SHA-1," *Advances in Cryptology, Proceedings Crypto'05, Lecture Notes in Computer Science 3621,* V. Shoup, Ed., Springer-Verlag, 2005, pp. 17–36.

On NTRU and Its Performance

Ali Mersin, Mutlu Beyazıt

Izmir Institute of Technology, College of Engineering, Dept. of Computer
Engineering, Gülbahçe, Urla, 35430 Izmir, Turkey,
{alimersin, mutlubeyazit}@iyte.edu.tr

Abstract. Hardness of lattice problems has introduced a new candidate for public key cryptosystems. NTRU is one of such cryptosystems. The fact that it works with small integers and that the complexities of key generation, encryption and decryption are relatively small leads to good overall performance when compared to the other public key cryptosystems. The aim of this study is to make a theoretical to practical introduction to NTRU and investigate how it performs against popular public key cryptosystems such as RSA and, especially, ECC.

1 Introduction

Researches on complexity of lattice problems have raised a candidate for public key cryptography. Based on hardness of lattice problems, several cryptosystems have been developed. Such as Ajtai-Dwork [1], Goldreich-Goldwasser-Halevi [2] and NTRU [3] cryptosystems. With key complexity of $O(n \log n)$ instead of $\Omega(n^2)$, NTRU has the best performance among the other lattice based cryptosystems. Today, extensive researches have been going on concerning the NTRU and no crucial security issue has been found so far. On this paper we will concentrate on NTRU cryptosystem and its performance.

NTRU uses a special lattice called NTRU Lattice. Actually an NTRU lattice is a special version of a convolution modular lattice. If a convolution modular lattice contains a short vector, then it is called as NTRU Lattice.

There are several hard lattice problems which are shortest vector problem, closest vector problem, smallest basis problem and their variations. The security of NTRU cryptosystem is conjectured to be equivalent to the hardness of the shortest vector problem and the closest vector problem. Shortest vector problem is to find a vector, other than the zero vector, which has the smallest length or L^2 norm, defined in Subsection 2.1. It is known that the shortest vector problem is NP-Hard under randomized reduction hypothesis [4]. The closest vector problem is to find a lattice vector whose distance is minimum to the given vector or whose difference with the given vector has the smallest L^2 norm. It is also known that the closest vector problem is NP-hard and the solution for this problem is at least as hard as the solution of the shortest vector problem [5].

NTRU is not based on number theoretic hard problems, so any improvements on the solution of number theoretic hard problems won't affect the security of NTRU. On the other hand the usage of smaller integers and lower complexity of the operations provide the NTRU high performance. These properties strengthen the position of NTRU as a cryptosystem.

The rest of this paper is organized as follows. While the introductory theoretical background on NTRU cryptosystem is detailed in Section 2, parameter choices for practical implementations are covered in Section 3. General comparison and outcome of our implementation together with the obtained results are presented in Section 4. Finally, with Section 5 the study is concluded.

2   NTRU Cryptosystem

This section is intended to provide the theoretical background on the lattice-based public key cryptosystem NTRU (http://www.ntru.com).

2.1   Some Definitions

The main parameters of NTRU cryptosystem are integers n, p and q.[1] These values are used to define the following polynomial rings:

- $R = \mathbb{Z}[X]/(X^n - 1)$, which specifies the polynomials modulo $X^n - 1$ with integer coefficients.
- $R_p = (\mathbb{Z}/p\mathbb{Z})[X]/(X^n - 1)$, which specifies the polynomials modulo $X^n - 1$ whose coefficients are reduced modulo p.
- $R_q = (\mathbb{Z}/q\mathbb{Z})[X]/(X^n - 1)$, which specifies the polynomials modulo $X^n - 1$ whose coefficients are reduced modulo q.

For secure implementation of NTRU, the parameters should also satisfy $\gcd(p, q) = 1$ where $q > p$ and n should be chosen as a prime number due to the reasons discussed in [6].

$\mathcal{L}_f$, $\mathcal{L}_g$, $\mathcal{L}_m$ and $\mathcal{L}_r$ are also parameters which represent some special subsets of the polynomial ring R from which particular polynomials are chosen to be used in key generation, encryption and decryption.

In this study, all polynomials under our consideration have integer coefficients and generally belong to the ring R. In order to perform key generation, encryption and decryption operations, we need to specify some operations on polynomials.

Let u, v be arbitrary polynomials and m be a positive integer, then we can define following operations:

- $[u]_m$, or u (mod m), is reducing the coefficients of u to a specified interval of length m, generally $[0, m)$. However, we may take this interval to be $[A, A + m)$, for some integer A, in order to properly center the polynomial in some part of the decryption process.

---

[1] In fact, it is possible to choose p to be a polynomial if the parameters are properly defined. However, we shall slightly ignore this case since our forthcoming discussion makes use of p as a fixed integer value of 2.

- $u * v$ (mod m), or equivalently $u . v$ (mod m, $X^n - 1$), is called (cyclic) convolution product or star multiplication. Here the point [.] is the usual polynomial multiplication, and (mod m, $X^n - 1$) means reducing the polynomial modulo $X^n - 1$ and coefficients modulo m.
- For $u = u_0 + u_1 x + ... + u_{n-1} x^{n-1}$, we define

$$\text{Max}\, u = \max u_i, \quad \text{Min}\, u = \min u_i, \quad \text{and} \quad \text{Width}\, u = \text{Max}\, u - \text{Min}\, u.$$

- L^2-norm and centered L^2-norm of the polynomial u gives idea on the smallness or the length of u and are defined as

$$|u|_2 = \sqrt[2]{\sum_{i=0}^{n-1} u_i^2} \quad \text{and} \quad \|u\|_2 = \sqrt[2]{\sum_{i=0}^{n-1} (u_i - \bar{u})^2} \text{ where } \bar{u} = \frac{1}{n} \sum_{i=0}^{n-1} u_i$$

2.2 Key Generation

To create an NTRU public key, one chooses two polynomials such that $f \in \mathcal{L}_f$ and $g \in \mathcal{L}_g$. Here, polynomials in $\mathcal{L}_f$ and $\mathcal{L}_g$ have small widths. Also, the polynomial f should have inverses modulo p and q. In other words, one should be able to calculate f_p^{-1} and f_q^{-1} such that

$$f_p^{-1} * f \equiv 1 \pmod{p} \quad \text{and} \quad f_q^{-1} * f \equiv 1 \pmod{q}.$$

Private key is composed of the polynomials f and f_p^{-1}. After choosing the polynomials appropriately, public key can be computed as

$$h \equiv p f_q^{-1} * g \pmod{q}. \tag{1}$$

2.3 Encryption

In order to perform encryption, one chooses a polynomial m representing the message such that $m \in \mathcal{L}_m$, and a random polynomial $r \in \mathcal{L}_r$. Later the polynomial corresponding to the ciphertext is computed as

$$e \equiv r * h + m \pmod{q}. \tag{2}$$

As in key generation, $\mathcal{L}_r$ and $\mathcal{L}_m$ are special sets of the polynomials in R, having small widths.

2.4 Decryption

One can carry out the decryption by computing the polynomial

$$d \equiv f_p^{-1} * [f * e]_q \pmod{p}. \tag{3}$$

However, in some cases decryption may not be successful. The condition for successful decryption and its effects on the choice of parameters are briefly discussed in Subsection 2.5.

2.5   Conditions

Consider the polynomial

$$[f * e]_q \equiv p\,r * g + f * m \quad (\text{mod } q). \tag{4}$$

For different parameter sets $(n, p, q, \mathcal{L}_f, \mathcal{L}_g, \mathcal{L}_r, \mathcal{L}_m)$, it is probable that we will have the right hand side of Equation 4 in the interval $[A, A + q)$, $A \neq 0$. Therefore, we need to center the value of $[f * e]_q$ by reducing its coefficients into the correct interval in order to satisfy Equation 5.

$$[p\,r * g + f * m]_q = p\,r * g + f * m, \tag{5}$$

which guarantees the success of decryption.

Let $t = p\,r * g + f * m$. In some cases, the polynomial t may not be obtained easily due to the fact that it is not properly centered. This is called decryption failure. Although the probability of occurrence of decryption failure is significantly small for appropriately chosen parameter values, as discussed in [3] and [7], they should not be ignored [8].

[9] Discusses different types of wrap and gap failures which are different types of decryption failures. [7] Calculates the probability of failures, and discusses methods in order to correctly center the polynomial t to eliminate the wrapping failures. However, gap failures still remains untreated. On the other hand, [10] gives an algorithm to overcome all decryption failures. Furthermore, the same paper outlines an analysis relating the NTRU parameters to the decryption failures and presents the conditions for choosing the parameter values which prevents all decryption failures.

2.6   A Sample Digital Envelope

The original NTRU [11], which is mainly outlined so far, considers the plaintext directly as the polynomial m. However, this scheme is vulnerable to some certain types of attacks and in particular, if decryption failure occurs [12]. For example, if the attacker is allowed to send a large number of messages and observe which ones are accepted as valid he/she can easily recover the messages. Therefore, calculation of the polynomial m is modified as in [3] in order to improve the security of the cryptosystem.

Let $P_p(n - k)$ is the set of polynomials in R_p having degree at most $n - k - 1$, and let $m' \in P_p(n-k)$ be the plaintext polynomial. Then, during the encryption, one can compute the polynomial m as follows:

$$m = \left[m' + G([r * h]_p) + H(m', [r * h]_p)X^{n-k} \right]_p. \tag{6}$$

Here, $G : P_p(n) \to P_p(n)$ and $H : P_p(n) \times P_p(n) \to P_p(k)$ are generating function and hashing function respectively.

In order to obtain m' in the decryption process, after computing m, we need to calculate the values

$$x = [e - m]_p \quad \text{and} \quad y = [m - G(x)]_p.$$

It should be noted that in a valid decryption we expect the following equalities to hold

$$x = [r * h]_p \quad \text{and} \quad y = m' + H(m', [r * h]_p)X^{n-k}.$$

Later, we extract two polynomials $y' \in P_p(n - k)$ and $y'' \in P_p(k)$ from y as

$$y = y' + y''X^{n-k}. \tag{7}$$

If $y'' = H(y', x)$, it implies that $y' = H(m', [r * h]_p)$. Therefore, we conclude $y' = m'$ and decryption is valid.

Here, k is defined to be the security parameter of NTRU which provides resistance to some certain types of attacks and according to the chosen value of k, the probability of forging a valid ciphertext is p^{-k}.

Lastly, similar and more secure padding schemes like the one discussed above are also designated in [13], [14] and [8] for particular chosen set of parameters.

3   Instantiation of the Cryptosystem

In this section, we outline some conditions which vitally affect the way the parameters are chosen. Also, we briefly mention the latest recommended and the alternative choices of the parameters in order to provide efficient and secure realizations of the cryptosystem. However, we do not cover any of these in full detail. For a complete discussion, one should refer to [15], NAEP encryption scheme, and [16], SVES-3 an instantiation of NAEP.

3.1   Choosing Parameters

Since NTRU is first proposed, the recommended parameter values have been subject to changes. Many different parameter choices are discussed in the literature in order to provide different levels of efficiency and security, and in general, for each proposed set of parameters and defined security levels, the parameter p is fixed to be a small integer or polynomial value.

In order to realize efficient implementation of NTRU at least one polynomial in the convolution product should be binary, whose coefficients are in the set $\{0, 1\}$, or trinary, whose coefficients are in $\{-1, 0, 1\}$. Therefore, in the rest of our discussion we shall define d_z to be the number of coefficients in the polynomial binary or trinary polynomial z which are equal to 1.

3.2   Choosing n

If the message is binary, n is the number of bits that can be transported. In order to provide k bits of security and prevent some particular (birthday-like) attacks, $2k$ bits can be transported. In addition, SVES-3 uses k bits of random padding to gain security against enumeration attacks in case some low-entropy messages are transported. Therefore, we set n to be the first prime number greater than $3k$. It should be noted that n might need being changed if one cannot find appropriate values of the remaining parameters.

3.3   Choosing f, g, r and m

Let F, g and r to be binary polynomials with d_F, d_g and d_r number of 1s respectively. We take $f = 1 + pF$ so that the second convolution product in the decryption can be eliminated since $f_p^{-1} = 1$. Furthermore, since security increases when h is invertible, we also take g to be invertible and set $d_g = n/2$ to obtain the best lattice security, and choose smallest d_F, d_r and d_m such that

$$\frac{1}{\sqrt{n}} \binom{n/2}{d'/2} \geq 2^k$$

where $d' \in \{d_F, d_r, d_m\}$. Here, we can take $d_F = d_r = d$ in order to equalize the combinatorial security levels of F and r. Moreover, the message representative polynomial m is chosen in such a way that it does not contain very few 1s or very few 0s. Also, $\|m\|_2$ should be sufficiently large to provide resistance against attacks which stems from information leakage from the encrypted message, and we should have the probability of being rejected due to having insufficient security, P_{reject}, very small, for instance less than 2^{-40}.

3.4   Choosing p and q

It is already noted that p and q should be relatively prime. p is fixed to be the integer value of 2 so that we can work with binary polynomials. Also, q must have a higher order modulo n, i.e. the order of divisors of X^{n-1} modulo q should be high, for example $(n - 1)$ or $(n - 1)/2$. In addition, to achieve better lattice security we must keep f and g as large as possible relative to q. Though, for combinatorial security, it is better to increase p, it causes an increase in q and so decreases lattice security. As a result, we can select q as a prime number such that

$$q \leq p . \min(d_r, d_g) + 1 + p . \min(d_f, n/2)$$

and

$$\text{order of } q \text{ modulo } n \geq (n - 1)/2.$$

This choice gives us the best lattice security and zero probability of decryption failure.

3.5   Alternatives

It is possible to choose f not to be of form $1 + pF$ in order to decrease q. On the other hand, F and r can be chosen in the product form $f_1 * f_2 + f_3$ in order to obtain further performance benefits and slightly increased bandwidth.

We can also choose p and q values differently. Let s be the first power of 2 such that

$$s \geq p(1) . \min(d_r, d_g) + 1 + p(1) . \min(d_f, n/2).$$

Then for a small integer or polynomial value of $p = 2 + X$ or $p = 3$, in which cases we work with binary or trinary polynomials respectively, one can choose $q = s$. This speeds up the reductions modulo q. However, with the larger values of p lattice security worsens due to the fact that q gets larger. In addition, we can also speed up these reductions by choosing q to be the largest prime such that $q \leq s$ for $p = 2$ at the expense of lattice security.

As a last note, allowing the probability of decryption failures to be greater than 0 reduces q, thus improves the lattice security and the bandwidth.

3.6 NAEP Encryption Scheme

Let B_n be the set of binary polynomials whose degree is less than n, and $B_n(d)$ be the subset of B_n with polynomials having d number of 1s. Furthermore, let G and H be two hashing functions such that

$$G : B_{n-k} \times B_k \rightarrow B_n(d_r)$$
$$H : B_n \rightarrow B_n.$$

These functions should be chosen such that each of them has a very small probability of variation in running time, since the running time variations may causes leakage of information about the private key [17].

Encryption. During encryption we choose a random polynomial $b \in B_k$, and then we calculate the polynomial $r = G(m', b)$, where m' is the plaintext polynomial. Message representative polynomial m is given by

$$m \equiv (m' + b\,X^{n-k}) + H\left(\left[[r * h]_q\right]_p\right) \quad (mod\ p). \tag{8}$$

At this point, one should check whether m has the expected level of combinatorial security. If not the operation should be performed with a different and randomly chosen b.

For properly computed m, encryption is performed as defined before:

$$e \equiv r * h + m \quad (mod\ q).$$

Decryption. In decryption, one, first, calculates the polynomial m as described before:

$$m = \left[f_p^{-1} * [f * e]_q\right]_p.$$

Of course, the polynomial $[f * e]_q$ should be centered if decryption failure occurs.

In order to obtain m', we need to calculate the values

$$x = e - m \quad \text{and} \quad y = \left[m - H\left(\left[[x]_q\right]_p\right)\right]_p.$$

Later, we extract two polynomials $y' \in B_{n-k}$ and $y'' \in B_k$ from y as

$$y = y' + y'' X^{n-k}. \tag{9}$$

If the conditions

$$x = [G(y', y'') * h]_q \quad \text{and} \quad y' \in B_{n-k}(d_m)$$

are satisfied, the ciphertext is valid and $y' = m'$ is the plaintext.

4 Comparison of NTRU with Other Asymmetrical Cryptosystems

The underlying theory implies that NTRU can be yet another popular public key cryptosystem residing with ECC (http://www.certicom.com), RSA (http://www.rsa.co and the likes. Nevertheless, it is important to make detailed discussion of these cryptosystems in order to better comprehend how NTRU performs. For this reason, we shall make a brief comparison of NTRU, ECC and RSA in terms of key size, and key generation, encryption and decryption times.[2]

4.1 Parameters for NTRU

Using the conditions for recommended parameters in Section 3, one can obtain the following sets for the parameter values in Table 1.

Table 1. NTRU Parameter Sets.

k	$n\,(\geq 3k)$	p	q	$d\,(= d_F = d_r)$	$d_g\,(= \lfloor n/2 \rfloor)$	$d_{m_0}\,(\leq d_m)$
80	251	2	197	48	125	70
112	347	2	269	66	173	108
128	397	2	307	74	198	128
160	491	2	367	91	245	167
192	587	2	439	108	293	208
256	787	2	587	140	393	294

It should be noted that, besides complying the discussion we made so far, the parameter values in Table 1 are also recommended as SVES parameter choices

---

[2] A full fledged comparison of the cryptosystems requires taking into account different application areas, inclusion of different sets of parameters together with various data encoding and padding schemes etc. However, since this is slightly off-topic and a relatively large task, we shall just state basic space and performance measures. Nevertheless, we suggest the reader to refer to corresponding draft or completed standard specifications of IEEE (http://standards.ieee.org), NIST (http://csrc.nist.gov) and SEC (http://www.secg.org), and various other resources, in order to accomplish a more detailed study on this matter.

in [18], the latest IEEE draft standard (currently draft 9) for public key cryptosystems based on hard problems over lattices.

In the rest of our study, while making comparison of NTRU with other public key cryptosystems, we shall refer to NTRU instantiated with these sets of parameters.

4.2   Comparison of Key Sizes

In ECC and RSA, public and private keys can be chosen of approximately equal lengths, whereas NTRU public key size differs from private key size with a ratio of $\frac{n}{n-k} \log_p q$ -to- 1. The public key size of a cryptosystem gives useful insight on the bandwidth usage if the cryptosystem is intended to be used in key exchange schemes. Table 2 gives corresponding NTRU, ECC and RSA keys sizes for equivalent security levels (k) of 80 bits, 112 bits and 128 bits etc [19] [16].

Table 2. Public Key Sizes (in bits).

Security Level (bits)	NTRU	ECC	RSA
80	2008	160	1024
112	3033	224	2048
128	3501	256	3072
160	4383	320	4096
192	5193	384	7680
256	7690	521	15360

From Table 2, one can observe that ECC, among the three, makes the best use of bandwidth and NTRU's bandwidth usage becomes more efficient with respect to RSA as the security level increases.

4.3   Key Generation, Encryption and Decryption Times

Though RSA is the most studied, tested and scrutinized cryptosystem (among the three), the latest debates, such as in [19], point out that ECC gained significant trust over time, and now, many security vendors are including ECC modules in their own products.

Accordingly, we find it useful to give timing comparisons with ECC. On the other hand, preliminary timing comparisons with RSA can be found in [3].

While measuring key generation, encryption and decryption times of ECC, we make use of C code implemented in GMP (http://www.swox.com/gmp), which is the outcome of [20]. The curves used are NIST and/or SEC - Certicom recommended elliptic curves over prime fields [21] [22], and encryption and decryption measurements are taken in varying coordinate systems such as, affine, projective, Jacobian, Chudnovsky, and modified Jacobian.

Table 3 gives timing measurements for NTRU and ECC cryptosystems where the code is compiled (with no optimizations) and run on a personal computer with Windows XP Professional OS, P4 2.80 GHz CPU and 1 GB RAM.

Table 3. Key Generation, Encryption and Decryption Times.

Crytosystem	Security Level (bits)	Key Generation* (msec)	Encryption* (msec)	Decryption* (msec)
NTRU-251	80	75.65	1.68	8.22
ECC-192	between 80, 112	$57.87 - 152.73$	$37.81 - 116.39$	$19.15 - 57.68$
NTRU-347	112	144.16	3.11	15.70
ECC-224	112	$234.11 - 367.98$	$52.52 - 164.50$	$26.35 - 81.52$
NTRU-397	128	188.92	3.97	20.26
ECC-256	128	$478.22 - 656.63$	$68.72 - 223.29$	$35.00 - 111.16$
NTRU-491	160	288.31	5.97	30.96
NTRU-587	192	412.10	8.42	44.42
ECC-384	192	$947.43 - 1429.11$	$182.35 - 586.20$	$90.61 - 290.94$
NTRU-787	256	738.75	14.49	79.48
ECC-521	256	$2055.04 - 3175.87$	$423.25 - 1257.56$	$211.38 - 626.33$

*ECC timings are given as minimum - maximum of the values observed over all coordinate systems.

As one can derive from Table 3, NTRU seems faster than ECC with respect to all the security levels defined above. This mainly stems from the fact that NTRU operations are relatively simple and not as demanding as ECC's. For instance, they do not even require the use of multiprecision arithmetic in the sense ECC operations do. In addition, though the timing measurements can be affected by many things such as runtime environment, compiler options, code optimizations and, in case of ECC, the pros and cons of using a general purpose multiprecision arithmetic library etc., it is highly unlikely that ECC has better runtime performance.

5   Conclusions

Throughout the study we outlined NTRU and demonstrated, in basic terms, how it performs against RSA and ECC.

In conclusion, many researchers have been scrutinizing NTRU since the time it proposed for the first time. There has been serious amount of analysis on the security and performance issues, and NTRU seems to be quite a decent public key cryptosystem which will be useful in the field of security for many years to come.

6   Acknowledgement

We would like to thank to Serap Atay PhD. for lending us her source code to measure ECC timings, and Assoc. Prof. Ahmet Koltuksuz PhD. for his comments and corrections.

References

1. Ajtai, M., Dwork, C.: A public-key cryptosystem with worst-case/average-case equivalence. In: STOC. (1997) 284–293
2. Goldreich, O., Goldwasser, S., Halevi, S.: Public-key cryptosystems from lattice reduction problems. In: CRYPTO '97: Proceedings of the 17th Annual International Cryptology Conference on Advances in Cryptology, London, UK, Springer-Verlag (1997) 112–131
3. Hoffstein, J., Lieman, D., Pipher, J., Silverman, J.H.: NTRU: A public key cryptosystem. (1999) available at http://grouper.ieee.org/groups/1363/lattPK/submissions.html#NTRU1.
4. Ajtai, M.: The shortest vector problem in $L2$ is NP-hard for randomized reductions. In: Proceedings of the 30th Annual ACM Symposium on Theory of Computing (STOC-98), New York, ACM Press (1998) 10–19
5. Nguyen, P.Q., Stern, J.: The two faces of lattices in cryptology. In: CaLC '01: Revised Papers from the International Conference on Cryptography and Lattices, London, UK, Springer-Verlag (2001) 146–180
6. Gentry, C.: Key recovery and message attacks on NTRU-composite. In: EUROCRYPT '01: Proceedings of the International Conference on the Theory and Application of Cryptographic Techniques, London, UK, Springer-Verlag (2001) 182–194
7. Silverman, J.H., Whyte, W.: Estimating decryption failure probabilities for NTRUEncrypt (2003) NTRU Cryptosystems Technical Report 018, Version 1, June 20, 2003, available at http://www.ntru.com/cryptolab/tech_notes.htm#012.
8. Howgrave-Graham, N., Nguyen, P.Q., Pointcheval, D., Proos, J., Silverman, J.H., Singer, A., Whyte, W.: The impact of decryption failures on the security of NTRU encryption. Lecture Notes in Computer Science **2729** (2003) 226–246
9. Silverman, J.H.: Wraps, gaps and lattice constants (2001) NTRU Cryptosystems Technical Report Report 011, Version 2, March 15, 2001, available at http://www.ntru.com/cryptolab/tech_notes.htm#011.
10. Yu, W., He, D.: Study on NTRU decryption failures. In: ICITA '05: Proceedings of the Third International Conference on Information Technology and Applications (ICITA'05) Volume 2, Washington, DC, USA, IEEE Computer Society (2005) 454–459
11. Hoffstein, J., Pipher, J., Silverman, J.H.: NTRU: A ring-based public key cryptosystem. Lecture Notes in Computer Science **1423** (1998) 267–288
12. Silverman, J.H.: Plaintext awareness and the NTRU PKCS (2000) NTRU Cryptosystems Technical Report Report 007, Version 2, 2000, available at http://www.ntru.com/cryptolab/tech_notes.htm#007.
13. Hoffstein, J., Silverman, J.H.: Protecting NTRU against chosen ciphertext and reaction attacks (2000) NTRU Cryptosystems Technical Report Report 016, Version 1, June 9, 2000, available at http://www.ntru.com/cryptolab/tech_notes.htm#016.

14. Hoffstein, J., Silverman, J.: Optimizations for NTRU. In: Proceedings of Public Key Cryptography and Computational Number Theory, de Gruyter, Warsaw (2000)
15. Howgrave-Graham, N., Silverman, J.H., Singer, A., Whyte, W.: NAEP: Provable security in the presence of decryption failures (2003) available at http://www.ntru.com/cryptolab/articles.htm#2003_3.
16. Howgrave-Graham, N., Silverman, J.H., Whyte, W.: Choosing parameter sets for NTRUEncrypt with NAEP and SVES-3. Cryptology ePrint Archive, Report 2005/045 (2005) http://eprint.iacr.org.
17. Silverman, J.H., Whyte, W.: Timing attacks on NTRUEncrypt via variation in the number of hash calls (2006) NTRU Cryptosystems Technical Report 021, 2006, available at http://grouper.ieee.org/groups/1363/lattPK/submissions.html#2006-06.
18. IEEE 1363 Working Group: IEEE P1363.1 /D9 draft standard for public-key cryptographic techniques based on hard problems over lattices (2007) available at http://grouper.ieee.org/groups/1363/lattPK/draft.html.
19. NSA: (The case for elliptic curve cryptography) http://www.nsa.gov/ia/industry/crypto_elliptic_curve.cfm.
20. Atay, S.: Performance issues of elliptic curve cryptographic implementations (2006) Unpublished Ph.D. Dissertation Thesis, Ege University, Graduate School of Natural and Applied Sciences.
21. NIST-CRCS: Fips 186-2 digital signature standard (2001) available at http://csrc.nist.gov/publications/fips/index.html.
22. SEC: Recommended elliptic curves domain parameters (2000) available at http://www.secg.org/index.php?action=secg,docs_secg.

Establishing the ISMS at Tusaş Aerospace Industries Inc. – TAI: Experiences and Recommendations

Bilge Yiğit[1]

[1] Tusaş Aerospace Industries Inc. - TAI, Information Support Department, Fethiye Mahallesi, Havacilik Bulvari No:17 06980 Kazan, Ankara, Turkey
byigit@tai.com.tr

Abstract. This experience report describes the followed road of establishing the ISMS (Information Security Management System) and getting certified for ISO/IEC 27001:2005 on a time scale of 1.5 years at TAI. As a part of the whole management system, ISMS plays an important role in corporate governance. Experiences and lessons learned, as well as some specifics of TAI implementation will be presented. Lessons learned should benefit others who might be considering certification for ISO/IEC 27001:2005 in the future.

Keywords: ISO 27001:2005, ISMS, Information Security

1 Introduction

Enterprises are increasingly realizing the value of information they own. Managing the information is one of the key elements for corporate governance. ISMS is a living system which includes policies, objectives, processes, activities and resources to ensure information security based on a business risk approach.

TAI first met with ISO 17799 and ISO/IEC 27001 standards during a business partner's visit. ISO 27001 certification was not a goal on itself. Our management recognized the opportunity to improve management and processes with the guidance of the ISO 27001 standard.

We are aware of the importance and criticality of information security in our industry. TAI is the first ISO/IEC 27001:2005 certified company in the defense industry in Turkey.

In parallel to ISMS implementation we were also addressing improvement on software engineering processes in accordance with CMMI (Capability Maturity Model® Integration). We benefited from the parallel implementation as these issues were also discussed.

2   Establishing the ISMS in TAI

The most critical activities for establishing and implementing ISMS are;
- o   Scope definition
- o   Policy definition
- o   Risk assessment and treatment
- o   Internal audits
- o   Management review
- o   Corrective and preventive actions
- o   Measuring the effectiveness of ISMS implementation
- o   Training and awareness

2.1   Scope Definition

This is the starting point for establishing the ISMS. We faced some difficulties in defining the scope of our ISMS which were mostly caused by the reason that our quality management system was not on focus during scope definition. The scope of the quality management system and the ISMS should not necessarily be the same but if you already have a management system (i.e. quality management system) and you want to establish another (i.e. ISMS) then, it might be a good idea to review the scope of the existing management system first.

The resulting scope definition is set as to include capabilities that are owned by two distinct sites of TAI in different locations taking the quality management system into consideration in detail.

Processes that are followed are to be known and defined for effective ISMS. If not, it could be difficult to define assets and ownership of these assets. Defining the critical assets is the requirement for risk assessment. In this sense, there are numerous advantages for a company/organization to build ISMS over an existing quality management system.

2.2   Policy Definition

TAI policies are approved by board of directors and they sit at the top of the hierarchical document structure of quality management system. Under the policies' directions there are procedures, work descriptions, manuals, plans, instructions and other controlled items.

TAI information security policy is prepared in guidance of the standard and reviewed by appropriate management levels and approved by the board of directors.

2.3   Risk Assessment and Risk Treatment

TAI ISMS organization is based on a team named "Security Work Group" whose members are representatives from each functional department. This team meets monthly or on emergency basis. Security Work Group is responsible for performing risk assessment. Each security representative defines and categorizes the assets, identifies the ownership of assets that are in the scope of their departmental processes in coordination with his/her department. Each security representative performs the risk assessment for these assets.

Final results of the risk assessment is reviewed by the steering committee of the Security Work Group and reviewed and approved by the management.

We defined the process for the creation of asset inventory and performing risk assessment. Periodic reassessment of the risks is also ensured by this defined process.

Our internal software development department developed an application to support the risk assessment process. The application has the following basic functionalities;

- o   Inventory of assets
- o   Ownership of assets
- o   Classification of assets
- o   Threat definition
- o   Vulnerability definition
- o   Risk computation
- o   Risk treatment
- o   SOA (Statement of Applicability) generation

2.4   Internal Audits

Internal audits are also a requirement for both ISMS and quality management systems. Our internal audit function was already established therefore we didn't need to spend too much effort. The internal audit team is trained for the ISMS implementation at the beginning of the studies. After the certification, the internal audit team received the internal audit training for ISO/IEC 27001.

Our implementation team consisted of representatives from industrial security and information technologies departments. The participation of the people from internal audit team and the quality assurance team in the implementation of ISMS is a critical success factor.

The most common misconception is that the implementation of the ISMS is the duty of the information technologies department. ISMS is a management system and its implementation should be a collective effort of the representatives from all departments of interest. A good example can be the participation of the subcontractor management department in the team where there are certain benefits in addressing information security issues in the subcontractor audits.

2.5   Management Review

In TAI, the executive board meets regularly under the chairmanship of the general manager. The executive board consists of department heads directly reporting to the general manager. Our management review meeting is conducted as an executive board meeting. All the requirements (e.g. inputs, outputs, approvals) are met during these meetings and minutes of meetings are kept in the records of the ISMS.

Management reviews are also a requirement for other management systems where they guarantee the commitment of the management and give management the opportunity to review the strategies and policies, implementations with respect to those policies. Information security should be a consideration in the strategic planning process.

2.6 Corrective and Preventive Actions

Corrective and preventive actions are the triggers for process improvement and they are common the requirements to management systems [1]. A generic process may be adopted to meet the requirements of all the management systems.

2.7 Measuring the Effectiveness of ISMS Implementation

Measuring the effectiveness of ISMS is a requirement of ISO/IEC 27001 but the standard does not provide sufficient guidance. Before starting to measure the effectiveness, it is better to implement an effective ISMS first.

Effectiveness of the ISMS depends on several factors such as; successful management of ISMS implementation project and management commitment. Successful management of the ISMS implementation project depends on actions relating to the activity sequencing and resource estimating. By performing activity sequencing and resource estimation, an overall insight for the project is obtained. Sponsorship and management commitment is a critical requirement for effectiveness. It could be a good idea for managers/sponsors to prepare a project charter at the beginning of the ISMS project in order to document the business needs, project justification, assumptions and constraints, resources, responsibilities, authorizations, scope, strategic alignment. This charter may be distributed among the appropriate management levels.

Measuring the effectiveness of the controls is another issue. Standard requires measuring the effectiveness of selected controls or control groups. At the very beginning, we chose to concentrate on control groups. We established information needs and measurement constructs for control groups. For establishing the measurement model, we are inspired by the measurement information model documented in the international standard ISO/IEC 15939, "Software Measurement Process." The ISO/IEC 15939 standard describes how data is collected and organized to satisfy defined information needs. [2]

An example information need and measurement construct to measure the effectiveness of intrusion prevention systems are given in Table 1 and Table 2.

Table 1. Information Need for intrusion prevention system.

Information Need #	1
Description	Successful attacks over a period of time
Reporting Period	Monthly
Rationale	To measure and improve the effectiveness of intrusion prevention systems
Responsible	Bilge Yiğit

Table 2. Measurement Construct for intrusion prevention system (IPS).

Measurement Construct #	Base Measure(s)	Unit of Measure	Measurement Function	Indicator	Period	Where?	How?
1	Successful attacks	Each	Count successful attacks	>0	Monthly	IPS	Automatic

An example criterion for measuring the effectiveness of the security organization may be the attendance percentage of the departmental security representatives to the monthly scheduled meetings. There is limited guidance in the literature for measuring the effectiveness of ISMS. The only guide we could find was a British Standards Institution (BSI) publication [3].

We prepared a technical document which describes the information needs and measurement constructs for each control group. This TAI document is approved by the management and published [4].

2.8   Training and Awareness

In the implementation phase, all departmental security representatives are trained for the ISO 27001 requirements and all decisions are coordinated among the representatives. Decisions are made in the Security Work Group Meetings and documented in the meeting minutes.

TAI has over 2500 employees. Nearly all of them are trained on information security. Orientation training is also extended to include information security policies and procedures.

3   ISO 27001 and CMMI Level 3

ISMS implementation was in parallel with studies for CMMI Level 3 certification. ISO 27001 mainly addresses verification and validation process areas of CMMI in controls and control objectives.

Change management which is included in configuration management process area in CMMI is also addresses in ISO 27001. Both changes to the information processing facilities and systems and application software are addressed in ISO 27001.

In the software requirements development process, among other requirements, the identification of security requirements is explicitly stated. The traceability of the security requirements into design, coding, testing and the final product is also established.

A standard set of organizational security requirements are defined for in-house developed software. They are considered as an input to requirements engineering process in every software development project.

Project managers in software development projects are expected to consider the security related risks and classify their project's security level.

We had defined a process management process to manage our software engineering processes before ISMS implementation started. All the process assets (i.e. process definitions, guidelines, templates, etc.) are stored in a controlled process asset library in a configuration management system. Our process asset library is extended to include all new process assets that are created during the ISMS implementation.

Experiences gained from implementing the measurement and analysis process area in CMMI are used for defining the process of measuring the effectiveness of ISMS.

4 Experiences and Lessons Learned

Based on the establishment of ISMS and certifying for ISO 27001, we have come up with a list of experiences, benefits and lessons learned.

1. Senior management involvement
Senior management involvement must be established at the beginning of the implementation and continue as the system lives. Resource allocation (both labor and non-labor) is ensured by the senior management involvement.

2. Never underestimate the power of a good plan
Good planning may prevent future critical problems and may help saving effort, time and budget.

3. Start working on your quality management system
ISMS has lots of correspondence with other management systems. Effort and time may be saved by reviewing those correspondences. The integration of management systems will be more efficient then managing them separately.

4. Good tools help
Especially the risk assessment process can not be easily performed in large organizations without tool support. Tool selection is critical if it will be a commercial off-the-shelf one. The risk assessment process must be customizable in the tool according to the needs of the organization. TAI utilizes in-house developed software to define the information assets and implement risk assessment process.

5.   Do not forget that information security is not only the job of IT
Establishing the ISMS should be a collaborative project in which different disciplines must be involved. Information technologies, physical security, quality management, subcontractor management, program management are some of the most important disciplines.

6.   Proper risk assessment
The risk assessment helps to answer the question, 'How much security do we need for which assets?' The risk assessment should involve all asset owners. It is unlikely to be able to conduct an effective risk assessment without them. Risk management process should achieve a cost-effective balance between the implementation of security controls and the significant threats. The cost of a control should never exceed the expected benefit to be derived.

7.   Information security depends on mainly three information criteria:
Confidentiality, integrity and availability. Without any one of these criteria, we can not ensure information security.

8.   Make sure the roles and responsibilities are defined and approved for ISMS

9.   Benefit knowledge sharing between different process improvement initiatives
Different models, standards may have different focus areas and benefiting is possible if they are considered together.

Acknowledgments. ISMS project has been successful due to the contributions of many people. I would like to gratefully acknowledge the valuable efforts of project team members Pınar Altınten and Hasan Güngör, my manager Gülsen Bayramusta for her commitment and support. I would like to also acknowledge all other management levels that had supported the project.

References

1. ISO/IEC 27001:2005, Information technology – Security techniques – Information security management systems – Requirements, Annex C
2. ISO/IEC 15939:2002, Software engineering - Software measurement process
3. Humphreys T., Plate A.: Measuring the effectiveness of your ISMS implementations based on ISO/IEC 27001:2005, BSI, UK (2006)
4. Altınten P., Yiğit B.: Measuring the effectiveness of TAI ISMS in accordance with ISO 27001:2005, Version 1, TAI Internal Technical Document

A Memory Management Model For Cryptographic Software Libraries

Ali Mersin, Mutlu Beyazıt

Izmir Institute of Technology, College of Engineering, Dept. of Computer
Engineering, Gülbahçe, Urla, 35430 Izmir, Turkey,
{alimersin, mutlubeyazit}@iyte.edu.tr

Abstract. Cryptographic protocols are implemented on the abstraction
of multiple precision number libraries in which the dominant design criterion mostly turns out to be the maximization of the system performance.
In contrast, each protocol may have its own memory usage pattern. In
general case, the memory allocation and release routines are frequently
called during the runtime. For this reason, an improper memory management strategy may yield an inefficient implementation. In this paper, we
propose a memory management technique which is constructed under the
consideration of the context of high level cryptographic software running
on multi-programmed environments. Also, we show the implementation
results of our approach and discuss with respect to the common static
and dynamic memory allocation strategies.

1   Introduction

Memory management is a common task in software engineering. Each application has its own memory allocation and release pattern. Therefore, there is no
generally agreed upon management strategy. [1] The only common point for all
strategies is that the memory management can be abstracted from the higher
layers of the software. This approach provides the flexibility of using the software
with the proper management technique which is determined under the characteristics of the underlying hardware. The implementation environment of this
study, namely Crympix which is a fixed size multi-precision cryptographic software library, is developed having the ability to supply this opportunity to its
users [2].

Cryptographic software deals with computation intensive executions. Therefore, memory allocation and release operations are expected to be light-weight,
so that their effect on the overall runtime performance can be neglected. Cryptographic software has its own memory allocation pattern which is suitable to
be exploited to prevent memory fragmentation. More specifically, since the cryptographic protocols run with large numbers, roughly a hundred digits, they can
only grow up to a predefined length, i.e. all numbers are stored in memory blocks
of fixed size.

The presentation of a number x is given in radix representation in Equation 1.
[3] Radix representation is enforced by the underlying hardware which provides

binary operations on the fixed precision variables. Actually, this is a natural selection criterion to utilize the hardware at its peak.

$$x = \sum_{i=0}^{l-1} x_i \beta^i = (x_{l-1}, x_{l-2}, x_{l-3}, ..., x_0)_\beta \tag{1}$$

In Equation 1, β is the base and l is the length of the number in terms of the number of digits present. The bound is generally called the modulus. Hence, the same amount of memory space is necessary for a number unit for each time. This bound is determined by Equation 2.

$$L = 2l - 1 = \left\lfloor n^2 \right\rfloor_\beta = 1 + \left\lfloor \log_\beta n^2 \right\rfloor \tag{2}$$

where β is the base and n is the modulus. Note that L is the desired length of numbers to be fixed. Most cryptographic libraries take the advantage of this property.

In this paper, we show some further improvements which provide the user with a sight of determining the desired point in the trade off between memory utilization and performance. To guide the reader on the memory usage pattern of cryptographic software primitives, the memory allocations for some basic functions are given in Table 1.

Table 1. Unit memory requirement of basic arithmetic operations.

Operation	Input	Output	Inner	Total
Add / Subtract	$[1, 2]$	$[0, 1]$	$[0]$	$[1, 3]$
Mul-Comba	$[1, 2]$	$[0, 1]$	$[0]$	$[1, 3]$
Mul-Basecase	$[1, 2]$	$[0, 1]$	$[1]$	$[2, 4]$
Mul-Karatsuba	$[1, 2]$	$[0, 1]$	$[1]$	$[2, 3]$
Mul-Window	$[1, 2]$	$[0, 1]$	$[1, n]$	$[2, n]$
Mod	$[1, 2]$	$[0, 1]$	$[1]$	$[2, 4]$
ModExp	$[1, 3]$	$[0, 1]$	$[3, n]$	$[4, n]$
ECC Affine Double	$[1]$	$[0, 1]$	$[2]$	$[3, 4]$
ECC Affine Add	$[1, 2]$	$[0, 1]$	$[2]$	$[3, 5]$
ECC Projective Double	$[1]$	$[0, 1]$	$[17]$	$[18, 19]$
ECC Projective Add	$[1, 2]$	$[0, 1]$	$[14]$	$[15, 17]$
ECC Jacobian Double	$[1]$	$[0, 1]$	$[6]$	$[7, 8]$
ECC Jacobian Add	$[1, 2]$	$[0, 1]$	$[19]$	$[20, 22]$

In Table 1:

- Operation column lists the most common cryptographic primitives. Add / Subtract are the usual addition and subtraction operations. Mul-Comba, Mul-Basecase, Mul-Karatsuba and Mul-Window are primitives which use different techniques to accomplish the multiplication operation. Mod and

ModExp are modulo and modular exponentiation operations and the rest of the operations are elliptic curve cryptography primitives for point addition and doubling functions in affine, projective and jacobian planes respectively.

– Input column shows the minimum and the maximum number of instances, i.e. allocated memory slots, which should be input to the corresponding operation in order to perform the task.
– Output columns gives the minimum and the maximum number of additional instances to store the result of the operation.
– Total column is simply the sum of the values in the input and the output columns yielding the overall minimum and maximum number of instances used during one execution of the corresponding primitive.

Currently the concept of security in terms of trust and usability of technological infrastructure has attained a very critical level. For instance, the cryptographic applications are very important for mobile equipments but as such, also have many restrictions. Such as:

– Hardware scalability which calls for smaller platforms i.e. 8 bit or 16 bit processors,
– Restrictions on the memory size,
– Speed requirement which is another crucial indicator for large application areas such as giant mobile networks and smart cards,
– Restrictions on power consumption such as in smart cards, mobile phones and PDAs.
– Performance/Cost ratio that also calls for a minimum unit cost per equipment.

During the last couple of years it is noted that a lot of cryptographic applications have been implemented with specific processors like ASIC or FPGAs [4]. In this way those equipments could be optimized with one specific algorithm and thus fitted into a particular size market requirements. In the meantime, it is also noted that the most of software applications are slow. A hardware-software co-design approach is being undertaken to overcome this problem [5]. At this point, the efficiency of a memory management model while using a powerful cryptographic software library has become a critical point.

The rest of this paper is organized as follows: While common solutions as to the problem of memory management are detailed in Section 2; while our proposed solution is given in Section 3. Implementation details are covered in Section 4. And, finally the results and our contribution are presented in Section 5.

2   Common Solutions (Static Vs. Dynamic)

In this section, we discuss the possible effects of using two common approaches of memory management. All approaches herewith are discussed under the basis of manual memory management that well suits with the ANSI C programming language which is employed for this study as well.

Static memory model depends on the use of a pre-allocated memory block. Each allocation unit is an equally sized partition of a block-memory. Since partition lengths and the total size of block-memory are fixed and thus known by the initial state, there is no allocation and release overhead for the runtime. On the other hand, these parameters may not be present in some environments. For instance, the maximum number of memory units in user level operations is assumed to be unpredictable. From this point of view, static memory management may not be the optimum solution for environments with shortage on memory.

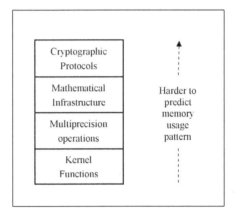

Fig. 1. The memory utilization for cryptographic software

Dynamic memory model provides the size independent memory allocation by instant allocation. Furthermore, instant releasing keeps memory utilization at the highest level. Despite the fact that being best solution in terms of memory, emergent runtime overhead may be unacceptable for some environments, i.e. the real time systems.

Figure 1 depicts the usual memory utilization model for a cryptographic software. Since, memory management can not be performed as isolated from environmental variables (or limitations); there is no unique optimum strategy. Decision should be made according to the knowledge about environment.

3 Proposed Solution (Hybrid)

The dynamic memory management ignores any assumptions by dynamically allocating and releasing each memory unit in an on demand fashion. Although this approach provides maximum memory utilization, it is computationally expensive to be utilized in cryptographic software.

The static memory management strategy relies on the fact that cryptographic applications, in general, do not need more than a certain number of memory units during their runtime. However, it is not easy to predict the behavior of the user level memory usage pattern. User level unit allocations may lead to the collapse of all predictions and/or assumptions on memory requirement. From this point of view, we propose a hybrid memory management technique which is the combination of static and dynamic memory management techniques.

Hybrid memory management assumes that all the numbers are in fixed size. It classifies the memory management routines into fast and slow memory management strategies. The fast memory management strategy is used for the allocation and the deallocation of numbers that are used inside the low level library functions which perform time critical tasks. It merely uses up the heap space. The algorithm is summarized in below Algorithm 1.

```
pseudocode Allocate()
    newUnit := NULL
    if isEmptyStack() then
        newUnit := dynamicAllocation()
    else
        newUnit := popStack()
    end
    memoryUtilizer()
return newUnit

pseudocode Release(oldUnit)
    pushStack(oldUnit)
    memoryUtilizer()
return
```

Algorithm 1: Memory Allocation and Release Operations.

Since, the core functions never use more than a certain number of memory units during the runtime, the technique is adequate for the allocation and deallocation of low level units. Nevertheless, the addition of an upper-bound-checking parameter will be helpful to detect memory leaks in the development phase.

The slow memory management strategy basically implements the same logic as in the fast management technique with some modifications. At this point, we relax the assumption that there is no need for more than a certain and relatively small number of runtime units in order to perform a cryptographic task. It only takes some basic parameters into account and performs the release of unused memory slots in order to establish the desired level of memory utilization lower bound.

The parameters used in the proposed model are as follows:

− x, Total number of allocated memory units,
− y, Number of memory units in use,
− d, Maximum difference between x and y,

- w, Number of actions before dynamic release is activated,
- c, Number of actions before a new crash occurs,
- h, Number of units to be dynamically released at a time,
- k, Number of actions between two dynamic release operation,
- u, Lower bound for the memory utilization; $0 \leq u \leq 1$.

Figure 2 depicts the crash intervals by the aforementioned parameters.

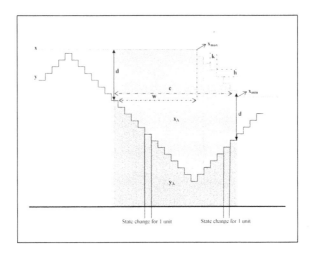

Fig. 2. Crash Interval

A state change is caused by an action which is defined to be either an allocation or a release request. The algorithm is designed to keep the memory utilization greater than u by satisfying the condition $x - y \leq d$. When the condition $x - y \leq d$ stays violated for w number of actions, the ratio x/y is decreased to increase the utilization level. There are more than one way to accomplish this task. For example, there can be a concurrent mechanism, a separate module, which periodically monitors the memory utilization level and take proper steps in order to keep the desired conditions. However, these kind of mechanisms are not likely to be light-weight and generally depends on some sort of feature or support in the underlying platform. The model we propose is very basic and performs memory management tasks only when the actions occur.

A crash occurs if the condition $x - y \leq d$ is violated and then satisfied again after some actions have elapsed. The interval in which the condition stays violated, is called the crash interval. When the crash occurs, d is updated by $(\bar{y}_c/u) - \bar{y}_c$ where $\bar{y}_c$ is the average of the y values observed at the end of the crash intervals. The new value of w is given by Equation 3.

$$w = \max \left(\frac{1}{\bar{x}_{\max}} \left(\frac{\bar{y}_A}{u} - \bar{x}_{A_{wc}} \right), 0 \right) \tag{3}$$

In Equation 3, $\bar{x}_{max}$ is the average of x values observed at the beginning of the crash interval, $\bar{y}_A$ is average of the sum of y values observed at the actions in the crash interval and $\bar{x}_{A_{wc}}$ is the average of the sum of x values observed at the actions between w and c in the crash interval. The values h and k are not updated. Note that the sum of x between the wth and c^{th} actions of the crash interval is given by the below Equation 4.

$$\sum_{w<i<c} x_i \cong \frac{1}{2}\left(2x_{\max} - \frac{h}{k}(c-w)\right)(c-w) \tag{4}$$

The basic algorithm for hybrid memory management is updated by Algorithm 2 as shown below.

---

pseudocode $memoryUtilizer()$
  if $x - y > d$ **then**
    if $xMaxSet = FALSE$ **then**
      $xMax := x$
      $xMaxSet := TRUE$
    end
    $c := c + 1$
    $y_A := y_A + y$
    if $c > w$ **then**
      if $k = 1$ **or** $c - w \equiv 0 \,(\mathrm{mod}\ k)$ **then**
        $dynamicRelease(h)$
        $x := x - h$
      end
      $x_{A_{wc}} := x_{A_{wc}} + x$
    end
  end
  if $xMaxSet$ **and** $x - y <= d$ **then**
    $n := n + 1$
    $\bar{x}_{\max} := (\bar{x}_{\max} \cdot (n-1) + x_{\max})/n$
    $\bar{x}_{A_{wc}} := (\bar{x}_{A_{wc}} \cdot (n-1) + x_{A_{wc}})/n$
    $\bar{y}_c := (\bar{y}_c \cdot (n-1) + y)/n$
    $\bar{y}_A := (\bar{y}_A \cdot (n-1) + y_A)/n$
    $w := \max\left(\dfrac{1}{\bar{x}_{\max}}\left(\dfrac{\bar{y}_A}{u} - \bar{x}_{A_{wc}}\right), 0\right)$
    $d := \bar{y}_c\left(\dfrac{1}{u} - 1\right)$
  end
  return

Algorithm 2: Memory Utilizer.

At this point, it might be useful to note how the formulas for w and d are derived. For this purpose, let x_i and y_i be the values observed in the crash interval as usual. It is obvious that for x_i and y_i values which are observed outside any crash interval we have the ratio y_i/x_i satisfies the minimum utilization level u, or more precisely $x_i - d \leq y_i \leq x_i$. Therefore, we need to employ a mechanism which updates the values of w and d in order to keep the ratio y_i/x_i as desired by evaluating the observations in the crash intervals.

In order to derive the new formula for d we use the boundary value for the condition $x - y \leq d$ when a crush occurs by substituting the x with y/u. In other words, remembering that y_c be the y values observed at the end of the crash interval. Thus, it can be expected that, at the next crash, y will be around the average of y_c, namely $\bar{y}_c$. Since $u = y/x$, if we use the boundary formula $d = x - y$ at the crash point we obtain Equation 5 which is the new formula for variable d.

$$d = \frac{\bar{y}_c}{u} - \bar{y}_c = \bar{y}_c\left(\frac{1}{u} - 1\right) \tag{5}$$

Deriving the formula for w is a little bit complicated. There might be many approaches to determine the new value of w. Since, we intend to keep things as simple as possible considering the time critical characteristics of the cryptographic tasks, we determine w by trying to keep the value of the ratio in Equation 6 close to u.

$$\frac{y_A}{x_A} = \frac{\sum\limits_{i<c} y_i}{\sum\limits_{i<c} x_i} \tag{6}$$

To do this, again we use the average of the values observed in the crash interval and obtain the Equation 7.

$$w = \frac{1}{\bar{x}_{\max}}\left(\frac{\bar{y}_A}{u} - \bar{x}_A\right) \tag{7}$$

The most visible downside of the model is that when the user is able to carry on without any allocate and release requests the memory slots stay non-managed due to the fact that one cannot carry out the memory management routine. This mainly stems from using actions to access the mechanism which manages the memory slots during runtime. On the other hand, if the implementer does not display such a deviant behavior this approach most probably succeeds at keeping the desired conditions with a little performance cost. As a last note, we should remark that if this logic is deactivated both fast and slow memory management routines employs exactly the same mechanism, which is, in fact, often what we expect from the time critical cryptographic tasks.

4   Implementation

The proposed algorithm keeps the overall memory utilization above the specified level. To observe the outcome of the designed memory utilizer, the algorithm is implemented in the kernel of Crympix Multiprecision Library which is written in ANSI-C Programming Language. Therefore, dynamic memory management is performed with malloc() and free() functions, assuming their response time to be stationary, and the stack memory is not used.

The test scenario consists of random allocate and release requests being successively performed for all utilization parameter values $0 \leq u \leq 1$. And its execution platform is a general use personal computer with Intel 2x3.00 GHz dual core processor and 1 GB of RAM.

The time versus utilization graph of the proposed algorithm is depicted in Figure 3. The 0% utilization corresponds to measured running time in static kernel management. 100% memory utilization represents the dynamic memory management scheme.

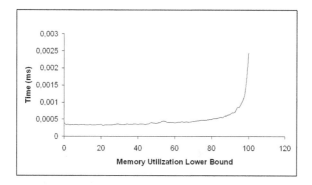

Fig. 3. Utilization vs. Time curve

In general the utilization vs time curves we obtain have significant increase in the slope after approximately 55% - 65% utilization levels. From Figure 3, one can derive the following results:

– Performance loss for $0 - 55\%$ or 65% can be considered as insignificant.
– Memory utilization lower bound specified in our model had better not exceed approximately 65% if the performance loss is intolerable.
– Static memory management technique seems not to be the optimum solution for multiprogrammed environments.
– Dynamic memory management technique leads to inefficiency in cryptographic applications.

As a last note, we also observed the runtime memory utilization values for the specified memory utilization lower bounds, u, versus the number of occurring actions. It has been noted that for the specified values $0\% \leq u \leq 65\%$, the change in runtime memory utilization levels show some similarities in the manner that they require relatively less number of dynamic allocate and release operations to achieve the desired level of utilization. On the other hand, in case of greater values of memory utilization lower bound, the number of required operations increase significantly as u approaches 100%.

5   Results and Contribution

Static kernel has the best runtime performance; however, since the maximum number of memory units in user level operations is assumed to be unpredictable, the memory utilization of static management strategy is the worst.

With the 100% utilization level the dynamic kernel has the best memory utilization level, but runtime performance greatly suffers under such a scheme which is obviously unacceptable in time critical cryptographic tasks.

Hybrid kernel is designed to find the balance between memory utilization and the runtime performance. Furthermore, if the user's behavior complies with the assumptions of what we expect from a cryptographic application then the hybrid approach gets closer to the static management technique in terms of performance.

In conclusion, we have proposed a self-tuning memory manager for cryptographic software. The measurements with proposed solution show that an approximately 55% - 65% memory utilization lower bound provides an approximate speedup factor of 6 over 100% memory utilization.

6   Acknowledgement

We would like to thank to Assoc. Prof. Ahmet Koltuksuz PhD., Serap Atay PhD., Hüseyin Hışıl, Sevgi Uslu and Evren Akalp for their continuous support and comments.

References

1. Bosselaers, A., Govaerts, R., Vandewalle, J.: A fast and flexible software library for large number arithmetic. In Macq, B., ed.: 15th Symp. on Information Theory in the Benelux, Louvain-la-Neuve (B), Werkgemeenschap Informatie- en Communicatietheorie, Enschede (NL) (1994)
2. Koltuksuz, A., Hisil, H.: Crympix: Cryptographic multiprecision library. In Yolum, P., Güngör, T., Gürgen, F.S., Özturan, C.C., eds.: ISCIS. Volume 3733 of Lecture Notes in Computer Science., Springer (2005) 884–893
3. Menezes, A.J., Vanstone, S.A., Oorschot, P.C.V.: Handbook of Applied Cryptography. CRC Press, Inc., Boca Raton, FL, USA (1996)
4. Dhem, J.F., Feyt, N.: Hardware and software symbiosis helps smart card evolution. IEEE Micro **21**(6) (2001) 14–25
5. Zambreno, J., Choudhary, A., Simha, R., Narahari, B.: Flexible software protection using hardware/software codesign techniques. In: DATE '04: Proceedings of the conference on Design, automation and test in Europe, Washington, DC, USA, IEEE Computer Society (2004) 10636

Information Security Policy: Positioning the Technological Components of Information Security Services under the Perspective of Electronic Business

Kaan Kurtel

Izmir University of Economics, Sakarya Cad. No.156, Balcova, 35330, Izmir, Turkey.
kaan.kurtel@ieu.edu.tr

Abstract. Emerging trends increase the role of security in modern business life. Thus, the role of security policy has become more significant. An information security policy defines the firm's security requirements also minimizes security risks. This objective is simply realized by good collaboration between the security policy and the existing cryptographic technologies. This paper includes a descriptive and analytical understanding of the role of information security services for developing security policy; in addition, the study also explains some details of possible effects an electronic document related to this concept. Security policy developers are the main beneficiaries of this work.

Keywords: Information Security Policy, and E-business.

1 Introduction

One of the most significant changes over the past few decades has been the rise of information as a strategically important, integral part of everyday economic and social life. The Internet, World Wide Web and related information technologies have accelerated to the importance of information. The growth of the Internet is strategically influenced by three main factors: *trust*, *privacy* and *security*. As a consequence, millions of firms face very costly threats from theft of information (intellectual property, customer data, etc.), financial fraud, or quite simply disruption of their information systems through targeted security breaches to more advanced forms like viruses or worms.

In fact, information security risks are the most important trade obstacles for online business life. Difficulties in making electronic payments, online banking and concerns over the security and privacy of transactions are limiting to the business itself and negatively impact customers purchasing decisions.

The factors afore mentioned make the planning for effective information security policies essential. An information security policy should protect the firm's electronic resources and operations. Another consideration of importance of the security policy is creating sustainable relationships between company and the members of its value chain, including customers.

Preparing an effective information security policy is only possible with a clear understanding of the existing problems, requests of customers and also to be familiar with the IT related to cryptography, the digital signature mechanism, and its implementation. Therefore, in this paper we intended to define the interrelation of security policy, and the technological components of security services from the perspective of e-business. The paper contains five sections including the introduction. The second section defines information security policy. The third section explains the components of security infrastructures. The fourth section details information security policies and security infrastructure, and the final section contains the conclusion.

2   Information Security Policy

A security policy is a high-level management document informing all users about the goals and constraints of using information system; it also must answer three questions: *who* can access *which resources* in *what manner*? [1] This document regulates how an organization will manage, protect and distribute its sensitive information and lays the framework for the company's physical and information assets. The security objectives of the organizations must be tempered with the organization's goals and situation, and determine how the organization will apply its security objectives. Security polices have evolved gradually and are based on a set of security principles. While these principles themselves are not necessarily technical, they do have implications for the technologies that are used to translate the policy into automated systems [2]. The typical policy goals are to [1]:

- Promote efficient business operations.
- Facilitate sharing of information throughout the organization.
- Safeguard business and personal information.
- Ensure that accurate information is available to support business processes.
- Ensure a safe and productive place of work.
- Comply with applicable laws and regulations.

A security policy should consider a managerial decision, for example, who has access to the corporate server, what the payment procedure is, who confirms to the order, who confirms the delivery method and transportation scheduling. Also, a security policy regulates the document flowing activities between the departments; describe responsibility levels, and the roles. The issues of who approves documents on the existing multi-communicative environment, which parts of documents are approved, and who approves them, and responsibility levels, are essential.

3   The Components of Security Infrastructures

All these type of managerial and role based organizational questions are related some well defined cryptographic concepts. The Fig. 1 represents the relationships between security policies and the technological infrastructures of secure online business [3].

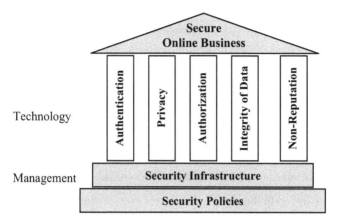

Fig. 1 Security policies and the technological infrastructures of secure e-business.

The technological components of information security services are listed and briefly defined in the following [3].

- **Authentication:** Customers must believe and be assured that they are doing business with, and sending private information to a real entity — not a "spoof" site masquerading as a legitimate bank or e-store.
- **Confidentiality:** Sensitive Internet communications and transactions, such as the transmission of credit card information, must be kept private.
- **Data integrity:** Communications must be protected from undetectable alteration by third parties during transmission on the Internet.
- **Non-repudiation:** It should not be possible for a sender to reasonably claim that he or she did not send a secured communication or did not make an online purchase.

The technologies to implement these well known services are as follows:

- Symmetric cryptography as an encryption tool and it establishes confidentiality.
- Asymmetric cryptography with the following components assures the authentication, data integrity and non-repudiation.
  - Hashing algorithms.
  - Digital signatures.
  - Certification authority.

The public key infrastructure (PKI) is the set of hardware, software, people, and procedures needed to create, manage, store, distribute, and revoke digital certificates based on asymmetric cryptography. The principal objective for developing a PKI is to enable secure, convenient, and efficient acquisition of public's keys for the electronic business transactions. It relies on public key technology in which the hash (or message digest) of a message is cryptographically signed. Because of the nature of public key signatures, anyone with the signer's digital certificate (or public key) can validate that the signer indeed signed the message, providing legal proof that the signer cannot refute. The Certification Authority (CA) as a trusted third party certifies public keys by embedding the owner identity information and public key info into a certificate. A CA plays a critical role in this process since, the public key and the owner of this key should be bind together.

A digital signature provides a mechanism for assuring *integrity* of data, the *authentication* of its signer, and the *non-repudiation* of the entire signature to an external party [4].

The Secure Electronic Transaction (SET) is a protocol designed for protecting credit card transactions over the Internet. It is an industry-backed standard that was formed by MasterCard and Visa (acting as the governing body) in February 1996. The SET standards provide the major business requirements for credit card transactions by means of secure payment processing over the Internet and, are listed below: [5, 6]

- Confidentiality of information.
- Integrity of data.
- Cardholder account authentication.
- Merchant authentication.
- Security techniques.
- Creation of brand-new protocol.
- Interoperability.

4 Information Security Policies and the Security Infrastructure

An e-business system is an open system and the system interacts with external entities such as e-mails, web based applications, and the data transfers. For this reason an e-business company needs an information security policy that defines the firm's security requirements plus the means to minimize the security risks. This objective is realized simply by good collaboration between the security policy and the existing cryptographic technologies.

Security technologies will potentially affect many aspects of the IT environment. For example, PKI itself must operate within a well-defined set of security practices and policies. Existing structures of an enterprise will need to be re-define and modified such as additional naming structures may need to be defined; employee desktops will need to be upgraded with certificate-enabled software; smart cards may need to be issued and managed. The relationship between the information security policy, the security technology, and digital signature is defined below.

4.1 PKI Policy

The cryptographic technology and protocols of PKI requires an infrastructure, and that infrastructure encompasses much more than cryptographic technologies. It includes the policies governing the use of PKI, the risk management controls and business processes needed to enable PKI-supported systems and the applications that serve the newly emerging digital analogues replacing and extending our traditional business, government, and inter-personal transactional relationships [7].

A properly developed PKI policy is a fundamental part of the organization's trust model. For that reason, an organization's overall security policy should depend on

PKI policy. For example, while the company PKI policy accepts a PKI-authenticated order up to $1 million in value for specific customer, the general security policy prohibit electronic order transactions above $100.000.

4.2 Certificate Policy

Certification authority (CA) plays a major role in security policy as an issuer or canceller of certificates can also produce many administrative functions such as initialization, key pair recovery, key pair updates, and cross certification. For example, a CA may work with a group of companies in the same supply chain in creating specialized and unified policy rules. In other situation, CA may have to set de facto policy, for example, to authenticate Web sites using SSL (Secure Sockets Layer) and may have established the policies and processes under those rules [7].

A certification policy is defined "a named set of rules that indicates the applicability of a certificate to a particular community and/or class of application with common security requirements." [8]

4.3 Digital Signature Policy

Basically, a digital signature provides the ability to authenticate the document's originator and to verify that it is unaltered. Business dynamics generally more complex than that and most of the times it is necessary to digitally sign a portion of a document rather then to sign of the complete document. This might be useful if the information is dynamic and changes too often for a signature to be meaningful for the whole document, but for parts of it, it is important to maintain a signature. Another possibility is that the file is dynamic but different portions change at different times [9]. For instance, the two or three departments in a firm may update a transportation order independently. Every updating needs some unique authorization by the department authority or information creator. As a business reality, during the transportation process, parts of the document can be changed and every change needs a different signing procedure in spite of a single signing to the final document.

These concepts have been summarized by different signature validations such as *partial document ownership, soft signatures*, and *hard signatures* [10]. Partial document ownership involves assigning ownership of the content to an individual when that individual changes part of a document. Clearly, partial document ownership requires that a document be recognized as a collection of objects (for example, paragraphs). Soft signatures are analogous to initials: when an individual chooses to sign a document with a soft signature, the signature is not invalidated with a modification to the document. Hard signatures are the traditional digital signatures that are invalidated when the document is changed. Thus, the signing procedure moves to the document ownership from document level to information level.

5 Conclusion

In this paper we have explained how an information security developer can profit from the existing cryptographic based security concerns. And, from this perspective the major attributes that a successful information security policy should have:

- In the e-business arena, firms must have information security policy.
- Information security policy;
    - Should solve the existing security risks.
    - Should be suitable in regard to the general security policy.
    - Must respond to contemporary business needs.
    - Should be developed by security professionals who understand the significance of cryptographic infrastructures.
- There is a very close interaction between the information security services and the policy.

Good knowledge of cryptography and security abilities and practices coupled by experiences, planning, plus managerial leadership yield an enterprise level success for the company. The concept of information security policy will possibly provide suitable solutions to the existing security problems.

Acknowledgments. I am indebted to Dr. Ahmet Koltuksuz, Dr. Serap Atay and Selma Tekir from Izmir Institute of Technology for their guidance, thoughtful feedback, and other contributions to my work.

References

1. Pfleeger, P. C., Pfleeger S.L.: Security in Computing. Pearson Education.3rd edn. (2003)
2. Tipton, H.F., Krause N. (ed.): Information Security Management Handbook. Auerbach Publishers (2003)
3. VeriSign: Security for the Internet. http://www.verisign.com.au (2003)
4. Daum, B., Merten U.: System Architecture with XML. Morgan Kaufmann Publishers (2003)
5. Rhee, M.: Internet Security Cryptographic Principles, Algorithms & Protocols. Wiley (2003)
6. Stallings, W.: Cryptography and Network Security. 4th edn. Pearson Prentice Hall (2006)
7. Sabo, John, et.al.: PKI Policy White Paper. http://www.pkiforum.org (2001)
8. Chokhani S., et.al.: Internet X.509 Public Key Infrastructure Certificate Policy and Certification Practices Framework. Internet Engineering Task Force (IETF). http://www.ietf.org/rfc/rfc3647.txt (2003)
9. Loeb L. (ed.): Hack Proofing XML. Syngress Publishing (2002)
10. Gupta A., Tung Y., Marsden J.R.: Digital signature: use and modification to achieve success in next generational e-business processes: Information & Management, 41 (2004) 561-575

Applying ParseKey+ as a New Multi-Way Client and Server Authentication Approach to Resolve Imperfect Counter Utilization in IEEE802.11i for Impersonation Avoidance

Behnam Rahnama, Atilla Elci

Eastern Mediterranean University, Department of Computer Engineering and Internet Technologies Research Center, Gazimagusa, Mersin 10, Turkey
{behnam.rahnama; atilla.elci}@emu.edu.tr

Abstract. This paper presents the ParseKey+, an approach to a new highly-secure and safe authentication service. The scheme includes authentication process for both client and server sides in addition to passwords of each side. ParseKey+ employs transposition to hide the encrypted key in a key file retrievable only by the other side knowing the indices and lengths of sub-keys inside ParseKey+ encrypted file. ParseKey+ avoids client and server impersonations in addition to mutual client/server authentication. Therefore, the ParseKey+ scheme avoids counter-reply attack. The key file is changed at each signing on procedure. The key file is an additional security beyond the login password which avoids client and server impersonation.

1 Introduction

The importance of authentication service was highlighted by a high level experts committee looking into cyber security issues (President's Information Technology Advisory Committee, 2005): the committee listed authentication research as the top priority among a list of 10 priority areas. This problem has been with us since long time ago. The prevailing applications are distributed and tend to be realized through connectionless services where authentication of both parties (that is, client & server) has gained exceptional importance. Such systems must include new security mechanisms to guarantee the safety of private information. (Elci, Rahnama, & Amintabar, TEHOSS06, 2006). Accordingly, new software architectures are required to provide safety for large amount of processed data in ontology files. (Elci & Rahnama, IWSC 2005, 2005) Cipher analysis of authentication services through insecure networks shows possibility of misappropriation of protected information (Van, 1989). An authentication service should serve for safe access to securely protected information on insecure networks and platforms by controlling and managing interaction (Litwin, 2001).

Authentication service checks login information such as username, password, and perhaps some other static data (Haller & Atkinson, 1994). It will grant permission in such cases as user-supplied data matches server-side data. Cipher analysis discovers flaws in authentication service and is used to devise ways of attacking a system through knowledge of encryption algorithms and of the static data being held. A case in point is the three-way authentication flaw discovered in X.509 version 3 (ITU-T Recommendation X.509, 2005). As the transferred information is not in encrypted form, consequently, it requires a secure communication channel for transferring the messages between client and server. Therefore, it gives way to further possibility of attacks. Moreover, it suffers from requiring extra computation among the information kept in elements of X.509 certificate and certificate revocation list.

Yeh-Shen-Hwang's One-Time Password (OTP) (Yeh, Shen, & Hwang, 2002) supplies authentication procedure through insecure channels as an enhanced version of OTP (Lamport, 1981). However, the service does not revoke faking servers (Haller N. , 1995). Yeh-Shen-Hwang OTP authentication also suffers from eavesdropping, client and server impersonations (Yum & Lee, 2005). The strategy used in ParseKey+ follows the idea behind OTP where the key for next login is changed. Therefore it counters the reply attack (Haller & Atkinson, 1994).

The vulnerabilities of wireless 802.11i (Robust Security Networks: IEEE standard 802.11i, IEEE Standards) are found where it suffers of imperfect counter mode utilization for exchanging keys to AES blocks (Hussain, Mufti, & Ilyas, 2006).

Therefore securer and stronger authentication approaches are required to gaurantee the safety of online servises such as online banking applications, etc. For this purpose, we introduce ParseKey+ approach, a five-way strong authentication approach to client/server impersonation avoidance using Steganography for key encryption to authenticate legitimate clients and servers to each other.

ParseKey+ represents a less complex but more secure procedure than those mentioned above for authentication service hence the lifetime of a session is not our concern. The lifetime of a ParseKey+ file for future logging in to the server or in other words, ticket to ticket-granting server, might be set by the server in addition to the client's profile. However this issue is supported by X.509 certificate (ITU-T Recommendation X.509, 2005).

At the next section, cryptanalysis of the mentioned authentication approaches is covered. Then, ParseKey+ as a new approach is explained followed by its cryptanalysis against most prevalent attacks. Subsequently, results are compared against other approaches, followed by conclusions on practical examples of using ParseKey+.

2   Vulnerabilities of Existing Approaches

Many approaches in authentication services require a hash function to transpose subkeys. Moreover, the password itself is kept as a hash stream in database. However, in February 2005, an analysis to compromise the full SHA1 by Xiaoyun Wang, Yiqun Lisa Yin, and Hongbo Yu was reported (Wang, Yin, & Yu, 2005). The attacks can find collisions in the full version of SHA1, requiring fewer than 2^{69} operations. Likewise, MD5 hash streams are vulnerable by use of similar attacks. Furthermore, there are many online databases with millions of SHA1 and MD5 records to lookup the reverse stream (Reverse MD5 Hash Lookup).

Existing approaches have inherent vulnerability cases which we take up below. Subsequently we summarize our findings.

2.1   Cryptanalysis of X.509:

X.509 certificate verification is vulnerable to resource exhaustion. Included in X.509 certificates are public keys used for digital signature verification. Choosing very large values for the public exponent and public modulus associated with an RSA public key causes the verification of that key to require large amounts of system resources. Therefore, a remote, unauthenticated attacker could consume large amounts of system resources on an affected device, thereby creating a denial of service (NISCC, 2006). It's clear that server impersonation is possible wherever the messages for authentication steps are not sent encrypted. Moreover, in authentication procedure, nonce, as an identification element at certificate message can be signed by any attacker instead of server. This is because of having no previously registered information about the server at clientside. Additionally, the system is vulnerable to attacks in schemes when synchronised clocks are not available. In this situation, a middle point for exchanging messages among parties can impersonate one side to the other one. (Chefranov, 2006)

2.2   Lamport OTP:

Original One-Time-Password scheme (Lamport, 1981) suffers of client and server impersonation (Haller N. , 1995). The security of the OTP system is based on the non-invertability of a secure hash function. Due to the infrastructure of Lamport OTP and use of MD5 or SHA1 for encrypting the initial stream, the method is no longer secure as security flaws have been found in both hash functions. In addition, the result of OTP at each authentication time is based on a counter and the hash function; the method is no longer secure if a third party logs copies of previously entered values and results of hash.

2.3   Yeh-Shen-Hwang OTP:

Hwang-OTP (Yeh, Shen, & Hwang, 2002) as an enhancement on standard OTP is also vulnerable. The cryptanalysis consists of three cases: eavesdropping, impersonation of Server, impersonation of User. In the eavesdropping base case, the attacker records messages between Server and User in the t'th and t+1th login stages. By using the recorded messages, the attacker then impersonates Server. After receiving a response from User, the attacker can successfully impersonate User in the t+2th login stage.

In step (1) the attacker records communication tokens in the t'th and t+1th login stages. In step (2) as impersonation of server case, by using the tokens recorded in Step (1), the attacker computes the fake tokens for the $t+2$th login stage. With these tokens, the attacker plays the server's role for the $t+2$th login stage. The tokens are valid for the

$t+2$th stage except that D_{t+1} is the previously used random number in the $t+1$th stage. Since the user does not store the previous random numbers, he is unable to find out that the tokens are forged and thinks the attacker is a valid server. And finally at step (3) as impersonation of user case, the attacker tries to login to the server in the t+2th stage. After receiving the server's challenge, the attacker computes response.

Since the attacker's response is exactly the same as the expected response to the server's challenge, the attacker is accepted as a valid user. Hence, the attacker succeeds in this so called re-play attack (An adversary makes a replay attack when he performs an action before the valid application does so.) To counter this scenario of attack, the user could store all the previous random numbers received from the server. (Chefranov, 2006) (Yum & Lee, 2005)

2.4 WPA2:

WPA2 (Robust Security Networks: IEEE standard 802.11i, IEEE Standards) or IEEE 802.11i amendment suffers from imperfect counter mode initialization which results in the collapse of whole security mechanism (Hussain, Mufti, & Ilyas, 2006). The AES blocks are not transferred in a secure way anymore. The counter mode utilisation can be regenerated. Therefore, not only the transferred information can be decrypted but also client and server impersonation is possible.

Table 1. Summary table of comparative cipher analysis.

Method	Complexity	Decryption of Payload	Client Impersonation	Server Impersonation	Reply Attack
X.509	Very high	Possible	Possible	Possible	Not Possible
OTP	High for large N	Not Possible	Possible	Possible	Possible
Hwang-OTP	High	Not Possible	Possible	Possible	Not Possible
WPA2	Low	Possible	Possible	Possible	Not Possible

3 ParseKey+ Approach

The ParseKey+ approach provides authentication mechanism for both client and server sides in addition to ordinary ParseKey approach which is an authentication mechanism for client side only (Rahnama & Elci, 2005), whereby client is assured of server's genuineness; and at the same time the protocol guaranties that a legitimate client is communicating with the server. The ParseKey+ guaranties trusted communication between a client and a server each being certain of the other's authenticity. In this approach, a dual side mechanism is used to meet such secure authentication requirements.

In explaining the ParseKey+ encryption and decryption algorithm, its database requirements and key generation process are discussed. Notice that term "key" is used to mean the string of hexadecimal characters output by a hash function. In addition, the term "ParseKey+ file" means the file that hides the key among random noise sequences of hexadecimal characters produced by the same hash function. This practice assures uniform distribution of characters in the file. The key is chopped into sub-key pieces which are then scattered around in the ParseKey+ file. The position indices of the sub-keys in ParseKey+ file in addition to consequent sub-key lengths are kept at the server data-base or in a memory device at the user-/server-side together with its corresponding password.

ParseKey+ authentication process involves both user and server actively. Both parties maintain their original copies of ParseKey+ file, own and other side's User ID/password. At the beginning, client interacts with the server and the login screen appears prompting for username and password. The authentication service tries to match client supplied data against that saved in DB; if successful, the server sends a uniquely generated server-side password and the server-name to the client. If the client can verify the server data then an acknowledgement message is sent to the server. Having received acknowledgement, the server returns the client's ParseKey+ file. This latter is used to retrieve the key and compared against the key kept in client's memory device. If these match, then the client returns server's ParseKey+ file. If the key extracted from this file matches the one saved in DB, then server allows entry into the system. This proves to either party that the other side is authenticated. At the culmination of this process, a new ParseKey+ file is downloaded to the client for use in future logon.

The process is introduced algorithmically starting from the client's first move as follows:

1. Client first registers with the system by providing its unique username and a password. The password is kept encrypted (using a one-way encryption hashing such as SHA (National Institute of Standards and Technology (NIST), 1995) or MD5 (Rivest, 1992) at the server side in order to avoid unauthorized system-side use. However, there is no restriction in the type of hash function as long as it provides uniform distribution of output characters. Our concern here is to use a hash function to distribute the characters uniformly rather than strength of the hash function as security holes were found of late in both algorithms (Wang, Yin, & Yu, 2005) (Reverse MD5 Hash Lookup).

2. After the server approves the validity and availability of username and password, at server side, a server password is generated, saved locally and it's one-way encrypted string is downloaded to the client for future use. This and the server name is saved by the client.

3. A process at the client's side is run to generate client ParseKey+ file (ParseKey$_{(C)}$) as explained below. ParseKey$_{(C)}$ is used for authenticating the server at the next login.

    a. A random seed (RS$_{(C)}$) is generated.
    b. The key string (Key$_{(C)}$) is generated using (RS$_{(C)}$) by applying a one way encryption algorithm such as SHA.
    c. The fixed size ParseKey+ file is created as follows:
        i. Key$_{(C)}$ is chopped into random length and number of sub-keys.
        ii. For each sub-key, an index is randomly generated determining the position of it in the ParseKey+.
        iii. Each sub-key is then inserted in to the ParseKey+ file starting at its designated index location.
        iv. Sub-keys themselves and their corresponding lengths and indices to their respective positions are kept in IL$_{(C)}$ stream.
        v. Sections of the ParseKey+ file unoccupied by sub-keys are then filled by noise text. The noise text may be created em-

ploying the same encryption algorithm that is used to generate $Key_{(C)}$.

    d. $IL_{(C)}$ stream and $Key_{(C)}$ are saved with the client.

    e. $ParseKey_{(C)}$ is uploaded to server to be used for server authentication later. $ParseKey_{(C)}$ is discarded. Execution continues at the server side.

4. Comparable operation to that of the client is also done at the server. A process is run to generate ParseKey+ file for authentication of client at next login as follows:

    a.   A random seed $(RS_{(S)})$ is generated;

    b.   The key string $(Key_{(S)})$ is generated using $(RS_{(S)})$ by applying a one way encryption algorithm such as SHA;

    c.   Random lengths for each sub-key of $Key_{(S)}$ and indices to their position $(IL_{(S)})$ within a ParseKey+ file is computed. ParseKey+ file is pre-filled with random noise;

    d.   Using $IL_{(S)}$ and $Key_{(S)}$ the ParseKey+ file (to be used for client authentication next time) is generated $(ParseKey_{(S)})$;

    e. $ParseKey_{(S)}$ is downloaded to the client;

    f. $IL_{(S)}$ and $Key_{(S)}$ are kept at server's database. But $ParseKey_{(S)}$ is discarded.

5. The client is admitted to the secure member's area allowing access to the secured services.

6. Client can logout or close the communication with the server at any time.

7. At a later time when client decides to utilize the server, it must provide the correct username and password (same as that kept in server's database).

8. If client is authenticated by username/password pair above, then the server supplies its own name and password to client side process. Server's name/password is compared against that kept in the step 2 by the client.

9. If the server name and password is correct, then the server is requested to download the $ParseKey_{(C)}$ file (that was uploaded in step 3).

10. At the client side, $ParseKey_{(C)}$ is checked for verification as follows:

    a.   Client decrypts $ParseKey_{(C)}$ using $IL_{(C)}$ by fetching the sub-keys out of ParseKey+ file and then combining them to generate Key_T.

    b. Key_T, the output of decryption above, is compared against the original $Key_{(C)}$ as generated at step 3.b.

11.   If Key_T does not match $Key_{(C)}$ then Fake-Server Exception is raised. Otherwise, client uploads to the server the $ParseKey_{(S)}$ file provided in step 4.

12. Then at the server side, client authentication process is carried out as follows:

    a.   Server decrypts $ParseKey_{(S)}$ using $IL_{(S)}$ by fetching the sub-keys out of ParseKey+ file and then concatenating them together to produce Key_T.

    b. Key_T is compared against the original $Key_{(S)}$ as generated at step 4.b.

1. If match occurs, the algorithm continues from step 3. Otherwise, Fake Client Exception is raised.

The above process is depicted pictorially in the following UML swim lane shown as Figure 1:

First Login:

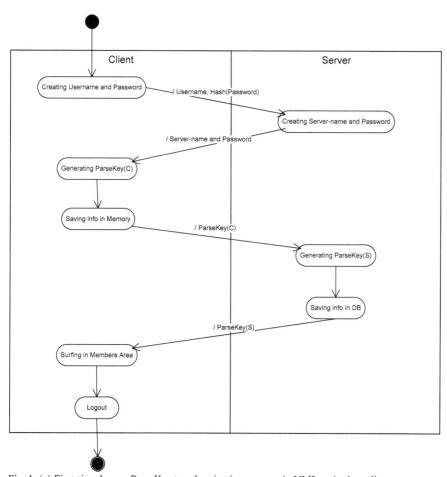

Fig. 1. (a) First-time logon: ParseKey+ authentication process in UML swim lane diagrams.

Next Logins:

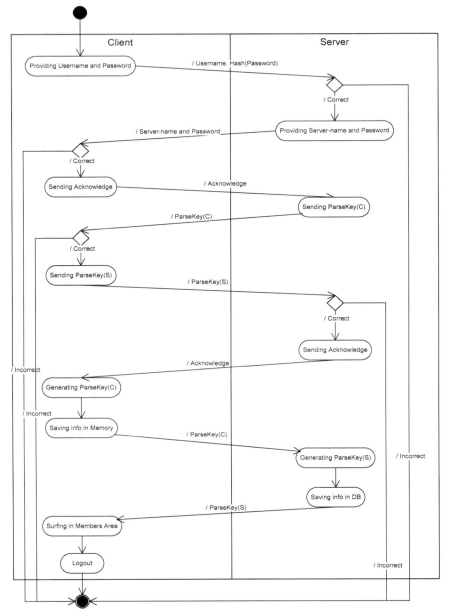

Fig. 1. (b) Subsequent logons: ParseKey+ authentication process in UML swim lane diagrams.

Underlying operations of the ParseKey+ protocol may be precisely formulated in a new representation scheme that will be introduced below. The nomenclature of this scheme is highlighted in Table 2 below.

Table 2. Legends for formulation of ParseKey+ approach

S	Server side
C	Client side
→	Transferring control flow
Hash()	A hash function such as MD5 or SHA1
Subscript $_{(S)}$	Points to server's indices
Subscript $_{(C)}$	Points to client's indices
Encrypt()	Scattering of a key and distributing its sub-keys among random noise in ParseKey+ file
Decrypt()	Merging all sub-keys retrieved from ParseKey+ file using indices and lengths elements to obtain the key.
Sizeof()	Size of a stream in terms of bytes
Save()	Inserting or updating the information as a record in DB or a Memory device
{}	Message
Random()	Generating an integer random number between 0 and the input value
Noise-file	Preferred by user (i.e. 1024 bytes)
Acknowledge	A message to say that the sent data was received

Using the nomenclature of Table 2, ParseKey+ process for the first time and subsequent logon are given below:

First time login:

C→ S: {username, Hash (password$_{(C)}$)}

S→ C: {server-name, Hash (password$_{(S)}$)} // password$_{(S)}$ is unique for each client

$RS_{(C)}$ = Random (Timestamp)

$Key_{(C)}$ = Hash $(RS_{(C)})$

$IL_{(C) I}$ = Vector [Random (EOF), Random (Sizeof $(Key_{(C)})$)]

$IL_{(C)}$ = Array $[IL_{(C) i}]$ for all i

$ParseKey_{(C)}$ = Encrypt $(Key_{(C)}, IL_{(C)})$

Save $(Key_{(C)}, IL_{(C)}, ParseKey_{(S)})$

C→S: {ParseKey$_{(C)}$}

$RS_{(S)}$ = Random(Timestamp)

$Key_{(S)}$ = Hash $(RS_{(S)})$

$IL_{(S) i}$ = Vector [Random (Sizeof (noise-file)), Random (Sizeof $(Key_{(S)})$)]

$IL_{(S)}$ = Array $[IL_{(S) i}]$ for all i

$ParseKey_{(S)}$ = Encrypt $(Key_{(S)}, IL_{(S)})$

Save $(Key_{(S)}, IL_{(S),} ParseKey_{(C)})$

S →C: {ParseKey$_{(S)}$}

Figure 2 below depicts an overview of exchanges durşng the first time logon.

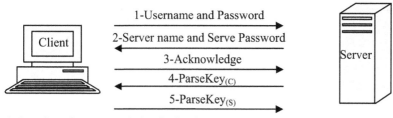

Fig. 2. Overview of ParseKey+ during the first logon.

Next Logins:

C→S: {username, Hash (password$_{(C)}$)}
S→C: {server-name, Hash (password$_{(S)}$)}
C→S: {Acknowledge}
S→C: {ParseKey$_{(C)}$}
Key$_T$ = Decrypt (ParseKey$_{(C)}$, IL$_{(C)}$)
Compare (Key$_T$, Key$_{(C)}$)
C→S: {ParseKey$_{(S)}$}
Key$_T$ = Decrypt (ParseKey$_{(S)}$, IL$_{(S)}$)
Compare (Key$_T$, Key$_{(S)}$)
RS$_{(C)}$ = Random (Timestamp)
Key$_{(C)}$ = Hash (RS$_{(C)}$)
IL$_{(C) i}$ = Vector [Random (EOF), Random (Sizeof (Key$_{(C)}$)]
IL$_{(C)}$ = Array [IL$_{(C)i}$] for all i
ParseKey$_{(C)}$ = Encrypt (Key$_{(C)}$, IL$_{(C)}$)
Save (Key$_{(C)}$, IL$_{(C)}$, ParseKey$_{(S)}$)
C→S: {ParseKey$_{(C)}$}
RS$_{(S)}$ = Random (Timestamp)
Key$_{(S)}$ = Hash (RS$_{(S)}$)
IL$_{(S) i}$ = Vector [Random (Sizeof (noise-file)), Random (Sizeof (Key$_{(S)}$)]
IL$_{(S)}$ = Array [IL$_{(S) i}$] for all i
ParseKey$_{(S)}$ = Encrypt (Key$_{(S)}$, IL$_{(S)}$)
Save (Key$_{(S)}$, IL$_{(S),}$ ParseKey$_{(C)}$)
S →C: {ParseKey$_{(S)}$}

The following table represents a sample list of client login records at the server. Similar record structure is used at the client side. Note that the sequence of IL contains set of pairs as sub-key's file pointer and its length.

Table 3. Login table

#	Username	SHA1(pass)	Key$_{(S)}$	IL$_{(S)}$	ParseKey$_{(C)}$
1	Elci	356a192...	0760c...	5,4;173,7;260,5;475,13;599,9	913b04c...
2	Chefranov	da4b923...	46e6a...	27,2;312,9;729,15;900,14	cab7aec...
3	Rahnama	77de68d...	ecd81...	527,12;690,2;739,18;37,8	c8e14d7...
...	...	...	...	...	...

An improvement in creating the fixed-size ParseKey+ file is to randomly distribute sub-key packets in the file in any order, and treating the file cyclically (i.e. sub-keys in file do not follow their real sequence as it is in the key itself) and not necessarily in a forward linking chain (as in Amin, Salleh, Ibrahim, Katmin, & Shamsuddin, 2003) (Artz, 2001). Figure 3 displays the structure of a sample ParseKey+ file.

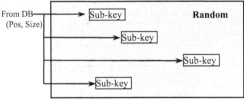

Fig. 3. ParseKey+ general file representation – the position and size variables are in DB and only sub-keys are in the file spread over ParseKey+ file pre-filled with noise.

4   Cryptanalysis of ParseKey+

We first study the possible circumstances under which attacks may be organized in order to show the strength of ParseKey+ in such cases. Subsequently, salient features of Parse-Key+ are discussed in order to display how strong it is.

The cryptanalysis is done over three circumstances: Eavesdropping for reply attack, impersonation of server, and impersonation of client. In the eavesdropping case, the attacker copies a transmitted ParseKey+ file and then tries to impersonate as client or server.

4.1   Reply Attack Case

Having obtained a copy of the last used ParseKey+ file has no use since the user/server name and its corresponding password are required to be supplied at the first step of authentication procedure. Moreover due to ParseKey+ OTP nature, the compromised ParseKey+ file is not valid while the new ParseKey+ file has been generated for the next time use (Haller, 1995)

4.2   Server Impersonation Case

At the beginning of logging in procedure, client supplies its username and password. Fake server acknowledges the correctness of data. The user data can be used later for impersonating the client. The next step is to provide the server name and the generated server password for that particular client. This step avoids the server impersonation. As a legitimate client having access to the server may later become an attacker, the server's password and ParseKey+ file are individually created corresponding to each user, server impersonation is not possible.

4.3   Client Impersonation Case

The server has to supply the Server's ParseKey+ file corresponding to the valid client in addition to the unique password created for an individual client. This double check procedure requires a valid registered user's data to let the server send client's ParseKey+ file. Furthermore, a fake server might have obtained a client's username and password at the first step. This username and password has no more use because the fake server has to also produce the client's ParseKey+ file in order to impersonate a legitimate client. That's why the acknowledgement message is used in this approach to force the server to send its ParseKey+ file in advance.

As may be judged by the above cryptanalysis study, ParseKey+ is strong in precluding the above mentioned penetration cases. The strength of ParseKey+ approach partly stems from its use of a transposition technique to confuse data stream. For this purpose, some extra information for the ParseKey+ file (index elements and lengths) is needed. The minimum file size may be as small as just 40 bytes. This would support just transposition of sub-keys and would not leave room for using Random Noise for Steganography. It would supply only 16^{40} different answers that are possible for brute force cipher attacking processes, but is indeed still much more than 168 bit security. Of course that is not a small value but having the larger file size would improve confusion factor nonlinearly. For example, 1 KB ParseKey+ file size gives $16^{1024} = 1.044388881E+1233$ different situations and it designates a fairly unbreakable key.

On the other hand, the ParseKey+ data stream necessitates employing an indexing array data as well. In the worst situation DB should save totally $40 \times 1 + 40 \times 2 = 120$ bytes for sub-key sizes and sub-key positions (binary representation). Then the cost efficiency in terms of usage space will be $40/ (40+120) = 25\%$. While this level of efficiency is not suitable for data transfer purposes, it is pretty well suited for key distribution for use with other algorithms like AES and etc. Finally, the best usage of this algorithm is for authentication services.

To compare the ParseKey+ approach with existing ones we summarize its features in the following table.

Table 4. Summary table of comparative cipher analysis including ParseKey+

Method	Complexity	Decryption of Payload	Client Impersonation	Server Impersonation	Reply Attack
X.509	Very high	Possible	Possible	Possible	Not Possible
OTP	High	Not Possible	Possible	Possible	Possible
Hwang-OTP	High	Not Possible	Possible	Possible	Not Possible
WPA2	Low	Possible	Possible	Possible	Not Possible
ParseKey+	Low	Not Possible	Not Possible	Not Possible	Not Possible

5 Conclusions

This study describes a five-way strong authentication procedure as an approach to client/server impersonation avoidance using steganography for key encryption. To increase security of authentication service, other security steps in addition to exchanging username/password are needed. ParseKey+ uses steganography and transposition to hide a scattered key within a file of similarly produced random noise (Stallings, 2001) (Wrixon, 1998). This file is changed at each user session. Therefore, it also guarantees unique login in addition to countering the reply attack.

ParseKey+ Algorithm uses SHA1 hash function to create both key and random noises. These hexadecimal random noise characters are used for hiding the scattered sub-keys among them. The hidden key is recoverable by using the indices of sub-key file pointer stored in DB and sub-key size in key file.

In order to further enhance the security of ParseKey+, an improvement is applied whereby the positions of sub-keys are cyclically distributed in the ParseKey file. In this form, cipher analysis indicates that ParseKey+ performs better than X.509, OTP, Hwang-OTP and WPA2 schemes.

6   References

Amin, M. M., Salleh, M., Ibrahim, S., Katmin, M., & Shamsuddin, M. (2003). Information hiding using steganography. *Proc. 4th National Conf. on Telecommunication Technology*, (pp. 21-25).

Artz, D. (2001). Digital steganography: hiding data within data. *International Journal of Internet Computing*, 5 (3), 75-80.

Chefranov, A. G. (2006). *Lecture Notes of CmpE552, Database and File Security*. Retrieved October 5, 2006, from http://cmpe.emu.edu.tr/chefranov/cmpe552-06/index.htm

Computer Emergency Response Team (CERT). (1995, January). IP Spoofing and Hijacked Terminal Connections. *CA-95:01* .

Haller, N. (1995, February). The S/KEY one-time password system. *RFC 1760* .

Haller, N., & Atkinson, R. (1994). *On Internet Authentication*. Retrieved from RFC 1704: http://www.ietf.org/rfc/rfc1704.txt

Hussain, M. J., Mufti, M. U., & Ilyas, U. (2006). Vulnerabilities of IEEE802.11i Wireless LAN CCMP Protocol. *Transaction on Engineering, Computing and Technology*, 11.

ITU-T Recommendation X.509. (2005, 8). *Information Technology - Open Systems Interconnection - The Directory: Authentication Framework* . ITU-T.

Lamport, L. (1981). Password Authentication with Insecure Communication. *Communications of the ACM*, 24 (11), 770-772.

Litwin, E. (2001). Cryptography. *International Journal of Potentials*, 20 (1), 36-38.

National Institute of Standards and Technology (NIST). (1995, April). Announcing the Secure Hash Standard. *U.S. Department of Commerce*, FIPS 180-1.

NISCC. (2006). *X.509 Certificate Vulnerabilities*. Retrieved January 12, 2007, from http://www.niscc.gov.uk/niscc/docs/re-20060928-00661.pdf?lang=en

President's Information Technology Advisory Committee. (2005, February). *Cyber Security: A Crisis of Prioritization*. Retrieved April 10, 2006, from http://www.itrd.gov/pitac/reports/20050301_cybersecurity/cybersecurity.pdf

Rahnama, B., & Elci, A. (2005). PARSEKEY: A New Approach in Secure Authentication Service. *Ag ve Bilgi Güvenligi Ulusal Sempozyumu* (pp. 108-113). Istanbul: Istanbul Technical University and the Electrical Engineers Chamber of Turkey.

Reverse MD5 Hash Lookup. (n.d.). Retrieved January 12, 2007, from http://md5.benramsey.com/

Rivest, R. (1992, April). *The MD5 Message-Digest Algorithm*. Retrieved July 3, 2006, from RFC 1321: http://www.ietf.org/rfc/rfc1321.txt

Robust Security Networks: IEEE standard 802.11i. (IEEE Standards). Retrieved December 25, 2006, from http://standards.ieee.org/getieee802/download/802.11i-2004.pdf

Stallings, W. (2001). *Cryptography and Network Security Principles and practices* (3rd Edition ed.). Pearson Education: Prentice Hall.

Van, R. G. (1989). Authentication services: Theory and practice. *Technische Hogeschool Eindhoven* .

Wang, X., Yin, Y., & Yu, H. (2005). Finding Collisions in the Full SHA-1. *Proc. CRYPTO '05, Lecture Notes in Computer Science*. 3621, pp. 17-36. Springer Verlag.

Wrixon, F. (1998). *Codes & Ciphers*. New York: Barnes & Noble Books.

Yeh, T., Shen, H., & Hwang, J. (2002). A secure one-time password authentication scheme using smart cards. *IEICE Transaction on Communication*, E85-B (11), 2515-2518.

Yum, D. H., & Lee, P. J. (2005). Cryptanalysis of Yeh-Shen-Hwang's One-Time Password Authentication Scheme. *IEICE Transactions on Communication*, E88-B (4).

Centralized Role-Based Access Control (RBAC) Framework for Critical Web Applications

Alireza Goudarzi

Department of Computer Studies and Information Technology
Eastern Mediterranean University (EMU)
Gazimagusa , Turkey via Mersin-10
alireza.g@prodigy.net
+90 533 879 6832

Abstract. Development and enhancement of the web technologies have made them a reliable alternative for regular desktop applications, especially after the Web 2.0 wave. The only problem that IT managers still resist – and for good reason – is the high potential of security breaches in these applications. If web applications are going to replace desktop applications, there should be a solid security framework which can also work the same way desktop security mechanism work to make them completely ready for corporate and sensitive situations, that is using a flexible administrative role based access control. The current implementations to tackle this issue are based on enterprise development frameworks like J2EE or .NET. However popularity of open-source platforms in enterprise environment calls for open-source alternatives. This paper presents and implementation of a centralized role based access control using popular open-source products PHP and MySQL.

Keywords: role based access control, web security, access control in web based systems, secure web access, centralized security in web systems, PHP security framework, high security websites, sensitive web applications, authentication and access control in web applications, access control in enterprise open-source web application.

1 Introduction

During my job experience in governmental and financial organizations as a Microsoft system and security engineer I encountered several occasions in which custom applications that were caused a variety of problems for the organizations from minor security holes to major breaches that could potentially expose all the resources of the organization to a malicious user. These kinds of problems were very common in, but not limited to, the web applications which were in use. This made them even more sever because of the accessibility of web resources from the Internet. When we interrogated these issues most of the times the root led to incompatibility of products with the security protocols and policies of the host network. The unfortunate customers of these applications had to allocate enough man power to go through the source codes and ensure the compliance of the applications with the protocols. Alternatively, they could outsource the job to another company. Both ways it was so expensive that they usually refused to do it at all and the job rolled to administrator to set as many restrictions as possible on the servers to ensure the security of the IT assets. Anyone with a little experience of network and system administration knows that this will never work out as expected. What I gained from this experience, all came back to me during design of my senior project, a portable web desktop.

The main aim of my senior project is to give easy and reliable access to information assets of an organization. It didn't take long until I figured if this software is ever going to be deployed in an organization it needs to be secure enough in order to replace the common desktop applications. This means there should be a way to enforce the security defined by the administrator throughout all the modules of the software regardless of the company number of modules and companies who develop them. Therefore we started to design and develop a security and access control framework that ensures all the resources that are being accessed in our system are fully compliant with the rules defined by the policies. In this paper we will focus on the challenges and the countermeasures we used to cope with them during design of the security modules of my software.

2   Web Application Security

Security of web application comprise of many different parts. Some of these are general an applicable to any kind of web applications. They are also considered a foundation on which more sophisticated security countermeasures can be built. Of the most important and vital ones I can name input validation, as a fundamental and crucial to all the others. All the common attacks on web application are initiated using not validated or poorly validated inputs. Cross-site scripting, code injection, SQL injection and buffer overrun are all of these sorts of attacks. Authentication and access control are of other security controls that don't apply to all web applications. Although a successful authentication and authorization strategy needs to be backed up by a good validation or the results could be catastrophic. Organization and relation of these controls with other parts of the application is another important issue to consider.

A significant architectural issue in designing any application, and web application for the matter of our discussion, is reflected in a way security controls are enforced throughout the application, regardless of the type of control. In most of the web applications today, all the security related functions are grouped in different libraries and developers use these libraries whenever applicable to implement a specific security feature. A noticeable example is access control mechanism: if the developer determines that a page or portion of the page requires privileged access, the functions that restrict access to that portion of the page is included to the page plus the required access level so the page can determine if the visitor's access token contains that privilege. In other words the enforcement of the security is done on page by page basis. The problem with this method is that first, the change in the access control can be hard to apply and requires a bit of programming knowledge. In addition, the developer may forget or don't put the necessary controls in the page, perhaps because of miss-communication between customers and developers. This latter one happens a lot because of the extent that agile development methodologies depend on communication between customer and developer. Another issue would arise in collaborative development between multiple companies. It is quite challenging to ensure that all of the involving groups are exactly using the protocols stated by the customer. Event after the completion of the project the challenges still stand.

Custom made applications requires maintenance very often. Given the distributed approach to enforcing security it is obvious that modification to the policies or security controls could be very expensive for the customer and very cumbersome for the developers to achieve. However the simplicity of development in this fashion has made it very popular, and all web application frameworks and almost all the developers use this method to achieve common security enforcements. but due to current waves in web application development, which is using Web 2.0 to simulate desktop applications, it is time to develop a more solid security framework that gives assurance of a centralized security control in the web environment as well.

3   Centralized Schema for Security Enforcement in Web Applications

The functionality of model-view-controller (MVC) in modern application development has opened its way to the web world. Almost all of the common frameworks that support web application development present a model to implement the same approach in web applications. However among these frameworks only the those which are build upon enterprise solutions such as J2EE and .NET gained the ability to implement true centralized schema for access control.

Current MVC frameworks for web applications doesn't include integrated centralized security schema. The security is enforcement is in the hands of programmer for each module and there is a chance that mistakenly the security code of the enterprise is neglected and cause a breach in one of the modules which usually happens to opens an exploit in the system. I've examined several different well-known frameworks such as Joomla, Mambo, ActiveRBAC for Ruby on Rails, Zend Framework and several others, to see if I can find a security schema that is enforced centrally. Unfortunately all of them create an ACL which is enforced by the programming for those parts of codes that needs it.

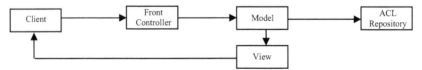

Figure 1: Enforcing ACL by the programmer

To overcome the issues mentioned earlier, we developed a centralized security model based on four *principles* and four *layers*. Principles of the frameworks are the objects of the framework which we are using to control, and layers are the protections that this model is going to be built upon. Each layer has a set of protocols which should be complied while installing and using the framework during the development and later in the production environment or the frameworks won't work as expected and blocks access to resources to ensure the security. We first start to elaborate on the principles and then we expand our discussion to the layers to see how these layers help us implement this framework.

The four principles of the system are *resource, access, source* and *user*. *Resource* is an object that we are controlling access to, *access* is the action that is going to be accomplished on the resource, *source* is where this request is coming from and *user* is the identity of the user who initiated this request. Using this model will give us the ability to control over the finest grain security aspect of the system. For every given request that is sent to the system all these principles should be determined if the request is going to be fulfilled otherwise it will be discarded. So each resource or, which is also called a webpart, in the system should be first registered in the system to become accessible. So is applied to the access, source and user. Interfaces of the system will give ability to developers and administrator to register new instances of these principles to fit their particular requirements. Among these principles, the resources have special importance to system because they are the links between principles and layers of the system.

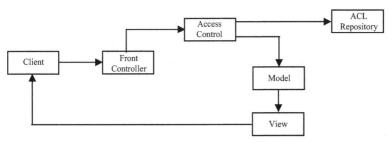

Figure 2: Enforcing ACL using a centralized schema

Layers of the system are like the skeleton that holds the structure of the framework together. They are considered external factor of the application logic that we develop using the framework, but they ensure that the overall application works as defined by the framework and all the controls are applied to all the requests the way they should. These layers from the bottom to top are *file system, web server, interpreter* and *security module*. File system and web server are two building blocks of the layers that work hand in hand. While the file system ensures that there can be no resources available and accessible directly, the web server will direct all the requests to the security module of the framework. In order to achieve the best result we have established best practices to help administrator and developers quickly and securely configure web server and file system of the host environment for this framework.

Interpreter is the application programming model that you are using. In our case we have specifically worked on PHP but the concepts we are using may be easily adapted to other programming languages as well. Finally the security module, "index.php" which is acting as the window to the application checks all the incoming requests and outgoing response to enforce whatever rules that have been put into effect for restricting or monitoring the user activity in the web application.

4   Implementation, Practices and Drawbacks

To describe the implementation of the framework let's start from the bottom, which is the file system and web server as two fundamental layers, and work our way up to the security module. The role of file system and web server layers is to ensure that all the resources are accessed through the security model. To achieve this there are 2 configurations. First we have to confirm that all requests are forced through our framework controller which is "index.php" in the document root. This'd better be the only file that exists under this root even though we make sure that no other files under the document root can be accessed directly. The below is the configuration line we use to reroute everything to our front controller.

```
AliasMatch /.* [path to document root]/index.php
```

Parameters that are passed to the "index.php" will be used later to route the request to the destination module that expects this request. But before forwarding the request to the module the request will be authenticated, validated and authorized. During this process if the resource or actions to be taken on that are not valid for the user, the request will be routed to the authentication page which asks the user to provide the username and password that has privilege for the requested resource. If the resource is not valid, that is not registered in the system; the request will be routed to the error page to show the user a proper error message. If the request passes all the controls it enters the redirector module to be routed to the proper destination module. The redirector module can be modified by developers to reflect their specific needs.

During this process all of the details of the request will be logged to the database according the auditing rules to be used later by administrators or developers for monitoring application access or advanced troubleshooting purpose.

Access control lists are stored in the database. We are aware that this might slow down the process of the authorization but there are several reasons that we considered to vote for the database-at least for now.

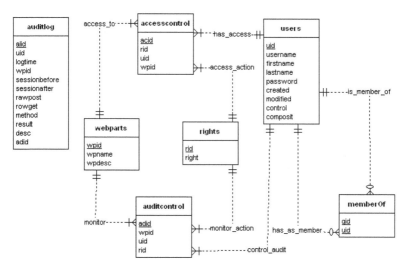

Figure 3: Data model for centralized role based access control

Storing access control lists in the database give us the chance to easily make the system scalable by deploying it on a load balanced cluster, whereas if we had used files it might have given us performance but the system was very difficult to scale and required special replication techniques for the files. But we have also developed a cache for the authorization module that stores the authorization of the user for the frequently accessed page and the current page. This will improve the performance very much. Although activating this cache will result in real-time access modification not to reflect the changes on currently logged in users. Administrator can configure this feature according to the requirements of their environment.

Developers who start to use this system might find it difficult at first but they have to note that usability and security have always stood against each other. Using the distributed model is much easier than a centralized framework but there is no escape from mistakes that might lead the application vulnerable to hackers and information thieves.

6 Developments

There are specific areas that are under still to be developed for this framework to make it easier to user for others. We are working to develop software that can make migration to this framework easier for existing and new websites [1]. Also we are trying to integrate it with our AJAX framework [2]. Algorithms and other processing parts are still other constant change and therefore we could not provide any statistics about the performance of the framework. But using the links to the project website you can easily access these statistics. We hope that this framework can increase security for web applications and help quicker adoption of the web systems.

References

1. Forecast: Open-Source Enterprise Application Software, Worldwide and Regional, 2006-2011. (http://www.gartner.com/DisplayDocument?ref=g_search&id=501649) (2007)
2. Open Source Software Closer to Commercial Enterprise. Strani Izdelava (2007) http://www.webpronews.com/expertarticles/2007/01/09/open-source-software-closer-to-commercial-enterprise-software
3. ActiveRBAC https://activerbac.turingstudio.com/003_arbac_schema.html#sample_implementation
4. RBAC in Ruby on Rails.http://www.yup.com/articles/2006/04/22/role-based-access-control-for-ruby-on-rails
5. phpShop http://www.phpshop.org/trac/wiki/TracPermissions
6. Guide   more   secure   Components/Modules/Plugins. http://forum.joomla.org/index.php/topic,78781.0.html
7. Zend   framework   manual. http://framework.zend.com/manual/en/zend.acl.html#zend.acl.introduction.querying
8. Context-Dependent Access Control for Web-Based Collaboration Environments with Role-Based Approach. Ruben Wolf and Markus Schneider (2005)
9. Integrative Security Management for Web-Based Enterprise Applications- Chen Zhao, Yang Chen, Dawei Xu, NuerMaimaiti Heilili, and Zuoquan Lin (2005)
10. Towards Dynamically Administered Role-Based Access Control - Andreas K. Mattas 1, Ioannis K. Mavridis 2 and George I. Pangalos (2003)
11. Role Based Access Control in Enterprise Application – Security Administration and User Management -Vinith Bindiganavale and Dr. Jinsong Ouyang,

Enforcing Security & Privacy Measures on Semantic Networks

Pooyan Balouchian
Marjaneh Safaei
Alireza Goudarzi

Department of Computer Studies & Information Technology
Eastern Mediterranean University (EMU)
Gazimagusa – Turkey via Mersin 10
pooyan1982@yahoo.com

Abstract. Privacy and security measures of information on the web are of utmost importance in the world today. Knowledge management is, now, taking advantage of semantic network principles. This paper deals with enforcement of security and privacy measures on a software package, the aim of which is to partly emulate the human brain memory. The software equips the human with an artificial memory, capable of storing and retrieving events occurred. Embedding security and privacy features in the associations among these events on the basis of semantic network principles on the web is considered as the core of our discussion.

Keywords: Semantic Networks; Security and Privacy Measures; Human Memory; Computer-Based Memory; Human Memory Simulation; Knowledge-Management Systems; Social Networking; Privacy and Semantic Social Netweforks

1 Introduction

Technology is pushed further whenever limitations are encountered by mankind. One of the most tangible limitations we are encountering with is loss of memory, especially long-term memory. You may have experienced moments when you want to recall an event, but you are unsuccessful. We all try to keep our memories somewhere out of our brain for further retrievals. It seems that we do not fully trust our brain, because of the possible memory-loss pointed out above. That is why we always try to write our memories on paper or in files on the computer. This approach has not yet fulfilled our desire for integrating our memories though. The reason is that enforcement of privacy measures on pieces of paper as well as computer files can not be as successful as enforcement of such privacy measures in the brain. Human brain may be considered as the most secure and private component throughout the world. The reason is that the only person, who has access to your brain, is you yourself.

We have come up with a software package called *Mental Network* [1], which is capable of being installed on Personal Digital Assistants, cell phones, laptops as well as desktop computers, with the aim of equipping the individuals with a secure and private artificial memory, through which they can easily store events of their interest in the form of sound, image, video as well as text. What the system does is relieving the brain memory from tensions of keeping an event in mind and giving the chance to the individual to keep his/her memories on the system.

During the day, events of interest are captured by the individual's handheld device, on which *Mental Network* is installed. Whenever the device is connected to the server-side of the application on the web, the data captured will be synchronized with the database server, where all the memories captured by the users exist. By applying semantic network principles, a social network is formed, where not only the people get connected to each other, but also their memories, experiences, events and locations they have been in are shared. It should be now obvious why the name of the software is *Mental Network*.

2   Security of Information Access in Semantic Data Models

Semantic data model is distilled in association with different data pieces than the data it contains. Associations or relations in a Semantic network are established and repealed dynamically between entities according to concepts and contexts. The result of these dynamic associations is what makes the data model able to capture vast information in proportion to the entities it contains. If we think of information as a chain of orderly connected entities, assuming that all chains made by connecting these entities have acceptable meaning, volume of information derived from a fixed set of data items can clearly show the importance of controlling access to this network if this model is to be used in a real world scenario. This model is used in *Mental Network* which partly emulates human brain to store and recall memories. In a way this system is like a social network but using Semantic models we are associating not only people but also locations and events. Since the data model is being populated by different users the privacy of the information is of utmost importance.

This model is analogous to a cube in which different users are observing this cube from different angles. Components of the cube visible to the observer who are next to each other are shared; their relative position with respect to each other is what makes each observer get a different perspective image. This is what we are going to implement in our Semantic social data model. The mechanics of this security model is very simple, though it is effective enough to make such a complex model work securely. The first elements of this model determine who has access to see an association. The second determines if the latter privilege is transitive. For example, Does a permission given to a friend pass on to his friends who have access to his information or not? State of this transition is being saved as an attribute of the privileges that are being saved in a separate data model. You will find out how exactly this model works later in this article.

3   Software Architecture

Mental Network software package is consisted of two sub-packages - client-side and server-side. The client-side is deployed on J2ME, the intended platform for *Mental Network*. J2ME was selected due to the fact that its architecture defines configurations, profiles and optional packages as elements for building complete Java runtime environment that meet the requirements for a wide range of devices and target markets. The independency of Java to the environment, on which it operates, accompanies *Mental Network* to be accessible on various devices. The user interface, with which the users come into contact, helps the user to capture the events into the local data model. Whenever the user feels there is an event of his/her interest, *Mental Network* is activated on the system by a simple keystroke. The user would be then able to store the event, including the location, date, time, people existing in the event, sound, image, video as well as textual description attached to that specific event. All the mentioned data are captured with as lowest amount of user intervention as possible to maximize simplicity of use and minimize probable errors.

The collection of these events forms the daily memory of the user. By combining all these memories, a life-time memory of an individual is formed, where the individual would be able to recall his/her memory, while benefiting from sound, image, video as well as details of an event captured by him/her at a long time ago. You would be able to recall your memory in the form of a report generated at companies. You are now capable of paging your integrated memories sorted by type of events, locations, times, dates and people, who have played roles in your events.

The server-side sub-package is a web-based database server. All events captured by the users are roomed in the database by synchronizing the local data model with the global one existing on the server. This chain of memories formed on the server is interrelated with respect to the events, locations, relationships and etc. However the availability of the memories to users has restrictions. The social impacts of memory sharing on the web are discussed in the upcoming section.

4   Social Impacts & Existing Social Networks

"Our privacy seems to be under siege on all fronts: the government, corporations, our employers, and of course even our neighbors seek to obtain more and more information about our lives, our preferences, our habits, and so on. Sometimes it seems as though our last refuge of privacy and security may be our own minds - but maybe not for long. [2]". It seems horrifying that other people have access to your very private memories.

Whereas *Mental Network* is aiming to partly emulate how human brain works, it is required to consider the privacy and security measures on the software as much as possible. A back up reliable memory, which is private and secure at the same time, would be of use to many people. The privacy and security will be achieved through requirement of a combination of access control mechanisms provided by all involving parties present in the event under review. Just like pieces of jigsaw puzzle.

Today, there are several social networks which have gained a high number of users throughout the world. So many of people who had always dreamt of finding their childhood friends or old friends are now able to find their lost friends, taking advantage of existing social networks. The heart of today's social networks is connecting people to each other. Of course, they are doing this in different ways. Some of them are connecting people to each other according to their spirits and behaviors, some others create this connection according to the location one is in and some others simply generate such connection based on no specific characterstic. What differentiates Mental Network with the existing social networks is that, the connection is not made among people, but among their memories. Such connection is made based on the involving parties in a given memory, date, location and event, for which that memory is stored. Connecting people's memories to each other, instead that of the people themselves can have different applications in long-term. The reason we think applications may get larger and larger in long-term is that knowledge-management systems gain experience after passage of a noticeable amount of time (For more details see Applications section).

5   Technical Measures

Mental Network uses a very easy-to-handle data model, though very efficient. Based on the fact that the number of users of such a network could be very high, storing and retrieving information in the database play a critical role. Using indexing technique on table columns, on which searching algorithms are working, provides faster response to the user. The heart of mental network lies on searching memories. Therefore fast response time to the searches requested by users is of utmost importance. Nevertheless the need for higher capacity is the trade-off, which is sacrificed to achieve better response time. Columns, for which indexing is needed include Location, Involving Parties in a given event, event titles and so on. Another measure which is taken into consideration to achieve a better response time is that there exists no joins in the search queries. All queries are generated using nested select statements instead of joining the tables. The only drawback is loosing simplicity of using joins in the algorithms. The logical ERD of the model is shown below:

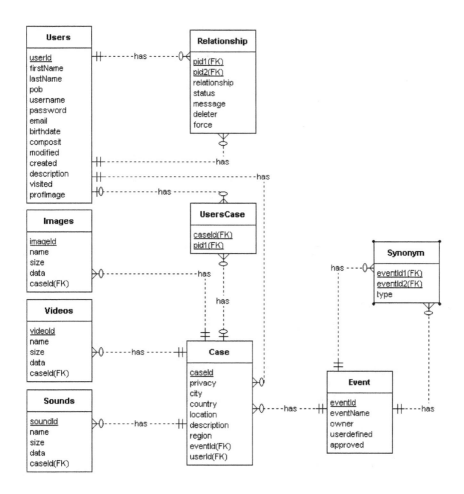

Figure 1: Data Model for Mental Network

"Representations of mental schema for personal or episodic knowledge include many different kinds of elements including concepts (ideas corresponding to things, actions, beliefs, judgments, etc) associations, relations, attributes, images, sensory perceptions, emotions, a sense of time and an awareness of context. These elements are linked in a complex web of verbal and non-verbal information (Anderson, 1990). [7]"

Some words represent different meanings when used in different contexts. The human brain analyzes the contexts in which words are used and therefore has a correct understanding of the events, but how can computers understand the correct meaning of an event. How can computers figure out in which context such events occur? Below you will find the logic to handle events. As you see in this specific example (in case of Ceremony), the system is dealing with four layers of information. The first three layers are system-defined and the last one (synonym layer) is dynamically generated by the users at the time of adding events. At the time of adding an event, when the user reaches the last   system-defined layer, he/she can either choose one of the choices available at that layer or he/she can add a synonym to one of them as deemed advisable. Notice that the added synonym is only visible to the adding party and no to third parties. The adding party can from now on search events of interest in the system based on the synonym he/she had added before and the result would be all leaves attached to its super-category.

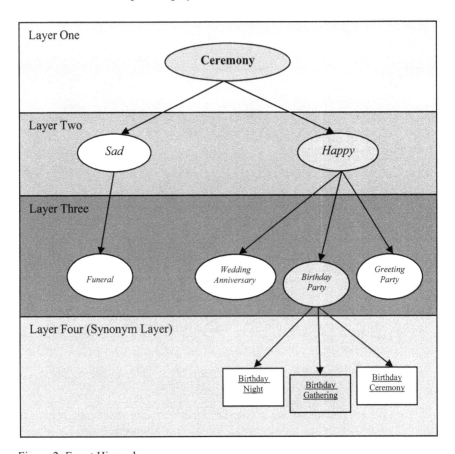

Figure 2: Event Hierarchy

The above model is representing a context-dependant environment for searching purposes. This model was resulted after we encountered a problem dealing with synonyms. Without such model, the system did not have any understanding of the context, in which a specific event occured. Using the above model, since the user is always involved in drilling down to events in a hierarchial fashion, there exists no way to misinterpret or misunderstand the context of an occurence.

The search functions are almost implemented using stored procedure in MySQL 5.0. The only problem is that mysql stored procedures do not support array data types, so to compensate that we created temporary tables using In Memory engine which has very high performance and used it to emulate arrays. We tag all the data with user sessions so in multiuser environment user's data do not mix with each other.

6   Applications

Mental Network can be used in different areas. In medicine, it can be applied for patients suffering from Alzheimer. From social point of view, it can be used for children monitoring by parents. The system would also be capable of gaining experience and guess what would happen in the future based on the experiences gained previously (Extrapolation). In marketing, business corporations can gain useful information of people's taste and that what group of people is more interested in what sort of products. This is what can revolutionize the industry in long-term. Finally Mental Network is able to generate a summary report of people's attitudes and tastes from a given point in time until a specific date.

7   Conclusion

The human brain is capable of more creativity and therefore it would be a good idea to reduce the memory load and reloading it on a computerized system. The concern for keeping the human memory on a computerized mechanism is privatization; that is keeping the memories secure. There is still a long way to achieve perfect privacy embedded in knowledge-management systems. In the future, a small tiny processor may be attached to newly-born babies at birth to accompany the baby during its whole life. Imagine to have a side memory, apart from that of the original one mankind is provided with by default at birth. There would be no need for writing biography of people; instead, the full story of one's life would be available on let's say a digital file.

Reference

1. http://exit.it-solutionz.net/mentalnetwork
2. *Brain Privacy. Are Your Thoughts Safe?* Agnosticism / Atheism From http://atheism.about.com/library/FAQs/phil/blphil_ethbio_brainpriv.htm?terms=medical+coding+employment
3. *Storing Memories.* (2006). The National Memory Test - 2006 from http://www.nationalmemorytest.net.au/storingmemories.aspx
4. *Types of Memory.* (2006). The National Memory Test - 2006 from http://www.nationalmemorytest.net.au/typesofmemory.aspx
5. Query Expansion Using an Intranet-Based Semantic Net, Dick Stenmark, Dept. of Informatics Göteborg University

6. Encyclopedia of Artificial Intelligence, edited by Stuart C. Shapiro, Wiley, 1987, second edition, 1992.
7. Knowledge and Semantic Network Theory, Kathleen M. Fisher - Robert Hoffman, Center for Research in Mathematics & Science Education, San Diego State University - San Diego, CA 92103

Simple Data Sharing Security Model Based On MLS

Muhammad Reza Fatemi, Shima Izadpanahi, Kiavash Bahreini

Computer Engineering, Eastern Mediterranean University, Magusa, TRNC
{muhammad.fatemi, shima.izadpanahi, kiavash.bahreini}@emu.edu.tr

Abstract. Multi-level security is a popular data sharing system. In this article we try to improve the security capability and flexibility of MLS. Our aim is to maximize security with respect to simplicity of system and keeping down the development cost. This system is based on MLS with minor modifications. The Group Security layer is used to add more flexibility and privacy of data. Also implementation of the system is considered with respect to secure communication techniques used.

Keywords: Multi-level security, trusted system, Multi-level, file sharing, secure communication, security model.

1 Introduction

The crucial characteristics of security are confidentiality (prevention of the unauthorized users from obtaining access to information), integrity (the prevention of the unauthorized modification or deletion of information), and availability (the prevention of the unauthorized withholding of information. [7]
Information needs to be granted access. Therefore, people can use and improve it; though there is always risk of information getting into the hand of wrong person. Security of data transmission is not a new topic and many algorithms and models are proposed to defeat the problem. What we consider in this paper is crucial aspect of IT systems nowadays. Colleagues need to inform each other of their achievements which may requires some levels of security from falling into wrong hands. Sharing documents are not just matter of IT systems; privacy of documents is needed in vast area of governmental and private institutions. The model we are proposing in this article is a simple and efficient of secure Object (Data) sharing system with low cost of implementation. To achieve a secure sharing system we have to consider aspects of that system which are: flexibility, granting secure connections, checking constraints on data requests and recording the exceptions and violation of the system.
The base model of this system is Multi level Security model (MLS) [1]. In this research we have added more flexibility and user integrity to MLS which is presented as Enhanced Multi-Level Security (EMLS). In section 2 brief explanations of secure system and multi-level security is provided. It is followed by definition of EMLS In section 3. In section 4 some information is shared about EMLS implementation.

2   Survey on Leveled Security

First aspect which should be considered in leveled security system is Trusted Systems [2,3]. In trusted systems we have classification of levels, *classification* and *clearance*. *Classification level* is in relation with level of sensitivity associated with some data in the system and *clearance level* is the level of trust granted to the users of the system.

Multi-Level Security (MLS) is the base model of our system. MLS is trusted system which at least has three levels of security. This system is mostly adapted in military systems as it is demonstrated in Fig.1. The most governing rule of MLS is No Read-up and No Write-down, which prohibits subject from accessing an object above its clearance level and writing into object below its clearance level. [1,2]

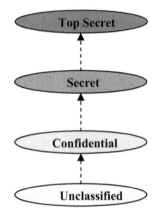

Fig. 1. Simple military classification level, which has four level of classification/clearance.

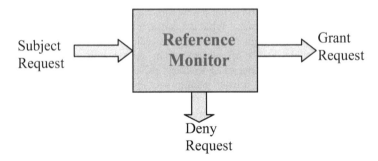

Fig. 2. Reference monitor (RM) block in the system, RM is a module in system which every request and response to them has to pass from it.

In MLS other than granting and revoking access to object, there are two other important issues which are *Reference Monitor* (RM) and *Auditing*. RM is recording every transaction which is happing on the system as demonstrated in Fig.2. In case of a transaction which access has been denied RM audits important data about that transaction like username, timestamp, object identity, reason of access denial and etc.

MLS system usually has intrusion detection. Using audit records which are available in the system intrusion detection tries to compare patterns of behavior which has been recorded with expected one.

3   Enhanced Multi-Level Security

Enhanced Multi-Level Security (EMLS) is a security model for sharing data. This system uses MLS base with more flexibility of securing the objects and with consideration secure data transaction implementation over the network. We present our enhanced multi-level security system, which is based on enforcing group security aspects on top of MLS in subsequent sections.

3.1 Group Security

Group Security (GS) is an object oriented perspective of data sharing over a secure system. GS gives the object creator the right of adding attributes to its object. We have considered three levels of accessibility on an object and one fixed property for owner of object as follows:

- *Private Level*
- *Group Level*
- *Public Level*
- *Ownership property*

If an object is set to "private level", only its owner can access the object. If GS is set to "group level" subjects having the same level of clearance as owner can access this object. If GS is set to "public level" any clearance level can access the object. Moreover, "ownership property" is giving the authority of modifying GS properties of an object only to its owner.

3.2 Enforcing GS over MLS

MLS applies the criteria of no read-up and no write-down on subjects and objects according to subjects' clearance level. GS is considered mostly for group management systems. By adding a layer of GS over MLS, more flexible and secure system is obtained. EMLS supports the properties and functionalities of MLS but in this system we have some more features as listed below:

- Owner of a subject can bind access modifiers according to his/her criteria as well as no read-up and no write-down.
- Owner can grant divided access levels to each object of his/her as follows:
  - o Read, Write, Delete for owner

    o   Read, Write, Delete for subject with same clearance level

    o   Read, Write, Delete for everyone

In EMLS flexibility of granting and revoking permissions on objects is pure dynamic. If owner revokes his/her own access by mistake, s/he can grant access to himself/herself by modifying the properties of that object. Only owners are granted to modify properties (ownership property). EMLS properties are demonstrated as a part of developed system in Fig.3.

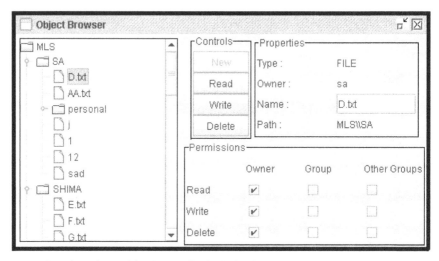

Fig. 3. Snapshot of EMLS implementation is depicted. Each user has the write to press read, write or delete buttons showed in "control" frame. But before triggering the audit system user can deduct if any access violation can be triggered by paying attention to permissions frame. In the "Permissions" frame owner, group and other groups are names assigned to three different levels of GS security levels.

4   Implementation of EMLS

As a testing phase of our research we have conducted series of tests during our implementation time, presented in this section.

4.1 Secure communication over network

Secure communication is a primary issue for any secure sharing system. This topic has been in research and development for years. What we used for our implementation was achievement of other researchers. During development phase we have implemented three different Encryption algorithms.

- Data Encryption Standard

- RSA
- Java Serialization

After comparison between time costs of each algorithm we have chosen java serialization because of its simplicity. Java has ability of storing object and retrieving them through object definitions and using XML technology. In our EMLS model we have considered simplicity and efficiency two most important issue. Therefore, if intruders over the network try to tap into our communication, they can not understand the information which is being transmitted. Data which is being transmitted over network is being protected by its object nature and architecture.

5   CONCLUSION

Secure data sharing is one the most considerable aspects of IT systems; considering the old fashion multi-level security systems which are almost expired these days. To achieve a secure sharing system we have to consider aspects of that system. In this research we have added more flexibility and user integrity to legendry multi-level security systems. Moreover, confidential communication over public networks demands end-to-end secure connections to ensure data authentication, privacy, and integrity. In our approach we have implemented DES[4] and RSA[4] algorithms. We used Java Script because it is computationally inexpensive and it supports acceptable level of security during data transmission. Security issues of this system will be undergoing research. In future we will consider implementing more realistic and proper version of this topic.

References

1. O'Halloran, C.: Trusted system construction. Computer Security Foundations Workshop, 1999. Proceedings of the 12th IEEE (1999) 124–135
2. Kelly, R.J.; LaBerge, E.F.C: MLS: a total system approach. Aerospace and Electronic Systems Magazine, IEEE, Vol 5, Issue 5, (1990) 27 - 39
3. Garuba, M.; Appiah, E.; Burge, L., III.: Performance study of a MLS/DBMS implemented as a kernelized architecture. Information Technology: Coding and Computing, 2004. Proceedings. ITCC 2004. International Conference on (2004) Vol 1, 566-570
4. Stallings, W. : Cryptography and Network Security, Prentice Hall, 4th Ed. (2005)
5. Stallings, W.: Operating Systems, Prentice Hall, 4th Ed. (2001)
6. Afyouni, H.A.: Database Security and Auditing, Thomson, (2006)
7. Proctor, N.E. and Neumann, P.G.:Architectural Implications of Covert Channels, Proc. 15th Nat'l Computer Security Conf., (1992) 28–43.

Examining the Resistance of the Developed Program Using LSB Methods Called Stego_LSB against RS Steganalysis

Andaç Şahin[1], Ercan Buluş[1], M. Tolga Sakallı[1], H. Nusret Buluş[1]

[1] Trakya University, Engineering and Architecture Faculty, Computer Engineering Department, Edirne -Turkey
{andacs, ercanb, tolga, nusretb}@trakya.edu.tr

Abstract. Steganography is information hiding method which has been used since ancient times. Due to the developing technology, to perceive the data in digital media is very difficult. Therefore, steganography is especially is used in the making plan over terrorist attacks nowadays. One of the common methods used for hiding information on image files is LSB (Least Significant Bit Insertion) method. On the other hand, RS steganalysis is used in LSB method to find out hidden information on 8-bit and 24-bit colored images. In the presented paper, a computer program which hides information on 24 bit BMP formatted image file has been developed and we have examined resistance of our program against RS Steganalysis. Moreover, a detailed comparison with other programs exist in the literature is given combined with our results.

1 Introduction

Steganography is very old information technique which takes its roots from the antique Greek and Herodotus [1]. Nowadays, steganographic techniques are applied on digital objects. Due to the developing technology, to protect the data or information from opponents are becoming more important matter.

The aim in the use of the steganography is to hide the existence of the secret message or information. The information transmitted is hidden on a seemingly innocent media to prevent opponent from becoming the aware of the existence of the information. It is possible to hide information on sound, digital images and graphics and this hidden information may be plain text, cipher text, moreover, images hidden on other images. From the point of the mentioned approach, the media on which the information is hidden is called cover data or cover object and the media which is formed after hiding the information on it is called stego text or stego object.

In this study, we have developed a computer program called Stego_LSB using LSB Insertion Method. In addition, we have developed a RS Steganalysis application to evaluate the degree of the resistance of our program and other programs exist in the literature against this steganalysis method. To determine the degree of the resistance of our developed program we have carried out different mask values to realize RS Steganalysis and evaluated the obtained results.

2   Image Steganography

Distribution of digital images on the Internet is very easy task. Therefore, we can face with these images almost on every page. Although, digital images may be different according to the format they use. They are small in size and the data or information hidden on these images may contain redundancy. Due to this fact, digital images are commonly used in steganographic applications. Therefore, the majority of the studies and developed techniques on steganography take part in the view of the image steganography.

Digital image is represented by a string formed by n rows and m columns. Binary images are formed by pixels which are the elements of a digital image. Digital images which are colorful images may have 8 or 24-bit value. On the other hand, grey level images may have 1,2,4,6 or 8-bit value. While each pixel in grey level images is represented by one byte which takes values between 0 (black) and 255 (white) [2], 24-bit images use three bytes for each pixel and the color of pixels is obtained from three basic colors: red, green, and blue. That is called the RGB value of a pixel [3].

LSB (Least Significant Bit) insertion method which is commonly used technique and spatial/image domain technique [4] is very easy method to put into practice. But, the method may cause some data loss in case of being careless while applying it [5].

3   RS Steganalysis Application

Steganalysis is attack methods to deduce information whether it is on cover data or not if there is information on cover data, then it aims to obtain hidden information and extract it from cover data.

In general, it assumed that steganalyst knows the steganographic system that will be attacked. This assumption is known as Kerchoffs's principle. If the steganalyst does not know the steganographic system that will be attacked, then to perceive hidden information will be more difficult. [7]. From attacker's point of view, it is necessary that he should have the information available to mount his attack. There are six main attack models on cryptosystems according to the information available for the attacker like stego-only attack, known cover attack, known message attack, etc.[7] [8]

Detection attacks are used to deduce information whether it is on cover data or not. These attacks are χ^2 Test, Histogram Analysis, RS Steganalysis (Dual Statistic Method), RQP Method, Visual Attacks and JPEG Steganalysis.

RS steganalysis is used in 24 bit colored image and 8 bit grayscale image and image pixels in RS Steganalysis can be categorized into three groups. These are as follows: Regular (R), Singular (S) and Unused (U). This steganalysis method has been developed by Fridrich [10].

We have developed a steganalysis application in Microsoft Visual Basic 6.0 to evaluate effectiveness of Stego_LSB program [6] from the point of RS steganalysis [11].

A 24-bit formatted image is chosen by the developed computer program and the information whether it is hidden or not on this image is examined. Pseudo-code for the application can be given as follows:

Step 1: Choose the image.
Step 2: Enter the mask values.
Step 3: For all color channels (Red, Green, and Blue);
    i. Divide the image to the groups which contain four elements.
    ii. Calculate the output value of discrimination function.
    iii. Calculate $f(F(G))$ value using appropriate flipping functions according to the given mask values (M).
    iv. Determine the number of regular, singular and unused groups using the result of the comparison of values obtained by discrimination function with values obtained by flipping functions.
    v. Repeat Step 3i, 3ii, 3iii, 3iv for $-M$ (negative mask values).
Step 4: Change the least significant bit of each byte of all pixels and Repeat Step 3.
Step 5: Calculate the difference of R_M , S_M , U_M values obtained from original image and image obtained by changing least significant bits for all color channels.

If the difference values obtained by executing the program are close to 0, then we can say that there is no information hidden on image.

4 Experimental Results

To examine the reliability of Stego_LSB from the point of RS Steganalysis, we applied RS Steganalysis attack on images hiding information according to the LSB method. Therefore, an image which is 260 × 320 pixels in size shown in Fig. 1 and Stools–4 [12]", "TurkSteg [13]", "Hermetic Stego [14]", "InfoStego [15]" ve "Stego_LSB [6] programs have been used to hide 5,4 KB text data .

Fig. 1. Original picture (ornek.bmp) used for the RS Steganalysis

Difference values which are obtained from various chosen mask values are shown in Table1, Table2 and Table3.

Table 1. Results obtained by choosing $M = (0,1,0,-1)$ as a mask value.

		Original picture	STools_4	TurkSteg	InfoStego	Hermetic Stego	Stego_LSB
R	R	1	2	29	0	5	1
	S	1	2	29	0	5	1
	U	0	0	0	0	0	0
G	R	3	13	8	7	14	6
	S	3	13	8	7	14	6
	U	0	0	0	0	0	0
B	R	21	23	16	21	8	11
	S	21	23	16	21	8	11
	U	0	0	0	0	0	0

Table 2. Results obtained by choosing $M = (1,-1,1,0)$ as a mask value.

		Original picture	STools_4	TurkSteg	InfoStego	Hermetic Stego	Stego_LSB
R	R	7	6	10	7	1	5
	S	7	6	10	7	1	5
	U	0	0	0	0	0	0
G	R	1	0	3	2	5	1
	S	1	0	3	2	5	1
	U	0	0	0	0	0	0
B	R	11	11	10	8	8	7
	S	11	11	10	8	8	7
	U	0	0	0	0	0	0

Table 3. Results obtained by choosing $M = (-1,0,-1,1)$ as a mask value.

		Original picture	STools_4	TurkSteg	InfoStego	Hermetic Stego	Stego_LSB
R	R	44	29	18	30	3	20
	S	55	45	32	47	11	32
	U	99	74	50	77	8	52
G	R	34	37	10	40	48	26
	S	53	67	24	64	29	35
	U	87	104	34	104	77	61
B	R	41	43	31	42	19	23
	S	52	46	43	54	20	38
	U	93	89	74	96	39	61

In Table 1 and Table 2, it is shown that the difference values of original image are close to zero according to the chosen mask values. In addition, it is also shown that the difference values of images on which the information is hidden using Stego_LSB are close to zero. As a result, it can be said that perceiving the information hidden on images using Stego_LSB by attacking with different mask values in RS Steganalysis is very difficult and Stego_LSB gives better results than other programs from the point of RS Steganalysis.

5 Conclusion

In the presented study, we have examined the resistance of Stego_LSB program developed by us against RS Steganalysis. Moreover, a detailed comparison with other programs using LSB method exists in the literature is given combined with our results. As a result, Stego_LSB gives better results with chosen right mask values from the point of RS Steganalysis than other programs exist in the literature. If wrong mask values are chosen, then Stego_LSB gives small difference values which confirms that Stego_LSB is more resistant against RS Steganalysis attack than the other programs exist in the literature

References

1. Petitcolas F.A.P., Anderson R.J., Kuhn M.G.: Information Hiding–A Survey, Proceedings of he IEEE, Special Issue on Protection of Multimedia Content, 87(7):1062-1078, July (1999)
2. Farid H.: Steganography: the art and mathematics of hiding information, Teaching Notes, Summer (2003). http://www.cs.dartmouth.edu/~farid/teaching/cs4/notes/steg.pdf
3. Johnson N. F., Jajodia S.: Exploring Steganography: Seeing the Unseen, February (1998). http://www.jjtc.com/pub/r2026.pdf
4. Johnson N. F., Jajodia S.: Steganalysis of Images Created Using Current Steganography Software, Second Information Hiding Workshop held in Portland, Oregon, USA, April 15-17, 1998. Proceedings LNCS 1525, 273-289, Springer-Verlag (1998)
5. Kessler G.C.: Steganography: Hiding Data Within Data, September (2001). http://www.garykessler.net/library/steganography.html
6. Şahin A., Buluş E., Sakallı M.T., "24-Bit Renkli Resimler Üzerinde En Önemsiz Bite Ekleme Yöntemini Kullanarak Bilgi Gizleme", Trakya Üniversitesi Fen Bilimleri Dergisi, Edirne-Türkiye, (2006).
7. Rijmen V., "Cryptanalysis and Design of Iterated Block Ciphers", PHd Thesis, October (1997).
8. Stinson D.R., "Cryptography: Theory and Practice, Second Edition", CRC Press, (2002).
9. Katzenbeisser S., Petitcolas F.A.P., "Information Hiding Techniques for Steganography and Digital Watermarking", pp. 81. Artech House, INC. 685 Canton Street Norwood, MA 02062, 2000.
10. Fridrich J., Goljan M., "Practical Steganalysis of Digital Images – State of the Art", Proc. SPIE Photonics West, Vol. 4675, Electronic Imaging 2002, Security and Watermarking of Multimedia Contents, pp. 1-13, San Jose, California, January 2002.
11. Şahin A., "Görüntü Steganografide Kullanılan Yeni Metodlar ve Bu Metodların Güvenilirlikleri", Doktora Tezi, (2007).
12. Brown A., "S-Tools for Windows", ftp://idea.sec.dsi.unimi.it/pub/security/ crypt/code/s-tools4.zip, (1996).
13. Tunçkanat M., Sağıroğlu Ş., "Güvenli İletişim İçin Yeni Bir Yaklaşım: Resim İçerisine Döküman Gizleme", GAP IV. Mühendislik Kongresi (Uluslararası Katılımlı), Şanlıurfa, vol.1, s.665-668, 6-8 Haziran (2002).
14. Hermetic Systems, "Hermetic Stego", http://www.hermetic.ch/hst/intro.htm, Program: http://www.hermetic.ch/hst/hst_setup.zip, 2006.
15. Antiy Labs, InfoStego, http://www.antiy.net/infostego/

Secure Routing in Ad Hoc Networks and Model Checking*

Evren Önem**, A. Burak Gürdağ*** and M. Ufuk Çağlayan

Computer Networks Research Laboratory (NETLAB),
Department of Computer Engineering,
Boğaziçi University, Istanbul, Turkey
{evren.onem,gurdag,caglayan}@boun.edu.tr

Abstract. Although secure routing issues in mobile ad hoc networks (MANETs) have always been a major focus in the recent years, the success of delivering a guarantee of secure communication has never been entirely achieved by any secure routing protocol. In this work, we give a survey on secure routing issues in MANETs, with a brief summary of the current state of the art in secure routing protocols and their resistance to known attacks. We describe formal specification and verification methods that are applicable in security property verification, especially having our focus on model checking. We also present current state of our research on using model checking to analyze security properties of secure routing protocols for MANETs. A formal security analysis of SAODV and ARIADNE by using a well-known model checker, SPIN, is provided. By modeling the SAODV protocol and formally specifying the security properties, we present two attacks automatically found by SPIN, in the presence of an external attacker. SPIN is also used to flag a sequence of possible events in the ARIADNE protocol, leading to an attack where one compromised node and one external node collaborate to remove the intermediate nodes from the route discovery process.

1 Introduction

An ad hoc network is a self-configuring network of terminals freely moving within an area, connected by wireless links. Such a network at any time forms an arbitrary topology by its unpredictably changing presence or absence of links. Following the widespread use of wireless communication technologies, concerns on secure communication in mobile ad hoc networks have gained significance

---

* This work is supported by the Turkish State Planning Organization (DPT) under the project number 2007K120610.
** The author is supported by the Turkish State Planning Organization (DPT) under the project number 2003K120250.
*** The author is with Argela Technologies and supported by the Directorate of Human Resources and Development (BAYG) of Turkish Scientific and Technological Research Council (TÜBİTAK).

in the research area by providing many challenges to the research community. Secure routing is one of them.

There are many proposals in the literature that claim to provide certain levels of security in routing functionality [1, 2]. However, there are few works that formally analyze the security of these proposals. The proposed secure routing protocols mostly come with an informal security analysis performed by their designers, which makes it doubtful to trust the protocol.

It is now widely accepted that secure protocols need a proof of security before being accepted as secure. Here comes the main flavor of formal methods, they help us either to prove or to refute the correctness of a design. Through the use of proper formal methods, it is possible to achieve provable correctness and reliability in any system design and to analyze a system for desired properties. In this work, we use model checking as our proper formal method. More specifically, we use PROMELA (PROcess MEta-LAnguage) as the specification and modeling language, and SPIN (Simple PROMELA INterpreter) as the model checker [3]. Our primary aim is to demonstrate design flaws that lead to violations of security requirements using model checking. Model checkers are good at finding design errors and they provide error traces representing a sequence of events that lead to a possible violation of the property that is being examined. In our case, an error trace is an abstract demonstration of a successful attack on the secure routing protocol that we are trying to verify.

There are several studies involving the modeling and verification of some routing properties of MANET routing protocols. Bhargavan et al. [4] model AODV with PROMELA and analyze it with SPIN to demonstrate that the protocol does not possess loop-freedom. In this study, the authors suggest an improvement to the AODV protocol and provide a formal proof showing that the improved version is loop-free. They use SPIN and HOL together for this purpose. In another study, Wibling et al. [5] analyze functional correctness of a hybrid MANET routing protocol called LUNAR with SPIN. There are five to seven nodes in their network models and their verification scenarios involves couple of topology changes. Renesse and Aghvami [6] used SPIN to verify certain properties of WARP, which is a hybrid MANET routing protocol. Their verification model includes five nodes with predetermined roles for sender and receiver. The model also includes a limited degree of mobility.

There are very few works in the literature that apply formal methods to analyze security of secure routing protocols for MANETs. In [7], Ács et al. developed a formal framework based on simulation paradigm to analyze the security of distance vector routing protocols for MANETs such as SAODV [8, 9] and ARAN [10]. The authors defined two attacks for SAODV, one of which involves deceiving the destination to think that the route is one hop shorter, and the other involves a spoofing attack.

In a recent work, Nanz and Hankin applied their calculus and static analysis technique to model and analyze a basic version of SAODV [11]. They demonstrate their technique to show the same spoofing attack as in [7].

To the best of our knowledge, our previous work [12] is the only one in the literature that applies model checking approach for the purpose of verifying a security property of a secure routing protocol for MANETs. In this work, PROMELA and SPIN are used to make a formal security analysis of ARIADNE [13], a secure on-demand source routing protocol for MANETs.

The rest of this paper is organized as follows: in Section 2, we give a discussion about routing in mobile networks with both general routing issues and secure routing issues. Here, a broad overview of current secure routing protocols is presented as possible prevention mechanisms, with some other approaches as possible detection mechanisms. Then, in Section 3, we discuss formal methods for system specification and verification, specifically focusing on the model checking approach. We describe various specification methodologies and verification tools, along with a shallow introduction to our tool of choice, SPIN. In Section 4, we give a summary of two secure ad hoc routing protocols, SAODV and ARIADNE. Then, we give a discussion about the security analysis of SAODV and ARIADNE using model checking. Here, we present our protocol models, security properties to verify and verification results. In the last section, the conclusions and future directions are presented.

2   Secure Routing in Ad Hoc Networks

First, we give an overview of general routing issues and possible attacks on routing protocols for MANETs. Then, we discuss secure routing by examining both prevention and detection schemes. We especially have our focus on the state of the art in secure routing protocols and their resistance to known attacks.

2.1   Routing in MANETs

MANET routing protocols are commonly classified in three categories, namely, proactive, reactive and hybrid protocols [14].

In *proactive protocols*, the routing information is exchanged between the neighbor terminals and is kept in local tables to use later while making routing decisions. Such behavior may provide a minimum routing delay as a result of its readiness. However, as the rate of change in topology increases, it becomes harder and harder for a terminal to be up-to-date so the performance degrades dramatically. OLSR [15] and TBRPF [16] are two examples of proactive protocols.

In *reactive protocols*, any routing information is queried only when a terminal needs it. Reactive protocols may react slower than proactive protocols for routing queries. However, they greatly reduce the amount of control packets exchanged in the network. AODV [17] and DSR [18] are two examples of reactive protocols.

In *hybrid approaches*, advantages of both classes of protocols above are used in specific cases. Using proactive routing for near terminals and reactive routing for distant ones may be an example for hybrid approaches. ZRP [19] is an example of this category.

Within the rapidly changing characteristics of a MANET topology, the reactive protocols are known to outperform the proactive ones with the exception of some certain cases [20, 21].

2.2   Attacks on Routing Protocols for MANETs

Attacks on MANET routing protocols are classified as *passive* and *active* attacks. A passive attacker, which is a threat only for privacy, is an eavesdropper on the network who may perform [22];

- *Sniffing*, where the attacker listens and records the traffic in the network.
- *Traffic analysis*, where the attacker intercepts and examines messages to deduce information from communication patterns [23]. Typically, the information that can be inferred from the traffic increases directly proportional to the number of messages observed. By traffic analysis, various advantages may be obtained, such as *location disclosure*, in which the attacker may discover the location of a node, or even the entire structure of the network.
- *Deliberate exposure*, where the attacker discloses the traffic information to the other terminals who do not have permission to see it.

Classification of active attacks is carried out in various ways in the literature, considering different points of views. We prefer to classify them in two categories both of which may be said to be instances of a denial-of-service (DoS) attack from an application-layer perspective [1, 24–26]:

- *Routing-disruption attacks*, where the attacker tries to disrupt the legitimate routes that packets in the network will follow and make them travel through unintended routes. By distributing forged false information in the network, various attacks can be realized:

  - *Route detours* may be created, where legitimate terminals are deceived to use suboptimal routes.
  - *Partitions* can be created, where one set of nodes within the network can be prevented from reaching another set.
  - *Gratuitous detours*, where the attacker makes a route seem longer than it actually is, by adding non-existent terminals to the route-reply.
  - *Blackmailing*, where the attacker might cause a selected terminal to be added to the blacklists of other terminals and make that terminal not appear in any route.
  - *Rushing attack* may be performed in protocols using duplicate route-request suppression, where the attacker broadcasts route-requests in account of other terminals thereby causing the legitimate route-requests in the future to be ignored.
  - *Wormholes* can be established, where a pair of attackers have a private physical connection and use this connection to disrupt routing, such as, one recording route-requests may forward them to the other to be rebroadcast thereby causing routes longer than one or two hops to be undiscovered.

- *Tunneling attack* may be performed, where two remote terminals collaborate to exchange legitimate messages of other terminals in an encapsulated form through existing message channels in order to show themselves as adjacent nodes. In this way, they may be able to attract and analyze certain network traffic.

– *Resource-consumption attacks*, where the attacker tries to acquire access to any resource in the network for which he/she does not have enough privilege. This type of attack may target bandwidth, memory or computation power of terminals. Some examples are as follows:

  - *Spoofing or masquerading*, where the attacker attempts to identify itself as some other terminal, which in turn will open the gate to perform further attacks such as creating routing loops.
  - *Hijacking*, where the attacker takes control of an on-going communication and masquerades as one of the communicating node.
  - *Misclaiming*, where the attacker advertises its authorized control of some network resources in a way that is not intended by the authoritative network administrator [22]. Compromised, unauthorized or masquerading nodes may misclaim network resources.
  - *Sleep deprivation torture attack* may be performed in which the attacker attempts to drain batteries of some other node by constantly keeping it busy in various ways.

Also, there are more attacks which must be classified under both of the above attack types, since they may target both routing-disruption and resource-consumption.

- *Blackholes* can be created by attacker terminals who attract packets and then drop all of them.
- *Grayholes* are the derivatives of blackholes, where not all but some packets are selectively dropped.
- *Routing loops* may be created by an attacker, thereby causing the packets to travel in cycles without reaching their intended destinations.
- *False route-errors* may be forged and disseminated in order to damage valid routes and initiate new route-discoveries within the network.

2.3   Secure Routing in MANETs

From the routing protocol point of view, there are two types of messages in an ad hoc network [8]: *routing messages* and *data messages*, both of which need totally different mechanisms to be secured. While any point-to-point security solution can provide *confidentiality* for data messages, a means of *differentiating between the legitimate nodes* and *per-hop authentication* are needed by intermediate terminals for securing the routing messages, and such a point-to-point mechanism alone can not provide such a means. Thus, the active focus on securing routing in MANETs is generally on securing the routing messages. Besides, it should be strongly underlined that a *prevention* mechanism is destined to be flawed if it is not perfect. Thus, to provide security in a wider sense, a proper *detection* mechanism should be deployed to react to attacks when they occur. One can find a very detailed survey in [27] about prevention and detection schemes at each distinct layer. We will mostly focus on network-layer prevention and detection schemes.

Prevention Mechanisms: These mechanisms include key-management and secure routing protocols:

- *Key-Management*: One of the most fundamental problems of a secure routing environment is the key-setup phase, which means disseminating the authentic key information to the mobile terminals. While in the case of simultaneous deployment, sharing the private keys before deployment is the most generic solution. In cases requiring incremental deployment, such as a MANET, establishing trust and keys between each two terminals becomes hard. F. Stajano and R. Anderson have proposed the *resurrecting duckling model* [28] and D. Balfanz *et al.* have given a more general shape to this approach by offering the use of privileged side channels [29]. S. Zhu *et al.* [30] described a secure protocol against a collusive attack by up to a certain number of compromised nodes, enabling any two nodes in the ad hoc network to establish a pairwise shared key on the fly, without requiring the use of an on-line key distribution center. Z. Haas and L. Zhu [31] described a distributed service in which the trust is divided into some number of shares using threshold cryptography and these shares are assigned to some predefined number of arbitrarily chosen nodes, called servers. Their mechanism can tolerate a certain number of compromised servers if a certain number of partial signatures are provided to compute a correct signature. G. Montenegro and C. Castelluccia have proposed *statistically unique cryptographically verifiable (SUCV) addresses* [32], that reduce the problem of distributing a list of *(node, public-key)* pairs to distributing a list of legitimate nodes. S. Capkun and J. P. Hubaux, assuming a source routing protocol, have proposed a mechanism to provide secure routing even with an incomplete set of security associations, provided that the percentage of security associations is sufficiently high [33]. Some other approaches can be examined in [34–36].

- *Secure Routing Protocols*: In order to cure the flaws of the general routing protocols defined in Section 2.2, different secure routing protocols are proposed, each with the ability of resisting some classes of attackers. Yih-Chun Hu *et al.* [13] introduced an attacker classification with the term *Active-y-x*, where y stands for the number of terminals that the attacker has compromised and x stands for the total number of terminals that the attacker owns within the network. In order to examine the current state of the art in routing protocols and their resistance-levels to known flaws, one should at least consider:

- *Secure Efficient Ad hoc Distance vector (SEAD)*: Based on DSDV and uses one-way hash chains to prevent multiple uncoordinated attackers from creating incorrect routing state in any other terminal [37, 1]. A node uses a specific single next element from its hash chain in each routing update that it sends about itself. By the use of this initial element, the hash chain can provide authentication for the lower bound of the metric in other routing updates for this destination. Any routing update can be authenticated using one of the previous authentic hash values from the same hash chain. However, SEAD cannot prevent the attack where a terminal re-advertises the same advertisement for a particular *sequence number* (freshness) and *metric* (the node's shortest known distance). This is because SEAD only secures the lower bound on the metric ensuring that the terminal does not reduce it [38]. Blackhole, location disclosure and wormhole attacks can still be performed.

- *Authenticated Routing for Ad hoc Networks (ARAN)*: Provides authentication and integrity through the use of cryptographic certificates [10]. Also provides non-repudiation. It consists of two stages of which the second is optional for a trade-off between power-saving and security. ARAN is able to detect the attackers in the ad hoc environment. However, as much as any mechanism using public key cryptography, it is vulnerable to DoS attacks launched by flooding the network with forged control messages for which signature verifications are required. Using this vulnerability, an attacker can force a terminal to discard a certain fraction of the control messages it receives [1]. Blackhole, location disclosure and wormhole attacks can still be performed.

- *Secure Routing Protocol (SRP)*: An extension to apply to the existing routing protocols so as to secure the route discovery phase [39], i.e., in order to accept only the legitimate RREP for a RREQ through the use of message authentication codes. Only an end-to-end shared key is required for the communicating pair of terminals. Flooding is limited since the neighbors are ranked inversely proportional to their send-rates and served accordingly. Since RERRs cannot be authenticated, a node on any route may forge a falsified RERR for that route. A node can freely modify the node-list of a RREQ packet that it forwards. SRP is also vulnerable to wormhole attack and attackers can at worst hide the routes they belong to [40]. An attacker may broadcast forged RREQs in the name of a legitimate terminal in order to reduce the effectiveness of the future legitimate RREQs of that terminal. A selfish node may not forward RREQs so that its own future RREQs will have higher priority. An attacker may also launch a masquerading attack using the security associations of the compromised terminal. Blackhole and location disclosure attacks can also be performed.

- *Secure Message Transmission (SMT)*: Different in the sense that it tries to ensure successful delivery of data packets forming an end-to-end security association, but not ever considering the security of route discovery and route maintenance phases [41]. Thus SMT should be deployed with some other protocol which can realize the route discovery phase and uses multiple routes each with a rating assigned by using feedback from the destination nodes. Messages are sent in a partial manner using secret sharing techniques so that if a certain number of the total packets are received, the message can be reconstructed at the destination. SMT handles link breakages and compromised routes by assigning them lower ratings.

- *Security-aware Ad hoc Routing (SAR)*: Introduces the notion of a *trust hierarchy* by distributing keys for each trust level [42]. Each terminal has a certain immutable trust level and while initiating a route discovery it declares the minimum value of a security metric (e.g. trust level) of the terminals that will participate on the route. Hence, no node with questionable trust level becomes a part of that route. The discovered route may not be optimal, but it is the most secure in terms of trust levels [38]. Scalability is a problem since it means distributing keys, but the protocol can prevent the malicious nodes from being in the discovered route. Blackhole, location disclosure and wormhole attacks may still be performed.

- *Secure Link-State Protocol (SLSP)*: Uses digital signatures and one-way hash chains to secure the link-state updates which are signed and propagated a limited number of hops [43]. Terminals are assigned priorities inversely proportional to the number of link-state updates they generate or forward. Therefore, the flooding attack described in SRP is also valid for SLSP. Blackhole, location disclosure and wormhole attacks may still be performed.

- *On-demand Secure Routing Protocol (OSRP)*: This protocol attempts to provide a fault-free path under situations where a group of nodes are possibly malicious, through assigning certain weights to each link between two adjacent nodes [44]. When a faulty link is found, its weight is multiplicatively increased so that the initiator of the route-request can avoid that link in the future by selecting from multiple routes the route whose sum of link-weights is the least, i.e., the route which has the least likelihood of having a faulty link inside. If there is one fault-free path to the desired destination, even in a highly adversarial environment, the protocol ensures the successful discovery after a bounded number of faults. Blackholes are prevented but grayholes are possible. Location disclosure and wormhole attacks may still be performed.

- *S-DSDV*: Relying on existing pair-wise secret-keys between each two terminals in the network, entity and message authentication is provided [45]. Data integrity protection and data origin authentication are realized. Besides, routes with falsified destination, advertised routes with falsified sequence numbers, advertised routes with falsified cost metrics, routing updates with falsified information, and advertised routes with falsified next hops are detected provided there is at most one bad node in the network. Blackhole, location disclosure and wormhole attacks may still be performed.

- *SAODV and ARIADNE*: Yih-Chun Hu *et al.* [13] design ARIADNE protocol to provide security against one compromised node and arbitrary active attackers. SAODV is the extended version of well-known AODV protocol with security enhancements. We will explain SAODV in Section 4.1 and ARIADNE in Section 4.2 in more detail.

Additional secure routing approaches are described in [46–51]. A much more detailed comparison between key-management schemes and secure routing protocols considering different metrics is provided in [2].

Detection Mechanisms: These mechanisms include *watchdog* and *pathrater* approaches [52], which target protection of the network from the misbehaving terminals by monitoring the neighbors for a future routing decision. Simulations show that, when used in combination, these two approaches provide considerable improvement in network throughput in an adversarial environment.

- *Watchdog*: Every terminal maintains a watchdog process monitoring their neighbors by listening promiscuously on the transmissions and checking whether the neighbor participates in forwarding process as expected. The misbehaviors are rated for each neighbor and after having exceeded a certain rating value the monitoring terminal notifies other nodes of the situation. The watchdog might not detect misbehaving nodes in presence of ambiguous collisions, receiver collisions, limited transmission power, false misbehavior, collusion between neighboring nodes and partial dropping [38]. Also the memory and computational resources are wasted for the actions performed while monitoring the neighbors, such as verifying the integrity of the sent packets.
- *Pathrater*: This approach enhances the knowledge of misbehaving nodes by considering also the link reliability data so as to pick the most reliable route. Terminals maintain a database storing link ratings of every other node. The optimal route is chosen considering an average link rating over the possible routes. If the watchdog reports a misbehaving terminal, its rating is set to a negative value. However, the recovery is possible for the malicious nodes

since they can increment their rating after a period of healthy participation in routing.

- *Packet Leashes*: This approach targets the detection of a wormhole which may severely disrupt routing if remains uncovered. A leash is the information included in a packet to limit the maximum distance that the packet is allowed to be transmitted. Two different types of leashes are defined [53] both requiring different mechanisms, but both with the same purpose: to determine the maximum distance a packet may travel so that if a wormhole exists then it can be detected since the packets through the wormhole travel more distance than feasible. *Geographical leashes* need geographical location information to send with each packet, for the packet to be verified at the receiving terminal. This type requires the nodes to be aware of the maximum speed of a terminal and also requires loosely-synchronized clocks. *Temporal leashes* require the nodes to have tightly synchronized clocks so that each packet has a timestamp showing the time when it was sent. The receiving node can thereby conclude if the packet has traveled too far, by using the fact that the upper bound on the distance the packet can travel is determined by the speed of light. Both leashes suffer from either exact positioning information or accurate time synchronization both of which are difficult to achieve.

Additional detection approaches are described in [54, 55].

3   Formal Methods and Model Checking

The term 'formal methods' is defined to be mathematical techniques for the specification, development and verification of software and hardware systems. Formal methods specially focus on requirements and specification phases of development. The main flavor of formal methods is that it is possible to achieve provable correctness and reliability in any system design and to analyze a system for desired properties. Here, it should be noted that the term 'system' refers to any application whose behavior can be described in a formal language.

It is common for any informal system specification to lead to ambiguous definitions of the desired features. Furthermore, there is no reliable way to prove the completeness and consistency of the design. In such systems, informal inspection is prosecuted with error-checking test suites of which the coverage gets weaker directly proportional to the system complexity. Instead, for the system not to be totally revised in case of any fault, mathematical techniques providing reliability and provability should be located.

Formal methods have already demonstrated success in specifying commercial and safety-critical software, and in verifying protocol standards and hardware designs [56].

Formal methods may be used for some distinct purposes [57] such as formal specification, formal verification, rapid prototyping, functional testing and performance testing, of which only the first two are in our interest.

3.1 Formal Specification

Formal specification is the unambiguous determination process of the system's main features and behaviors in a mathematical manner. The specification is meant to describe the 'what', not the 'how'. This formal description is generally used to guide further investigations, such as validation or verification. Recent advances have brought around the definition of formal description techniques (FDT), which can be classified according to their operational model as finite state machine, process algebra, Petri Nets and timed automata [57]. Some of the FDTs used in our field are Specification and Description Language (SDL), ESTELLE, LOTOS, UPPAAL, NuSMV, AVISPA, and the tool of our choice, SPIN, which is described in more detail in Section 3.3.

A detailed comparison between FDTs considering various metrics such as pertinence to protocols, modeling capabilities, validation capabilities, simulation capabilities, performance analysis, rapid-prototyping, tools availability, friendliness, main application domain and industrial applications, is given in [57].

3.2 Formal Verification

Formal verification is the act of proving or disproving the correctness of a system with respect to a certain formal specification or property, where the possible behavior of the system is checked against the desired behavior. Two well-studied formal verification approaches are *theorem proving* and *model checking*.

Theorem Proving: A technique where both the system and its desired properties are expressed as formulas in some mathematical formalism, which defines a set of axioms and a set of inference rules for deduction of further facts from the given axioms. Theorem proving is the process of finding a proof of a property from the axioms of the system [58]. Depending on the formalism used, the hardness of proving a theorem may be in some spectrum from trivial to impossible. The theorem provers can be categorized depending on the degree of how automated they are; from highly automated general-purpose provers to interactive provers dedicated to some special purpose. Interactive provers are generally more suitable to prove some property of a system since they require user-guidance at certain phases of the operation, meaning that in order to use these provers efficiently, the user must have a degree of proficiency in the field. The distinctive feature of theorem provers is that they can naturally handle infinite state spaces, using numerous proof procedures (e.g. mathematical induction, model elimination, tableaux method, superposition, higher-order unification and DPLL) in order to generate proofs about infinite sets.

Model Checking: A technique that relies on building a finite model of a system and checking that a desired property holds in that model [58]. *Temporal model checking*, in which the specifications are stated in a temporal logic and systems are modeled as finite state transition systems, is firstly developed by Clarke and Emerson [59] and by Queille and Sifakis [60] independently in the 80s. In a typical case, the user creates a high level representation of the system, i.e., the model, and the specification to be checked. The model-checker either terminates positively, meaning that the model satisfies the specification or give an execution trace showing how the property in question might not hold. The check is generally an exhaustive state space search, exploring all states and transitions in the model to see whether the required properties are valid in all reachable states and execution sequences of the model. Since the model is finite, this process is guaranteed to terminate. The size of the state space for such a model depends on the number of concurrent processes, the number and range of the internal variables, the type of the exchanged messages, and the nature of the communication [57]. A model being more abstract means having less of the parameters above. Therefore, it is usually much simpler to verify properties at a more abstract level.

Main flavors of model checking are:

− It is fully automated and fast in the sense that it takes as input a formally specified description of the system and a temporal logic formula to check the system against, and then automatically produce the answer without any user interaction.
− It is not needed to specify the whole system. Model checking can also produce handy results on a partially defined system and accommodates making decisions about functionality in the early stages of development.
− It comes up with counterexamples i.e., error traces, which demonstrate the deficiency clearly, greatly helping to cure the error.
− Various case studies have shown that the use of model checking has led to shorter development times [61].

Main problems are as follows:

− The state explosion problem, which arises due to high complexity of the described system. To defeat this common problem, techniques like ordered *binary decision diagrams (BDDs)* [62] to represent state transition systems efficiently; *partial order reduction* [63] to reduce the number of independent interleavings of concurrent processes; *localization reduction* [64] to automatically abstract a model relative to the property being checked; *equivalence reduction of identifiers* [65] to collapse two distinct identifiers that are semantically equivalent; and *semantic minimization* [66] to eliminate unnecessary states from a system model, are developed. Even in the light of these approaches, the state explosion problem may persist, for which the best solution is to find a more proper *abstraction* for the system under verification [3].
− In cases where a system can not be specified as a closed finite model, model checking loses its suitability.
− In communications protocol verification, model checking is generally run on a few concurrent processes (to the best of our knowledge, at most five), due to state explosion problem. Although it is possible to show under certain assumptions that this approach suffices to capture all possible behaviors of

a protocol [67], it may not be sufficient to generalize the verification results in some cases.

- It is hard to choose what in the real world to model and to decide on the right level of abstraction. Models only 'model' some aspects of the system in question. The correspondence between the formal description and the real world is limited since the real world is not a formal system. Therefore, a positive verification run for a system does not necessarily show that the system does not have deficiencies in the real world. However, this is a very general modeling problem [56].

3.3   The SPIN Model Checker

SPIN [3] is an automata-based temporal logic model checker. It is a widely accepted tool-box used for specification, simulation, validation and verification of asynchronous concurrent processes, such as the terminals in a communication protocol. Its specification and modeling language is called PROMELA.

Assertions and *never claims* are the most frequently used constructs for specifying correctness properties in PROMELA. An assertion has similar semantics as those in the C Language. When the expression to be asserted is false, the assertion fails and SPIN gives "assertion violated" error.

Never claims are used to specify the finite or infinite behavior that should never happen during the execution of a system. When we want to specify a property to be satisfied by the system, we formalize it in a logic formula and produce a *never claim* that corresponds to the negation of this formula. SPIN then tries to find a violation for this *never claim*. If it finds one, this means there is a case that the opposite of our property can occur in the system, which means our property cannot be satisfied by the system.

A *never claim* can be written by hand or can be translated from a linear temporal logic (LTL) formula. SPIN also includes a timeline property editor that helps users visually specify properties that are otherwise hard to formalize.

SPIN has two modes of operation: simulation and verification. In simulation mode, it runs the model and helps users get an impression on how their model behaves and debug their model. In verification mode, SPIN analyzes the model against the properties considering all possible executions performing an exhaustive search on the state space. It can also perform partial search on the state space, which is quite useful in case of very large models or insufficient computational resources.

If SPIN finds a violation, it produces an error trace. Using this error trace, a user can run a simulation of the execution that leads to the violation.

One can make efficient use of SPIN by making proper abstractions and thereby can keep the complexity of the model low enough to be analyzed by computers.

4   Formal Security Analysis of SAODV and ARIADNE

Our ongoing work is on formal security analysis of secure routing protocols for MANETs, through the use of SPIN. As our main target, we try to find security flaws in the design of secure routing protocols. We employ SPIN to check our protocol model against the security properties that we formally specify as *never claims* in PROMELA. We list any violations, if any, as security flaws.

We first give an overview on the two secure routing protocols that are currently in our focus: SAODV and ARIADNE. Then we give a discussion about the formal security analysis of these protocols conducted using SPIN.

4.1   Formal Security Analysis of SAODV

Secure AODV (SAODV) [8, 9] is an extended version of AODV protocol. It defines a set of message extensions to route request (RREQ), route reply (RREP) and route error (RERR) messages in AODV. There are also some new messages related to the detection of duplicate network addresses.

SAODV addresses integrity and authentication through digital signatures and hash chains. End-to-end digital signatures are used to sign non-mutable data in routing messages. This means only the initiators of RREQ and RREP messages (namely *originator* and *destination*, respectively) sign the packets and intermediate nodes verify the signatures before forwarding them. This mechanism provides the authentication of originator and destination nodes and the integrity of non-mutable information in routing messages.

SAODV utilizes hash chains to keep the integrity of distance information, namely the hop count field, which is supposed to be incremented at each hop.

A sample SAODV route discovery process is demonstrated in Fig. 1. Here, the originator node O initiates a route request for the destination node D by sending a RREQ message. Upon receiving RREQ message, the destination node initiates a route reply by sending a RREP message. Each node is supposed to check the signature and hash fields before processing an incoming message. If the message is to be forwarded, the hash field is updated after incrementing the hop count field.

Informal Security Analysis: SAODV has some weaknesses. First of all, there is nothing to prevent a node from leaving the hop count unchanged and also increasing the hop count arbitrarily. The first weakness is already defined in [8] since it helps a malicious node to attract traffic. We also think that the ability to increase the hop count arbitrarily can also lead to a situation that is desirable for an attacker. If there is at least one alternative to a route on which the attacker resides, it may consistently declare high hop counts so that this route is not selected. This can lead to a concentration of traffic on certain routes and to an excessive power consumption on certain nodes.

Besides the inability to detect incorrect hop count manipulation, SAODV has also another weakness. It does not do anything to protect the sender IP

m_1 (bcast) : { $RREQ$, O, $reqId$, D, SEQ_D, O, SEQ_O, d_{max}, h_{top}, K_O, SIG_O, h }
m_2 (bcast) : { $RREQ$, I, $reqId$, D, SEQ_D, O, SEQ_O, d_{max}, h_{top}, K_O, SIG_O, h }
m_3 (ucast) : { $RREP$, D, SEQ'_D, O, $lifetime$, d_{max}, h_{top}, K_D, SIG_D, h' }
m_4 (ucast) : { $RREP$, I, SEQ'_D, O, $lifetime$, d_{max}, h_{top}, K_D, SIG_D, h' }

Fig. 1. A Sample SAODV Messaging

address field, S, which is used as next hop information in routing tables. This weakness can be used to disrupt routing operation. For example, a malicious node can impersonate another node while forwarding a RREP, which causes incorrect routing information stored in the network. Such attacks potentially cause the nodes to consume excessive time and energy. The designers of SAODV deliberately left such denial of service (DoS) attacks out of scope [8].

SAODV Model: Abstraction is the key process of a model checking attempt. We need to make certain assumptions and simplifications on the protocol to produce a model that is not only relevant to the properties we specify, but also simple enough to allow a feasible analysis by the model checker.

We adopt all the assumptions that are specified by its designers in [9]. We also make some additional specific assumptions and simplifications to reduce the complexity of our SAODV model.

In this work, we used at most five nodes in our network model. Each node has only one network interface and a unique IP address. The network does not contain subnets and is not connected to external networks. Intermediate nodes are not allowed to reply to route requests[1]. There is a single external attacker in the network that does not have and cannot obtain any cryptographic properties that the honest nodes have. Topology does not change during the operation, i.e., mobility, link breakages, and node failures do not occur. The cryptographic functions and algorithms used in the protocol are assumed to be secure. Links and routes are symmetric. Since we are only interested in route discovery, we consider only mechanisms related to RREQ and RREP messages and omit those parts that involve RERR message, which is related to route maintenance.

In our model, there is only one originator, one destination, and one attacker node and there can be at most two intermediate nodes depending on the size of the network. The number of nodes does not change during the network lifetime.

---

[1] Normally, this is an option in SAODV to enhance protocol performance. Since we are interested in functionality, we can disregard this option.

The originator process non-deterministically checks whether it needs to send route request for the destination or not. If it has a valid route, it does nothing. If it does not have one, it initiates a route discovery. A route to the destination expires non-deterministically in the originator process. Route entries do not expire in the intermediate and destination nodes.

Security Properties to Verify: The requirement for the maintenance of correct distance information property can be informally stated as follows: "*It is always true that if there is a route from an originator node O to a destination node D, the length of this route known to node O is actually the shortest path from O to D.*"

We also define the loop freedom property: "*It is always true that if node O has a route entry for a destination node D then node D must be at most $n - 1$ hops away from node O, where n is the number of nodes in the network.*"

We negate the corresponding LTL formulas for these properties and convert them to corresponding *never claims* in PROMELA, by using the internal converter supplied with SPIN.

Verification Results: For each property that we defined, SPIN found a counterexample (i.e. violation) in milliseconds for networks with four and five nodes with linear topologies.

In the first scenario, the attacker increments hop count field in RREQ messages correctly but leaves it unchanged in RREP messages. It also does not spoof any other node.

Since SAODV does not have a mechanism to prevent or detect unchanged hop count values, SPIN flags this as a violation to the correctness of distance property.

In order to analyze the loop freedom property, we use a network with five nodes that are linearly arranged. In this analysis, the attacker is allowed to spoof the other nodes when forwarding RREP messages.

SPIN finds a counterexample that is demonstrated in Fig. 2. Here, attacker A sets the sender address field of the RREP message to node I_1 and sends it to node I_2. Node I_2 looks at the sender address field of the incoming RREP message and updates the next hop address for the destination to node I_1. Then, it forwards the RREP message to node I_1 which in turn updates the next hop address for the destination to the sender of this RREP message, namely node I_2. As a result, a routing loop is formed between nodes I_1 and I_2 as shown in Fig. 2.

4.2   Formal Security Analysis of ARIADNE

ARIADNE [13] is a secure routing protocol for ad hoc networks that aims for resilience against *Active-y-x*[2] attackers, relying only on efficient symmetric cryp-

---

[2] Yih-Chun Hu *et al.* [13] introduced an attacker classification with the term *Active-y-x*, where y stands for the number of terminals that the attacker has compromised

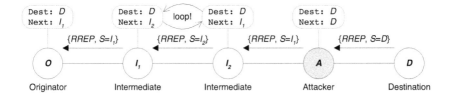

Fig. 2. Loop Formation Between Two Intermediate Nodes

tography. The protocol operates in an on-demand fashion, based on the well-known DSR [18] protocol. The protocol offers three different modes of operation in the sense that authentication of routing messages can be realized by one of three proposed schemes: shared secrets between each pair of nodes (ARIADNE with MACs), shared secrets between communicating nodes combined with broadcast authentication (ARIADNE with TESLA), or digital signatures. Our focus is on ARIADNE with MACs.

ARIADNE uses *hash-chains* and *MAC-lists* to provide security. A hash-chain is a list of hash values, linked together cryptographically. It is created by taking an initial seed and incrementally hashing it n times, where n is said to be the length of the chain. The usability of the hash-chain comes from the fact that given any element from the hash-chain, it is infeasible to calculate any previous elements from the chain. The target can check the authenticity of any RREQ through the use of hash-chains. The initiator initializes the first element of the hash-chain to a MAC computed over $(RREQ, S, D, id)$, using K_{SD}: the secret-key between the initiator and the target. Any intermediate node takes the current hash-chain value, append itself, and rehashes before rebroadcasting the packet. In this way, the target can make sure that the node-list in the RREQ is valid by reconstructing the MAC computed over $(RREQ, S, D, id)$ using K_{SD}, and then reconstructing the hash-chain from scratch by using the node-list, and finally checking whether it is equal to the hash-chain in the packet. The security here relies on the fact that any node does not hear a RREQ without itself listed.

A MAC-list is a list of MAC values that are stored separately. MAC values are used for both authentication and integrity purposes. Each terminal computing a MAC value in ARIADNE does so over the previous fields of the packet by using the shared key between itself and the target node. During a RREQ propagation, these MAC values are appended to the existing MAC-list, thereby increasing the packet-length while traversing the route. In this way, the target can make sure that the packet is not modified along the route by recalculating MACs for each hop over the route and by comparing each calculated value with the value in the RREQ packet. Furthermore, if someone over the route modifies the RREQ, the

and x stands for the total number of terminals that the attacker owns within the network.

target knows which node it is. Also, the initiator of the route discovery can make sure that the received RREP is valid by recalculating the MAC over the RREP using the shared key between itself and the target node and then by comparing it with the value in the received RREP packet.

Informal Security Analysis: Through the use of hash-chains and MAC-lists as described above, ARIADNE provides authentication of the initiator and the target of the route discovery, and makes sure that any modifications on a route discovery packet will be detected at the target or at the initiator. Furthermore, no node over the route can remove previous nodes from the node-list in a RREQ packet.

ARIADNE Model: We adopt all the assumptions originally made in the protocol specification [13]. We have not considered the route-maintenance mechanism to strengthen the focus on issues to be verified: the route discovery phase of the protocol. We assume the same setup defined in [7], where the authors present a previously unknown flaw on ARIADNE protocol. There are two compromised nodes which have one legitimate shared key to communicate within the network, so they use the same identity. We do not consider mobility since we aim to show a sequence of events in the protocol leading to a possible *Active-1-2* attack in a special configuration in which there are some legitimate nodes between two subverted terminals. By also including distinct sender and receiver nodes, this special configuration leads to a five-node linear topology in the most simple case. There are more complicated network topologies where it is still possible to have this configuration inside. But, we assume a linear topology for the sake of simplicity and verification performance. We assume that any terminal in our model may non-deterministically start a route discovery for a non-deterministically chosen destination. A RREQ-timeout may also take place non-deterministically.

Subverted nodes in our model act almost the same as the legitimate nodes. What they do more is similar to what Ács *et al.* describe in [7]. Basically, in the RREQ propagation phase, the compromised node tries to hide the *hash-chain* value in the place where it should have normally appended a MAC-value. By this way, the external attacker which is over the route and is aware of this possible attack understands that a hidden hash-chain element is sent when it sees the compromised node identifier in the *node-list* field of RREQ packet, thereby being able to remove the nodes between the two adversaries using this hidden hash-chain element. In the RREP forwarding phase, the external attacker adds the nodes it has removed to the *node-list* again, in order to provide a harmless forwarding process towards the initiator. Then the compromised node removes those added nodes from the *node-list* and renders the packet to the state on which the MAC-value was calculated. As a result, the source node cannot notice the node removals from the node-list, and accepts the RREP.

Security Properties to Verify: Our interest is that if the protocol can really prevent the intermediate nodes from removing other terminals from the node-list, in an *Active-1-2* adversary environment. So, the property that *"If some route is accepted by a legitimate node, then that route does really exist."* is our security claim to be verified.

We negate the corresponding LTL formula for this property and manually convert it to a corresponding *never claim* in PROMELA.

Verification Results: SPIN has generated an error trail which contains a sequence of events leading to a violation condition for our claim, namely, a security flaw. The logical representations of this attack can also be examined in Fig. 3 for the RREQ propagation phase and in Fig. 4 for the RREP forwarding phase.

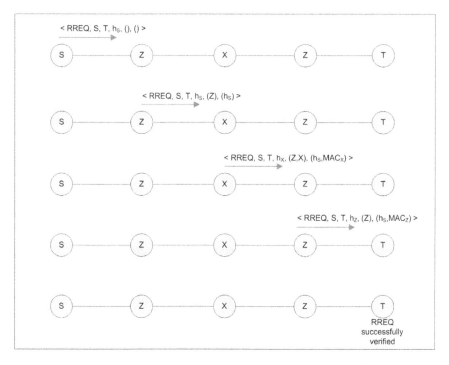

Fig. 3. Representation of the error-trail generated by SPIN, which is Ács' *Active-1-2* attack on ARIADNE; RREQ propagation phase.

It should be noted that the attack SPIN flags is a powerful one since it shows that the actual existing route can be shortened by the compromised nodes, resulting in a compromised route. Realization of such an attack forces the initiator to prefer such a compromised route, which in turn might possibly be used for routing-disruption or traffic analysis purposes.

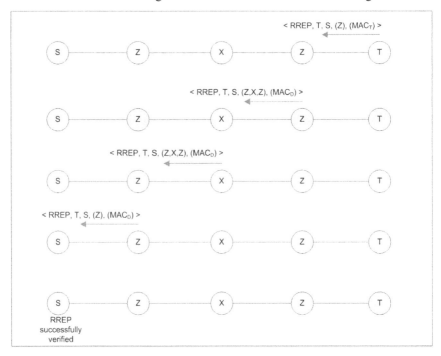

Fig. 4. Representation of the error-trail generated by SPIN, which is Ács' *Active-1-2* attack on ARIADNE; RREP forwarding phase.

5   Conclusions and Future Work

Although there are many proposals in the literature that claim to provide certain levels of security in routing functionality, there are few works that formally analyze the security of these proposals. However, through the use of proper formal methods, it is possible to achieve provable correctness and reliability in any system design and to analyze a system for desired properties. In this work, first we gave a broad overview of secure routing in MANETs and applicaple formal methods. Then, we presented the initial results of our research on using model checking to analyze the security of two secure routing protocols for MANETs. We used SPIN as our preferred model checker.

In our first study, we showed, by using model checking, the two important design deficiencies in SAODV which allow an external attacker to manipulate routing information in the network. Although, these deficiencies are already known, we think it is important to show them by using a well-established formal method such as model checking.

In our second study, again by model checking, we flagged a sequence of events in ARIADNE protocol that leads to a violation of the property that *no intermediate node is able to remove any other nodes from the route discovery process*.

Hence, we formally showed that one of the most powerful security properties of ARIADNE can be violated in an *Active-1-2* adversary environment.

Despite our assumptions and simplifications, our protocol models are nevertheless complicated as finite systems to be model-checked. However, SPIN can manage to reveal the attacks in a short time scale. Model checkers like SPIN are likely to be the most suitable tools to search for complicated security flaws that are hard to find without automation.

We currently focus on the analysis of security properties involving mobility and multiple attackers whose presence makes secure routing a challenge. We are planning to incorporate detection mechanisms such as neighborhood watching into our models. We are also considering to model-check other secure routing protocols against the same properties using the same attacker types in order to make a formal comparison.

References

1. Hu, Y.C., Perrig, A.: A survey of secure wireless ad hoc routing. IEEE Security and Privacy (2004)
2. Argyroudis, P.G., O'Mahony, D.: Secure routing for mobile ad hoc networks. IEEE Communications Surveys **7** (2005) 2–21
3. Holzmann, G.J.: Spin Model Checker, The: Primer and Reference Manual. Addison Wesley (2003)
4. Bhargavan, K., Obradovic, D., Gunter, C.A.: Formal verification of standards for distance vector routing protocols. Journal of the ACM (JACM) **49** (2002) 538–576
5. Wibling, O., Parrow, J., Pears, A.: Automatized verification of ad hoc routing protocols. In: Proceedings of 24th IFIP WG 6.1 International Conference on Formal Techniques for Networked and Distributed Systems (FORTE 2004). LNCS 3235, Madrid, Spain, Springer-Verlag (2004)
6. de Renesse, R., Aghvami, A.H.: Formal verification of ad-hoc routing protocols using SPIN model checker. In: Proceedings of The 12th IEEE Mediterranean Electrotechnical Conference(MELECON 2004), Dubrovnik, Croatia (2004) 1177–1182
7. Ács, G., Buttyán, L., Vajda, I.: Provably secure on-demand source routing in mobile ad hoc networks. IEEE Transactions on Mobile Computing **5** (2006)
8. Zapata, M.G., Asokan, N.: Securing ad hoc routing protocols. In: the ACM Workshop on Wireless Security (WiSe'02). (2002)
9. Zapata, M.G.: Secure ad hoc on-demand distance vector (SAODV) routing. Internet Draft, draft-guerrero-manet-saodv-06, work in progress (2006)
10. Sanzgiri, K., Dahill, B., Levine, B.N., Shields, C., Belding-Royer, E.M.: A secure routing protocol for ad hoc networks. In: Proceedings of the 10th IEEE International Conference on Network Protocols (ICNP'02), Paris, France (2002)
11. Nanz, S., Hankin, C.: A framework for security analysis of mobile wireless networks. Theoretical Computer Science **367** (2006) 203–227
12. Önem, E.: Formal security analysis of a secure on-demand routing protocol for ad hoc networks using model checking. Msc thesis, Bogazici University (2007)
13. Hu, Y.C., Johnson, D.B., Perrig, A.: Ariadne: A secure on-demand routing protocol for ad hoc networks. In: MOBICOM'02. (2002)

14. Abolhasan, M., Wysocki, T., Dutkiewicz, E.: A review of routing protocols for mobile ad hoc networks. Ad Hoc Networks **2** (2003) 1–22
15. Clausen, T.H., Jacquet, P.: Optimized link state routing protocol (OLSR). RFC 3626 (2003)
16. Ogier, R.G., Templin, F.L., Lewis, M.G.: Topology dissemination based on reverse-path forwarding (TBRPF). RFC 3684 (2004)
17. Perkins, C.E., Belding-Royer, E.M., Das, S.R.: Ad hoc on-demand distance vector (AODV) routing. RFC 3561 (2003)
18. Johnson, D.B., Maltz, D.A., Hu, Y.C.: The dynamic source routing protocol for mobile ad hoc networks (DSR). RFC 4728 (2007)
19. Hass, Z., Pearlman, M.: Zone routing protocol for ad hoc networks. Internet Draft, draft-ietf-manet-zrp-05.txt (1999)
20. Das, S., Castaneda, R., Yan, J., Sengupta, R.: Comparative performance evaluation of routing protocols for mobile (1998)
21. Lee, W.C., Lee, J., Huff, K.: On simulation modeling of information dissemination systems in mobile environments. Lecture Notes in Computer Science **1748** (1999) 45–57
22. Barbir, A., Murphy, S., Yang, Y.: Generic threats to routing protocols. Internet Draft (2004)
23. Raymond, J.F.: Traffic analysis: Protocols, attacks, design issues and open problems. In Federrath, H., ed.: Designing Privacy Enhancing Technologies: Proceedings of International Workshop on Design Issues in Anonymity and Unobservability. Volume 2009 of LNCS., Springer-Verlag (2001) 10–29
24. Djenouri, D., Khelladi, L., Badache, A.N.: A survey of security issues in mobile ad hoc and sensor networks. Communications Surveys & Tutorials, IEEE **7** (2005) 2–28
25. Patwardhan, A., Parker, J., Joshi, A., Karygiannis, A., Iorga, M.: Secure routing and intrusion detection in ad hoc networks. In: Proceedings of the Third IEEE International Conference on Pervasive Computing and Communications, Kauaii Island, Hawaii (2005)
26. Inkinen, K.: New secure routing in ad hoc networks: Study and evaluation of proposed schemes. HUT T-110.551 Seminar on Internetworking (2004)
27. B. Wu, J. Chen, J.W., Cardei, M.: A survey on attacks and countermeasures in mobile ad hoc networks. Wireless and Mobile Network Security (2006)
28. Stajano, F., Anderson, R.: The resurrecting duckling: Security issues for ad-hoc wireless networks. In Christianson, B., Crispo, B., Roe, M., eds.: Proceedings of 7th International Workshop on Security Protocols, Springer-Verlag (1999)
29. Balfanz, D., Smetters, D., Stewart, P., Wong, H.: Talking to strangers: Authentication in adhoc wireless networks (2002) In Symposium on Network and Distributed Systems Security (NDSS '02), San Diego, California.
30. Zhu, S., Xu, S., Setia, S., Jajodia, S.: Establishing pair-wise keys for secure communication in ad hoc networks: A probabilistic approach (2003)
31. Zhou, L., Haas, Z.J.: Securing ad hoc networks. IEEE Network Magazine **13** (1999) 24–30
32. Montenegro, G., Castelluccia, C.: Statistically unique and cryptographically verifiable (SUCV) identifiers and addresses. In: Proceedings of the 2002 Network and Distributed System Security Conference (NDSS'02), San Diego (2002)
33. Capkun, S., Hubaux, J.: Biss: Building secure routing out of an incomplete set of secure associations. In: Proceedings of 2nd ACM International Workshop on Wireless Security (WiSe'03). (2003)

34. Kong, J., Zerfos, P., Luo, H., Lu, S., Zhang, L.: Providing robust and ubiquitous security support for mobile ad hoc networks. In: Proceedings of the International Conference on Network Protocols (ICNP). (2001)

35. Asokan, N., Ginzboorg, P.: Key-agreement in ad-hoc networks. Computer Communications **23** (2000) 1627–1637

36. Capkun, S., Buttyan, L., Hubaux, J.: Self-organized public-key management for mobile ad hoc networks. In: Proceedings of 1st ACM International Workshop on Wireless Security (WiSe'02). (2002)

37. Hu, Y.C., Johnson, D.B., Perrig, A.: SEAD: Secure efficient distance vector routing for mobile wireless ad hoc networks. Ad Hoc Networks **1** (2003) 175–192

38. Gupte, S., Singhal, M.: Secure routing in mobile wireless ad hoc networks. Ad Hoc Networks **1** (2003) 151–174

39. Papadimitratos, P., Haas, Z.J.: Secure routing for mobile ad hoc networks. In: Proceedings of the SCS Communication Networks and Distributed Systems Modeling and Simulation Conference (CNDS 2002), San Antonio (2002)

40. Padoy, N.: Secure routing in ad hoc networks (2003)

41. Papadimitratos, P., Haas, Z.J.: Secure message transmission in mobile ad hoc networks. Ad Hoc Networks **1** (2003) 193–209

42. Yi, S., Naldurg, P., Kravets, R.: Security-aware ad hoc routing for wireless networks. In: The 6th World Multi-conference on Systemics, Cybernetics and Informatics (SCI 2002). (2002)

43. Papadimitratos, P., Haas, Z.J.: Secure link state routing for mobile ad hoc networks. In: Proceedings of the IEEE Workshop on Security and Assurance in Ad hoc Networks in conjunction with the 2003 International Symposium on Applications and the Internet, Orlando, FL (2003)

44. Awerbuch, B., Holmer, D., Nita-Rotaru, C., Rubens, H.: An on-demand secure routing protocol resilient to byzantine failures. In: Proceedings of the ACM Workshop on Wireless Security (WiSe'02), Atlanta, Georgia (2002)

45. Wan, T., Kranakis, E., van Oorschot, P.: Securing the Destination-Sequenced Distance Vector Routing Protocol (S-DSDV). LECTURE NOTES IN COMPUTER SCIENCE (2004) 358–374

46. Lu, B., Pooch, U.W.: Cooperative security-enforcement routing in mobile ad hoc networks. In: the 4th IEEE International Workshop on Mobile and Wireless Communications Networks. (2002)

47. Stephan Eichler, C.R.: Challenges of secure routing in manets: A simulative approach using AODV-SEC. In: In Proceedings of the 3rd IEEE International Conference on Mobile Ad-hoc and Sensor Systems (MASS), Vancouver, Canada (2006)

48. Ramanujan, R., Ahamad, A., Bonney, J., Hagelstrom, R., Thurber, K.: Techniques for intrusion-resistant ad hoc routing algorithms (TIARA). MILCOM 2000. 21st Century Military Communications Conference Proceedings **2** (2000)

49. Carter, S., Yasinsac, A.: Secure position aided ad hoc routing protocol. In: Proceedings of the IASTED International Conference on Communications and Computer Networks (CCN02). (2002)

50. Adjih, C., Clausen, T., Jacquet, P., Laouiti, A., Muhlethaler, P., Raffo, D.: Securing the OLSR protocol. Proceedings of Med-Hoc-Net (2003) 25–27

51. Buchegger, S., Le Boudec, J.: Performance Analysis of the CONFIDANT Protocol: Cooperation Of NodesFairness In Dynamic Ad-hoc NeTworks. Proceedings of IEEE/ACM Symposium on Mobile Ad Hoc Networking and Computing (Mobi-HOC) (2002) 226–236

52. Marti, S., Giuli, T.J., Lai, K., Baker, M.: Mitigating routing misbehavior in mobile ad hoc networks. In: Proceedings of the ACM Symposium on Mobile Ad Hoc Networking and Computing (MobiHoc'00). (2000) 255–265
53. Hu, Y., Perrig, A., Johnson, D.: Packet Leashes: A Defense against Wormhole Attacks in Wireless Ad Hoc Networks. Proceedings of INFOCOM **2003** (2003)
54. Hu, L., Evans, D.: Using Directional Antennas to Prevent Wormhole Attacks. Proceedings of the 11th Network and Distributed System Security Symposium (2003) 131–141
55. Čapkun, S., Buttyán, L., Hubaux, J.: SECTOR: secure tracking of node encounters in multi-hop wireless networks. Proceedings of the 1st ACM workshop on Security of ad hoc and sensor networks (2003) 21–32
56. Hall, A.: Seven myths of formal methods. IEEE Software **7** (1990) 11–19
57. Babich, F., Deotto, L.: Formal methods for specification and analysis of communication protocols. IEEE Communications Surveys (2002)
58. Clarke, E.M., Wing, J.M.: Formal methods: State of the art and future directions. ACM Computing Surveys **28** (1996) 626–643
59. Clarke, E., Emerson, E.: Synthesis of synchronization skeletons for branching time temporal logic. Logic of Programs: Workshop **131** (1981)
60. Queille, J., Sifakis, J.: Specification and verification of concurrent systems in CESAR. Proceedings of the 5th Colloquium on International Symposium on Programming (1982) 337–351
61. de Renesse, F., Aghvami, A.: Formal verification of ad-hoc routing protocols using SPIN model checker. Electrotechnical Conference, 2004. MELECON 2004. Proceedings of the 12th IEEE Mediterranean **3** (2004)
62. Bryant, R.: Graph-based algorithms for Boolean function manipulation. IEEE Transactions on Computers **35** (1986) 677–691
63. Alur, R., Brayton, R., Henzinger, T., Qadeer, S., Rajamani, S.: Partial-Order Reduction in Symbolic State-Space Exploration. Formal Methods in System Design **18** (2001) 97–116
64. Kurshan, R.P.: Computer-Aided Verification of Coordinating Processes. Princeton University Press (1994)
65. Larsen, K.G., Pettersson, P., Yi, W.: Compositional and Symbolic Model-Checking of Real-Time Systems. (1995)
66. Elseaidy, W.M., Cleaveland, R., Baugh, J.W.: Modeling and verifying active structural control systems. Science of Computer Programming **29** (1997) 99–122
67. Yang, S., Baras, J.S.: Modeling vulnerabilities of ad hoc routing protocols. In: Proceedings of the first ACM Workshop on Security of Ad Hoc and Sensor Networks. (2003) 12–20

Author Index

EDITORS BIOS

Atilla Elci (editor)

Internet Technology Research Center (President), and Dept. of Computer Engineering, Eastern Mediterranean University, Gazimagusa, North Cyprus.
e-mail: Atilla.Elci@emu.edu.tr; home site: http://cmpe.emu.edu.tr/aelci/.
Dr. Atilla Elci received M.Sc. and Ph.D. degrees at Purdue University in Computer Sciences with high honors citation. He was a faculty member in the Dept. of Computer Sciences & Engineering in Middle East Technical Univ., Turkey, during 1976-1985. He served in the positions of Chairman (2 years), Asst. Chairman (4 years), and Manager of the University Computer Center (3 years) besides

teaching. He was head of Systems and Languages Major for 5 years. He became Associate Professor in Software major in 1983. From 1985 till 1997, he was consultant to the International Telecommunication Union (ITU), serving in computerization development projects of UNDP. He held numerous positions as chief technical adviser, senior expert, project designer and project manager in more than 20 countries in five continents. Subsequently he founded and ran his own company in Turkey offering IT and communications solutions (1998-2003). In 2001 he established the Department of Computer Engineering in Haliç University, Istanbul, serving as its Chairman till 2003. He is currently with the Department of Computer Engineering, and the President of the Internet Technologies Research Center, Eastern Mediterranean University, North Cyprus. Dr. Elçi has extensive experience in semantic Web, Web technology, multi-agent systems, semantic robotics, Internet, systems and languages, software engineering, education and training, and information systems in the telecoms agencies. He published over 100 papers, edited several proceedings, a series of 17 computer-based coursewares, designed several UNDP/ITU development projects in telecoms computerization, and organized several conferences, such as ESAS 2006 & 2007 and SIN 2007. Dr. Elci is a member of ACM, IEEE, IEEE CS, and Turkish Informatics Association.

S. Berna Ors (editor)

S. Berna Ors received the Electronics and Communication Engineering Degree in 1995 and the M.Sc. Degree in Electronics and Communication Engineering in 1998 from the Istanbul Technical University, in Istanbul, Turkey. She received the Ph.D. Degree in applied sciences from the Katholieke Universiteit Leuven, in Leuven, Belgium in 2005. She is currently assistant professor at Istanbul Technical University. Her interests include circuits, processor architectures and embedded systems in application domains such as security, cryptography, digital signal processing and wireless applications.

Bart Preneel (editor)

Preneel received the Electrical Engineering degree and the Doctorate in Applied Sciences from the Katholieke Universiteit Leuven (Belgium). He is currently full professor at the Katholieke Universiteit Leuven. He has been visiting professors at the Technical University of Denmark (2007), Graz University of Technology in Austria (1997-2006), the University of Bergen in Norway (1997-2001), Ruhr-Universitaet Bochum in Germany (2001-2002) and at the University of Ghent (1994-2002). He is a scientific advisor of Philips

Research (the Netherlands). During the academic year 1993-1994, he was a research fellow of the EECS Department of the University of California at Berkeley. His main research interests are cryptology and information security.

He has authored and co-authored more than 200 scientific publications and is inventor of two patents. He is vice president of the IACR (International Association for Cryptologic Research) and a member of the Editorial Board of the Journal of Cryptology, the IEEE Transactions on Information Forensics and Security, and the International Journal of Information and Computer Security.

He is also a member of the Accreditation Board of the Computer and Communications Security Reviews (ANBAR, UK).

He has participated to more than 20 research projects sponsored by the European Commission, for four of these as project manager. He is currently project manager of the European Network of Excellence ECRYPT (http://www.ecrypt.eu.org) which groups more than 250 researchers in the area of cryptology and watermarking.

He was program (co-)chairman of 9 international conferences (including Eurocrypt 2000 and SAC 2005) and has been a member of more than 100 international program committees. He has been invited speaker at 25 international conferences and workshops. He has lectured at 30 international Summer Schools in 15 countries.

In 2003, he has received the European Information Security Award in the area of academic research, and he received an honorary Certified Information Security Manager (CISM) designation by the Information Systems Audit and Control Association (ISACA).

Since 1989, he is a Belgian expert in working group ISO/IEC JTC1/SC27/WG2 (Security Techniques and Mechanisms), where he has edited five international standards.

He is president of L-SEC vzw. (Leuven Security Excellence Consortium), an association of 50 companies and research institutions in the area of e-security.